A

SCHOOL COMPENDIUM

OF

NATURAL AND EXPERIMENTAL

PHILOSOPHY,

EMBRACING THE ELEMENTARY PRINCIPLES OF

MECHANICS, HYDROSTATICS, HYDRAULICS, PNEUMATICS, ACOUSTICS, PYRONOMICS, OPTICS, ELECTRICITY, GALVANISM, MAGNETISM, ELECTRO-MAGNETISM, MAGNETO-ELECTRICITY, AND ASTRONOMY.

CONTAINING ALSO A DESCRIPTION OF THE

STEAM AND LOCOMOTIVE ENGINES,

AND OF THE

ELECTRO-MAGNETIC TELEGRAPH.

BY

RICHARD GREEN PARKER, A.M.

LATE PRINCIPAL OF THE JOHNSON GRAMMAR SCHOOL, BOSTON; AUTHOR OF "AIDS TO ENGLISH COMPOSITION," A SERIES OF "SCHOOL READERS," "GEOGRAPHICAL QUESTIONS," ETC. ETC.

Delectando pariter que monendo
Prodesse quam conspici.

CORRECTED, ENLARGED AND IMPROVED.

NEW YORK:
A. S. BARNES & CO., 51 & 53 JOHN-STREET.
1856.

School Philosophy,

BY RICHARD G. PARKER, A.M.

I.

JUVENILE PHILOSOPHY, or PHILOSOPHY IN FAMILIAR CONVERSATIONS. Designed to teach Young Children to Think.

II.

FIRST LESSONS IN NATURAL PHILOSOPHY. Designed to teach the Elements of the Science. Arranged from the Compendium of School Philosophy.

III.

A SCHOOL COMPENDIUM OF NATURAL AND EXPERIMENTAL PHILOSOPHY. Embracing the Elementary Principles of Mechanics, Hydrostatics, Hydraulics, Pneumatics, Acoustics, Pyronomics, Optics, Electricity, Galvanism, Magnetism, Electro-Magnetism, Magneto-Electricity, and Astronomy. Containing also a Description of the Steam and Locomotive Engines, and of the Electro-Magnetic Telegraph. A new Edition, corrected, enlarged, and improved.

College Philosophy,

BY PROF. W. H. C. BARTLETT.

W. H. C. BARTLETT'S COLLEGE COURSE OF NATURAL PHILOSOPHY. Used at the United States Military Academy, and designed for Colleges and Universities.

    VOL. I. ELEMENTS OF MECHANICS.
    " II. ELEMENTS OF ACOUSTICS AND OPTICS.
    " III. ELEMENTS OF ASTRONOMY.

ANALYTICAL MECHANICS, treated by the Aid of the Calculus. In one volume 8vo. Designed for advanced Students.

School Chemistry,

BY JOHN A. PORTER,
Professor of Chemistry in Yale College.

Entered according to Act of Congress, in the year 1858,
BY A. S. BARNES & CO.,
In the Clerk's Office of the District Court of the United States for the Southern District of New York.

DAVIES'
COURSE OF MATHEMATICS.

DAVIES' FIRST LESSONS IN ARITHMETIC—For beginners.

DAVIES' ARITHMETIC.—Designed for the use of Academies and Schools.

KEY TO DAVIES' ARITHMETIC.

DAVIES' UNIVERSITY ARITHMETIC—Embracing the Science of Numbers, and their numerous applications.

KEY TO DAVIES' UNIVERSITY ARITHMETIC.

DAVIES' ELEMENTARY ALGEBRA—Being an Introduction to the Science, and forming a connecting link between ARITHMETIC and ALGEBRA.

KEY TO DAVIES' ELEMENTARY ALGEBRA.

DAVIES' ELEMENTARY GEOMETRY.—This work embraces the elementary principles of Geometry. The reasoning is plain and concise, but at the same time strictly rigorous.

DAVIES' ELEMENTS OF DRAWING AND MENSURATION—Applied to the Mechanic Arts.

DAVIES' BOURDON'S ALGEBRA—Including Sturms' Theorem,—Being an Abridgment of the work of M. Bourdon, with the addition of practical examples.

DAVIES' LEGENDRE'S GEOMETRY AND TRIGONOMETRY.—Being an Abridgment of the work of M. Legendre, with the addition of a Treatise on MENSURATION OF PLANES AND SOLIDS, and a Table of LOGARITHMS and LOGARITHMIC SINES.

DAVIES' SURVEYING—With a description and plates of the THEODOLITE, COMPASS, PLANE-TABLE, and LEVEL: also, Maps of the TOPOGRAPHICAL SIGNS adopted by the Engineer Department—an explanation of the method of surveying the Public Lands, and an Elementary Treatise on NAVIGATION.

DAVIES' ANALYTICAL GEOMETRY—Embracing the EQUATIONS OF THE POINT AND STRAIGHT LINE—of the CONIC SECTIONS—of the LINE AND PLANE IN SPACE—also, the discussion of the GENERAL EQUATION of the second degree, and of SURFACES of the second order.

DAVIES' DESCRIPTIVE GEOMETRY,—With its application to SPHERICAL PROJECTIONS.

DAVIES' SHADOWS AND LINEAR PERSPECTIVE.

DAVIES' DIFFERENTIAL AND INTEGRAL CALCULUS

National Series of Standard School Books,

DESIGNED AS

Class Books for the use of Schools, Academies, Colleges, Families, and Libraries.

PUBLISHED BY A. S. BARNES & CO., 51 JOHN-ST., N. Y.

The best talent that could be procured has been employed in the preparation of these works; and the high standing they have already attained, as Class Books for the institutions of our country, is gratifying evidence of their intrinsic merits, and, it is believed, fully entitles them to the name of the NATIONAL SERIES.

ENGLISH LANGUAGE.

ORTHOGRAPHY AND READING.—Price's English Speller, 12 cts.; Martin's Orthoepist, 50 cts.; Wright's Orthography, 25 cts.; Dictation Exercises, 25 cts.; Parker's Series of Readers; Parker and Zachos' Introductory Lessons, 37½ cts.; Parker's Rhetorical Reader, $1; High School Literature, $1.

ENGLISH GRAMMAR AND RHETORIC.—Clark's Analysis, 37½ cts.; Clark's Grammar, 50 cts.; Clark's Etymological Chart, $2 50; Day's Art of Rhetoric, 75 cts.

ELOCUTION.—Northend's Little Speaker, 30 cts.; Northend's American Speaker, 75 cts.; Northend's School Dialogues, 75 cts.; Zacho's New American Speaker, $1.

ENGLISH POETS, WITH NOTES BY BOYD.—Milton's Paradise Lost, Young's Night Thoughts, Thomson's Seasons, Pollok's Course of Time, $1 each. Cheap School Editions of the above, 62½ cts. each; also, Beautifully Illustrated Editions, in various styles of binding.

HISTORY AND GEOGRAPHY.

HISTORY.—Willard's History of the United States, $1 50; School History, 75 cts.; Spanish Translation, $2 00; Universal History, $1 50; Historic Guide, 75 cts.; Map of Time, 75 cts.; Last Leaves of American History, 50 cts., and Charts; Alison's History of Europe, Abridged by Edward S. Gould, $1 50.

GEOGRAPHY.—Monteith's Youth's Manual of Geography, 50 cts.; McNally's System of School Geography, in Press.

THE SCIENCES.

MATHEMATICS.—DAVIES' SYSTEM OF MATHEMATICS—*Arithmetical Course*—Davies' Table-Book, 10 cts.; First Lessons in Arithmetic, 20 cts.; Intellectual Arithmetic, 25 cts.; School Arithmetic, 37½ cts., and Key; Grammar of Arithmetic; University Arithmetic, 75 cts., and Key, 50 cts. *Academic Course*—Elementary Algebra, 75 cts.; Elementary Geometry, $1; Practical Mathematics, $1; Logic of Mathematics, $1 50. *Collegiate Course*—Davies' Bourdon, $1 50; Legendre, $1 50; Surveying, $1 50; Analytical Geometry, $1 25; Descriptive Geometry, $1 75; Shades, Shadows, &c., $2 50; Calculus, $1 25.

NATURAL PHILOSOPHY AND ASTRONOMY.—PARKER'S SCHOOL COURSE—Parker's Natural and Experimental Philosophy, $1; Juvenile Philosophy, 25 cts.; or, First Lessons in Philosophy, 37½ cts., Parts I. and II. McIntyre on the Globes, $1. *Bartlett's College Course*—Mechanics, $3; Optics and Acoustics, $2; Analytical Mechanics, $4.

ENGINEERING.—Gillespie on Roads and Railroads, $1 50; Lardner on the Steam Engine, $1 50.

CHAMBERS' SCIENTIFIC SERIES.—Introduction to the Sciences, 40 cts.; Treasury of Knowledge, 75 cts.; Drawing, 62 cts.; Natural Philosophy, 75 cts.; Chemistry and Electricity, 75 cts.; Vegetable and Animal Physiology, 75 cts.; Zoology, $1; Geology, 75 cts.

CHEMISTRY.—Reid and Bain, 75 cts.; Gregory's Outlines, $1 50; Porter's School Chemistry, in Press.

BOOKKEEPING AND PENMANSHIP.—Fulton and Eastman's System.

MUSIC.—School Song and Hymn Book, 37½ cts.; Kingsley's Juvenile Choir, 37½ cts.; Young Ladies' Harp; Cruikshank's S. S. Gems, 30 cts.; Kingsley's Harp of David, $1; Sacred Harmonist; Cheever's Christian Melodies, 37½ cts.

LATIN AND GREEK.—Brooks' Latin Lessons, 50 cts.; Ovid, $2; Greek Lessons, 50 cts., and Collectanea Evangelica, 50 cts.

PREFACE.

In the year 1837, the school-committee of the city of Boston ordered a few articles of philosophical apparatus to be furnished for each of the grammar-schools of that city; and the author of this work, who for many years had been at the head of one of those schools, finding no elementary work, unencumbered with extraneous matter, suitable to explain the apparatus, attempted to supply the deficiency. The result was the first edition of this work. A few years afterwards, the philosophical apparatus was exchanged for one of better construction, and much more extended application, and an enterprising publishing house in New York induced the author to revise and extend his work. This was done in the year 1848. Since that time the progress of science has been so great that another revision is imperatively demanded; and the author, anxious not to be "*behind the age*," has made another careful revision, in which he is conscious of no omission in the notices of the present state of science, in the departments embraced in this volume, suitable for a work designed to be strictly elementary, and designed for those only whose progress in "*the exact sciences*" must necessarily be limited. The "*Questions*" which have appeared in previous editions he had no hand in preparing. Indeed, in his opinion, such appendages to school-books, in the hands of experienced teachers, are of very questionable expediency. But, as it is a custom *most honored in "observance,"* he has, in this edition of 1854, complied with that custom, and prepared them with his own hands. If he is not deceived in the result of his labors, his work will commend itself by the following features:

1. It is adapted to the *present state* of natural science; embraces a wider field, and contains a greater amount of information on the respective subjects of which it treats, than any other elementary treatise of its size.

2. It contains engravings of *the Boston school set of philosophical apparatus;* a description of the instruments, and an account of many experiments which can be performed by means of the apparatus.

3. It is enriched by a representation and a description of the *Locomotive* and the *Stationary Steam Engines*, and the various forms of the *Electric Telegraph* now in operation in this country.

4. The subjects of Pyronomics, Electricity, Magnetism, Electro-Magnetism, and Magneto-Electricity, as well as Astronomy, have large space allotted to them. Most of the latest discoveries in physical science have also received their due share of attention.

5. It is peculiarly adapted to the convenience of study and of recitation, by the figures and diagrams being first placed side by side with the illustrations, and then repeated on separate leaves at the end of the volume. The number is also given, where each principle may be found to which allusion is made throughout the volume. Suitable questions, also prepared by the author himself, and obnoxious to no objection as "*leading questions*," have been placed in immediate connection with the most important principles contained in the volume.

6. It presents the most important principles of science in a larger type; while the deductions from these principles, and the illustrations, are contained in a smaller letter. Much useful and

interesting matter is also crowded into notes at the bottom of the page. By this arrangement, the pupil can never be at a loss to distinguish the parts of a lesson which are of primary importance; nor will he be in danger of mistaking theory and conjecture for fact.

7. It contains a number of original illustrations, which the author has found more intelligible to young students than those with which he has met elsewhere.

8. Nothing has been omitted which is usually contained in an elementary treatise.

A work of this kind, from its very nature, admits but little originality. The whole circle of the sciences consists of principles deduced from the discoveries of different individuals, in different ages, thrown into common stock. The whole, then, is common property, and belongs exclusively to no one. The merit, therefore, of an elementary treatise on natural science must rest solely on the judiciousness of its selections. In many of the works from which extracts have been taken for this volume, the author has found the *same language* and expressions without the usual marks of quotation. Being at a loss, therefore, whom to credit for some of the expressions which he has borrowed, he makes this general acknowledgment, in the hope that it may be said of him, as it was once said of the Mantuan bard, that "he has *adorned his thefts, and polished the diamonds which he has stolen.*"

BOSTON OCTOBER 1853.

ADVERTISEMENT TO THE NEW STEREOTYPE OF 1854

In the revision of this work the author has endeavored to present his materials under a better classification. The omission of seventy-five pages of Questions, prepared by another hand, found at the end of the book in previous editions, has given room for a large collection of new facts and principles which the present improved state of science has revealed, without materially enlarging the size of the volume. The author now gives it to the world, in confidence that it is much more deserving of the unexpected favor it has received. All changes in a text-book are necessarily attended with inconveniences to teachers; but they who would keep pace with the progress of science must submit to such inconvenience, or be behind the age. The present is emphatically the age of "*progress*," and they who profess to record the triumphs of science must keep a blank page in their journals for the record of new conquests. So much of apology seems to be due for the appearance of a new revision of this volume so soon after the former revision. The author indulges the belief that no advance has been made in fact, in principle, or in physical law, which has not received its due share of attention so far as is consistent with the plan of a work professing to be strictly elementary.

4 KNEELAND-PLACE, 1853.

LIST OF WORKS

WHICH HAVE BEEN CONSULTED, OR FROM WHICH EXTRACTS HAVE BEEN TAKEN, IN THE PREPARATION OF THIS VOLUME.

Annals of Philosophy; Arnott's Elements of Physics; Bartlett's Philosophy; Bigelow's Technology; Cambridge Physics; Chambers' Dictionary; Enfteld's, Olmsted's, Smith's, Blair's, Bakewell's, Draper's, Grund's, Johnson's, Jones', Comstock's, and Conversations on, Natural Philosophy; Davis' Manual of Magnetism; Encyclopedia Americana; Franklin's Philosophical Papers; Henry's Chemistry; King's Manual of Electricity; Lardner's Works; Library of Useful Knowledge; Orbs of Heaven; Paxton's Introduction to the Study of Anatomy; Pambour on Locomotive Engines on Railways; Penny Cyclopedia; Peschel's Elements of Physics; Philips' Astronomy; Sir John Herschel's Astronomy; Silliman's Journal of Science; Singer's Electricity; Scientific Class Book; Scientific Dialogues; Smith's Explanatory Key; The Year Book; Turner's Chemistry; Wilkins' Astronomy; Worcester's and the American School Geography; Lathrop, McIntire and Keith, on the Globes; World's Progress; Annual of Scientific Discovery; Webster's Dictionary; Treasury of Knowledge; Gregory's Chemistry; Science of Familiar Things; Loomis' Elements of Geology; Chambers' Educational Course; Brande's Encyclopedia; Ure's Dictionary; McCulloch's Commercial Dictionary; Patent Office Reports.

SCHEDULE OF PHILOSOPHICAL APPARATUS

USED IN THE GRAMMAR-SCHOOLS OF THE CITY OF BOSTON.*

Adopted by the School Committee, Aug. 1847.

LAWS OF MATTER.

Apparatus for illustrating Inertia.
Pair of Lead Hemispheres for Cohesion.
Pair of Glass Plates for Capillary Attraction.

LAWS OF MOTION.

Ivory Balls on Stand for Collision.
Set of eight Illustrations for Centre of Gravity.
Sliding Frame for Composition of Forces.
Apparatus for illustrating Central Forces.

MECHANICS.

Complete set of Mechanicals, consisting of Levers, Pulleys, Wheel and Axle, Capstan, Screw, Inclined Plane, Wedge.

HYDROSTATICS.

Bent Glass Tube for Fluid Level.
Mounted Spirit Level.
Hydrometer and Jar for Specific Gravity.
Scales and Weights for Specific Gravity.
Hydrostatic Bellows, and Paradox.

HYDRAULICS.

Lifting, or Common Water-pump.
Forcing Pump; illustrating the Fire-engine.
Glass Syphon-cup for illustrating intermittent Springs.
Glass and Metal Syphons.

PNEUMATICS.

Patent Lever Air-pump and Clamp.
Three Glass Bell Receivers, adapted to the Apparatus.
Condensing and Exhausting Syringe.
Copper Chamber for Condensed Air Fountain.
Revolving Jet and Glass Barrel.

---

* The cost of this apparatus is about two hundred and sixty dollars. It was made by Mr. Joseph M. Wightman, importer and manufacturer of Philosophical Apparatus, No. 33 Cornhill, Boston, and in an eminent degree unites beauty with durability. Messrs. Chamberlain & Ritchie, also, in Washington-street, excel in their manufacture of Philosophical Instruments of all kinds. In the department of Electricity and Magnetism, Messrs. Palmer & Hall, successors of Daniel Davis, 428 Washington-street, have many articles of excellent design and execution.

Fountain Glass, Cock, and Jet for Vacuum.
Brass Magdeburg Hemispheres.
Improved Weight-lifter for upward pressure.
Iron Weight of fifty-six pounds, and Strap, } for Weight-lifter.
Flexible Tube and Connectors,
Brass Plate and Sliding Rod.
Bolt Head and Jar.
Tall Jar and Balloon.
Hand and Bladder Glasses.
Wood Cylinder and Plate.
India-rubber Bag for expansion of air.
Guinea and Feather Apparatus.
Glass Flask and Stop-cock for weighing air.

ELECTRICITY.

Plate Electrical Machine.
Pith-ball Electrometer.
Electrical Battery of four Jars.
Electrical Discharger.
Image Plates and Figure.
Insulated Stool.
Chime of Bells.
Miser's Plate for shocks.
Tissue Figure, Ball and Point.
Electrical Flyer and Tellurian.
Electrical Sportsman, Jar and Birds.
Mahogany Thunder-house and Pistol.
Hydrogen Gas Generator.
Chains, Balls of Pith, and Amalgam.

OPTICS.

Glass Prism, and pair of Lenses.
Dissected Eyeball, showing its arrangement.

MAGNETISM.

Magnetic Needle on Stand.
Pair of Magnetic Swans.
Glass Vase for Magnetic Swans.
Horseshoe Magnet.

ASTRONOMY.

Improved School Orrery.
Tellurian, or Season Machine.

ARITHMETIC AND GEOMETRY.

Set of thirteen Geometrical Figures of Solids.
Box of sixty-four one-inch Cubes for Cube Root, &c.

AUXILIARIES.

Tin Oiler; Glass Funnel; Sulphuric Acid.
Set of Iron Weights for Hydrostatic Paradox.

CONTENTS

Divisions of the Subject, ... 17
Of Matter and its Properties, ... 19
Of Gravity, ... 33
Mechanics, or the Laws of Motion, ... 41
The Mechanical Powers, ... 70
Regulators of Motion, ... 100
Hydrostatics, ... 108
Hydraulics, ... 128
Pneumatics, ... 138
Acoustics, ... 173
Pyronomics, ... 185
The Steam-engine, ... 196
Optics, ... 210
Electricity, ... 258
Galvanism, or Voltaic Electricity, ... 283
Magnetism, ... 298
Electro-Magnetism, ... 308
The Electro-magnetic Telegraph, ... 319
The Electrotype Process, ... 331
Magneto-Electricity 332
Thermo-Electricity, ... 334
Astronomy, ... 335

The Index at the close of the volume, being full and comprehensive, will be found more convenient for reference.

INTRODUCTION.

The term Philosophy literally signifies, the love of wisdom; but, as a general term, it is used to denote an explanation of the reason of things, or an investigation of the causes of all phenomena, both of mind and of matter.

When applied to any particular department of knowledge, the word Philosophy implies the collection of general laws or principles, under which the subordinate facts or phenomena relating to that subject are comprehended. Thus that branch of Philosophy which treats of God, his attributes and perfections, is called Theology; that which treats of the material world is called Physics, or Natural Philosophy; that which treats of man as a rational being is called Ethics, or Moral Philosophy; and that which treats of the mind is called Intellectual Philosophy, or Metaphysics.

The word Theology is derived from two Greek words, the former of which (Θεος) signifies God, and the latter (λογος) means a discourse; and these two words, combined in the term *Theology*, literally imply a discourse about God. The latter of these two Greek words (λογος or logos) is changed into *logy* to form English compounds, and it enters into the composition of many scientific terms. Thus we have the words minera*logy*, the science of minerals; meteoro*logy*, the science which treats of meteors; ichthyo*logy*, the science of fishes; entomo*logy*, the science of insects; litho*logy*, of stones; concho*logy*, of shells, &c.

The word Metaphysics is composed of two Greek words, Meta (or μετα), which signifies *beyond*, and phusis (or φυσις), which signifies *nature*, and in composition these words imply something *beyond nature.* From the latter of these words, phusis (φυσις), we obtain the term *physics*, which in its most extended sense implies the science of nature and natural objects, comprehending the study or knowledge of whatever exists. The natural division of all things that exist is into body and mind — things material and immaterial, spiritual and corporeal. Physics relates to material things, Metaphysics to immaterial. Man, as a mere animal, is included in the science of Physics; but, as a being possessed of a soul, of intellect, of the powers of perception, consciousness, volition, reason, and judgment, he becomes a subject of consideration in the science of Metaphysics.

All material things are divided into two great classes, called organized and unorganized matter. Organized matter is that which is endowed with organs adapted to the discharge of appropriate functions, such as the mouth and stomach of animals, or the leaves of vegetables. By means of such organs they enjoy life. Unorganized matter, on the contrary, possesses no such organs, and is consequently incapable of life and voluntary action. Stones, the various kinds of earth, metals, and many minerals, are instances of unorganized matter. Fossils, that is, substances dug out of the earth, are frequently instances of a combination of organized and unorganized matter. Unorganized matter also enters into the composition of organized matter. Thus, the bones of animals contain lime, which by itself is unorganized matter.

Physical Science, or Physics, with its subdivisions of Natural History (including Zoology, Botany, Mineralogy, Conchology, Entomology, Ichthyology, &c.) and Natural Philosophy, including its own appropriate subdivisions, embraces the whole field of organized and unorganized matter.

The term Natural Philosophy is considered by some authors as embracing the whole extent of physical science, while others use it in a more restricted sense, including only the general properties of unorganized matter, the forces which act upon it, the laws which it obeys, the results of those laws, and all those external changes which leave the substance unaffected. It is in this sense that the term is employed in this work.

Chemistry, on the contrary, is the science which investigates the composition of material substances, the internal changes which they undergo, and the new properties which they acquire by such changes. The operations of chemistry may be described under the heads of Analysis or decomposition, and Synthesis or combination.

Natural Philosophy makes us acquainted with the condition and relations of bodies as they spontaneously arise, without any agency of our own. Chemistry teaches us how to alter the natural arrangement of elements to bring about some particular condition that we desire. To accomplish these objects in both of the departments of science to which we refer, we make use of appliances called philosophical and chemical apparatus, the proper use of which it is the office of Natural Philosophy and Chemistry respectively to explain. All philosophical knowledge proceeds either from observation or experiment, or from both. It is a matter of observation that water, by cold, is converted into ice; but if, by means of freezing mixtures, or evaporation, we actually cause water to freeze, we arrive at the same knowledge by experiment.

By repeated observations, and by calculations based on such observations, we discover certain uniform modes in which the powers of nature act. These uniform modes of operation are called *laws;* — and these laws are general or particular according to the extent of the subjects which they respectively embrace. Thus, it is a general law that all bodies attract each other in proportion to the quantity of matter which they contain. It is a particular law of electricity that similar kinds repel and dissimilar kinds attract each other.

The collection, combination, and proper arrangement of such general and particular laws, constitute what is called Science. Thus, we have the science of Chemistry, the science of Geometry, the science of Natural Philosophy, &c.

The terms art and science have not always been employed with proper discrimination. In general, an art is that which depends on practice or performance, while science is the examination of general laws, or of abstract and speculative principles. The theory of music is a science; the practice of it is an art.

Science differs from art in the same manner that knowledge differs from skill. An artist may enchant us with his skill, although he is ignorant of all scientific principles. A man of science may excite our admiration by the extent of his knowledge, though he have not the least skill to perform any operation of art. When we speak of the mechanic arts, we mean the practice of those vocations in which tools, instruments and machinery, are employed. But the science of Mechanics explains the principles on which tools and machines are constructed, and the effects which they produce. Science, therefore, may be defined, a collection and proper arrangement of the general principles or leading truths relating to any subject; and there is this connection between art and science, namely — "A principle in science is a rule of art."

NATURAL PHILOSOPHY.

DIVISIONS OF THE SUBJECT.

What is Natural Philosophy?
1. NATURAL PHILOSOPHY, or PHYSICS, is the science which treats of the powers, properties and mutual action of natural bodies, and the laws and operations of the material world.

2. Some of the principal branches of Natural Philosophy are

Mechanics,
Pneumatics,
Hydrostatics,
Hydraulics,
Acoustics,
Pyronomics,
Optics,

Electricity,
Galvanism,
Magnetism,
Electro-Magnetism,
Magneto-Electricity,
Astronomy.

NOTE.—This list of branches might be considerably enlarged, but perhaps a rigid classification would rather suggest the omission of some of them, as pertaining to the department of chemistry.

What is Mechanics?
3. MECHANICS.—*Mechanics* is that branch of Natural Philosophy which relates to motion and the moving powers, their nature and laws, with their effects in machines.

4. Mechanics is generally considered under two divisions, called **Statics** and **Dynamics**.

5. The word *Statics* is derived from a Greek word implying *rest*, and it is applied to that department of mechanics which treats of the properties and laws of bodies *at* rest.

6. *Dynamics*, from a Greek word signifying *power* or *force*, treats of the properties and laws of bodies *in motion*.

7. *Pneumatics* treats of the mechanical properties and effects of air and similar fluids, called elastic fluids or gases.

8. *Hydrostatics* treats of the gravity and pressure of fluids in a state of rest.

9. *Hydraulics* treats of fluids in motion, and of the instruments and machines by which their motion is guided or controlled.

10. *Acoustics* treats of the laws of sound.

11. *Pyronomics* treats of the laws and effects of heat.

12. *Optics* treats of light, color and vision.

13. *Electricity* treats of an exceedingly subtle agent, called the electric fluid.

14. *Galvanism* (sometimes called chemical electricity) is a branch of Electricity.

15. *Magnetism* treats of the properties and effects of the magnet or loadstone.

16. *Electro-Magnetism* treats of magnetism induced by electricity.

17. *Magneto-Electricity* treats of electricity induced by magnetism.

18. *Astronomy* treats of the heavenly bodies,— the sun, moon, stars, planets, comets.

19. The agents whose effects or operations are described in Natural Philosophy are divided into two classes, called respectively Ponderable and Imponderable Agents.

NOTE.—Some writers on Philosophy have suggested a different classification, into Bodies and Agents, calling bodies *ponderable*, and agents *imponderable*.

20. Ponderable agents are those which have weight, as water, air, steam.

21. Imponderable agents are those which have no weight such as light, heat, magnetism and electricity.

OF MATTER AND ITS PROPERTIES.

What is Matter?   22. MATTER.— Matter is the general name of everything that occupies space.

23. Matter exists in four different states or forms, namely, in the solid, liquid, gaseous and vesicular forms.

24. Matter exists in a solid form when the particles of which it is composed cohere together. The different degrees of cohesion which different bodies possess causes them to assume different degrees of hardness.

25. Matter exists in a liquid state when the component parts do not cohere with sufficient force to prevent their separation by the mere influence of their weight. The surface of a fluid at rest always conforms itself to the shape of the portion of the earth's surface over which it stands.

26. Matter exists in a gaseous or aëriform state when the particles of which it is composed have a repulsion towards each other which causes them to separate with a power of expansion to which there is no known limit. Of this, smoke presents a familiar instance. As it ascends it expands, the particles repelling each other until they become wholly invisible.

NOTE.— The word aëriform means, *in the form of air*.

27. The vesicular form of matter is the form in which we see it in clouds. It consists of very minute vesicles, resembling bubbles, and it is the state into which many vapors pass before they assume a fluid condition.

28. Some substances are capable, under certain conditions, of assuming all these different forms. Water, for instance, is solid in the form of ice, fluid **as** water, in the gaseous state when converted into steam, and vesicular in the form of clouds.

29. All matter, whether in the solid, liquid, gaseous, or vesicular form, is either simple or compound in its nature. But this consideration of matter pertains more properly to the science of chemistry. It is proper, however, here to explain what is meant by a simple or homogeneous and a compound or heterogeneous substance.

30. All matter is composed of very minute particles or atoms, united together by different degrees of cohesion. When all the atoms are of the same kind, the body is a simple or homogeneous substance. Thus, for instance, pure iron, pure gold, &c., consists of very minute particles or atoms, all of which are pure iron or pure gold. But water, and many other substances, are compound substances, composed of atoms of two or more different substances, combined by chemical affinity.

NOTE.— The ancient philosophers supposed that all material substances were composed of Fire, Air, Earth and Water, and these four substances were called the four elements, because they were supposed to be the simple substances of which all things are composed. But modern science has shown that not one of these is a simple substance. Water, for instance, is composed of two invisible gases, called Hydrogen and Oxygen, united in the proportion of one part, *in weight*, of hydrogen to eight of oxygen; or, by measure, one part of oxygen to two of hydrogen. In like manner air, or, rather, what the ancients understood by air, is composed of oxygen united with another invisible gas, called nitrogen or *azote*, in the proportion of seventy-two parts of the latter to twenty-eight of the former.

The enumeration of the elementary substances, which, either by themselves or in union with one another, make up the material world, properly belongs to the science of chemistry. As this work may fall into the hands of some who will not find the information elsewhere, a list of the simple substances or elements is here presented, so far as modern science has investigated them. They are sixty-one in number, forty-nine of which are metallic and twelve are non-metallic.

The forty-nine metals are

Gold,	Manganese,	Potassium,	*Didynium,*
Silver,	Cadmium,	Sodium,	*Erbium,*
Iron,	Uranium,	Lithium,	*Terbium,*
Copper,	Palladium,	Barium,	*Ruthenium,*
Tin,	Rhodium,	Strontium,	*Pelopium,*
Mercury,	Iridium,	Calcium,	*Niobium.*
Lead,	Osmium,	Magnesium,	Selenium.
Zinc,	Titanium,	Aluminum,	[This substance is of questionable nature, some of its properties indicating a metallic and some a non-metallic character.]
Nickel,	Columbium,	Glucinum,	
Cobalt,	Tellurium,	Yttrium,	
Bismuth,	Tungsten,	Zirconium,	
Platinum,	Molybdenum,	Thorium,	
Antimony,	Vanadium,	Cerium,	[The last seven, in Italic, have not yet been fully investigated.]
Arsenic,	Chromium,	Lantanium,	

The non-metallic elements are

Oxygen,	Sulphur,	Chlorine,	Fluorine,
Hydrogen,	Phosphorus,	Bromine,	Borax,
Nitrogen,	Carbon,	Iodine,	Silica.

Of the elementary substances now enumerated, about fourteen constitute the great mass of our earth and its atmosphere. The remainder occur only in comparatively small quantities, while nearly a third of the whole number is so rare that their uses in the great economy of nature are not understood, nor have they as yet admitted of any useful application.

The science of Geology reveals to us the fact that *granite* appears to be the foundation of the crust of the earth; and in the granite, either in its original formation, or in veins or seams which have been thrown up by subterranean forces into the granite, all of the elementary substances which have been enumerated are to be found. A chart is presented below in which the materials composing the strata of the crust of the earth are enumerated, together with a tabular view of the composition of these materials. It is not contended that this chart is perfectly accurate in all its details; but, as it affords an interesting and extensive subject of investigation, and as it is not to be found elsewhere in print, it is thought that it will be well worth the space which it occupies, although a rigid classification would exclude it from this work.

Dr. Boynton's Chart of Materials that enter into the Composition of Granite.

	Silica	Alumina	Potash	Soda	Lime	Magnesia	Ox. Iron	Ox. Manganese	Water	Carb. Acid
Quartz	100									
Feldspar	65	19	14		1		1			
Albite	70	20		10						
Mica	46	26	10		1	5	8	1	2	2 Fluor. Acid.
							Prot. Ox.			
Hornblende	48	12			14	19	7			
Augite	54	1			24	17	4			
Diallage	47	4			13	25	8		3	
							M.			
Chlorite	27	18	2			15	31		7	
							Prot.			
Talc	57	1			4	27	8		3	
Hypersthene	56	2			2	14	25		1	
Actynolite	56	2			12	13	17			
							M.			
Steatite	62	1			1	28	2		6	
Serpentine	42				5	33	7		13	
Schorl	36	36	1	2		5	14	2		4 B. Acid.
							Prot.			
Garnet	40	20			1		36	3		
							Prot.	Prot.		
M. Garnet	36	18					15	31		
Clay	75	10			5	2	3			
							Prot.			
Green Sand	48	7	8				26		11	
Carbonate of Lime					56					44 Carb. Acid.
Carbonate of Magnesia						48			2	50 " "

What are the essential Properties of Matter?

31. There are seven essential* properties belonging to matter, namely, 1. Impenetrability; 2. Extension; 3. Figure; 4. Divisibility; 5. Indestructibility; 6. Inertia; 7. Attraction.

What is Impenetrability?

32. IMPENETRABILITY.—Impenetrability is the power of occupying a certain portion of space, so

---

* An *essential* property of a body is that which is necessary to the absolute existence of the body. All matter in common possesses these essential properties, and no particle of matter can exist without any one of them. Different bodies possess other different properties which are not essential to their existence, such as color, weight, brittleness, hardness, &c. These are called accidental properties, as they depend on circumstances not essential to the very existence of a body.

that where one body is another cannot be without displacing it.

33. This property, Impenetrability, belongs to all bodies and forms of matter, whether solid, fluid, gaseous, or vesicular.

The impenetrability of common air may be shown by immersing an inverted tumbler in a vessel of water. The air prevents the water from rising into the tumbler. An empty bottle, also, forcibly held horizontally under the water, will exhibit the same property; for the bottle, apparently empty, is filled with air, which escapes in bubbles from the bottle as the water enters it. But, if the bottle be inverted, the water cannot enter the bottle, on account of the impenetrability of the air within.*

---

* This circumstance explains the reason why water, or any other liquid, poured into a tunnel closely inserted in the mouth of a decanter, will run over the sides of the decanter. The air filling the decanter, and having no means of escape, prevents the fluid from entering the decanter; but, if the tunnel be lifted from the decanter but a little, so as to afford the air an opportunity to escape, the water will then flow into the decanter in an uninterrupted stream.

When a nail is driven into wood or any other substances, it forces the particles asunder and makes its way between them.

An experiment was made at Florence, many years ago, to show the impenetrability of water. A hollow globe of gold was filled with water and subjected to great pressure. The water, having no other means of escape, was seen to exude from the pores of the gold.

The reason why fluids appear less impenetrable than solids is that the particles which compose the fluids move easily among themselves, on account of their slight degree of cohesion, and when any pressure is exerted upon a fluid the particles move readily into the unoccupied space to which they have access. But, if the fluid be surrounded on all sides, and have no means of escape, it will be found to possess the property of impenetrability in no less a degree than solid bodies.

A well-known fact seems, at first view, to be at variance with this statement. When a vessel is filled to the brim with water or other fluid, a considerable portion of salt may be dropped into the fluid without causing the vessel to overflow. And, when salt has been added until the water can hold no more in solution, a considerable quantity of sugar can be added in a similar manner. The explanation of this familiar fact is as follows. The particles of the sugar are smaller than the particles of the salt, and the particles of the salt are smaller than the particles which compose the water. Now, supposing all of these particles to be globular, they will arrange themselves as is represented in Fig. 1, in which the particles of the water are indicated by the largest circles, those of the salt by the next in size, and those of the sugar by the smallest.

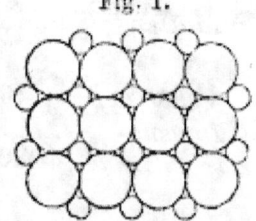

Fig. 1.

Familiar Experiment.— Fill a bowl or tumbler with peas, then pour on the peas mustard-seed or fine grain, shaking the vessel to cause it to fill the vacant spaces between the peas. In like manner add, successively, fine sand, water, salt and sugar. This will afford an illustration of the apparent paradox of two bodies occupying the same space, and show that it is only *apparent*.

OF MATTER AND ITS PROPERTIES.

What is Extension? 34. EXTENSION.— Extension is but another name for bulk or size, and it is expressed by the terms length, breadth or width, height, depth and thickness.

NOTE.— Length is the extent from end to end. Breadth or width is the extent from side to side. Height, depth or thickness, is the extent from the top to the bottom. The measure of a body from the bottom to the top is called height; from the top to the bottom, is called depth. Thus we speak of the depth of a well, the height of a house, &c.

What is Figure? 35. Figure is the form or shape of a body.

36. Figure and Extension are separate properties, although both may be represented by the same terms, length, breadth, &c. But they differ as the words shape and size differ. Two bodies may be of the same figure or shape, but of vastly different size. A grape and an orange resemble each other in shape, but differ widely in size. The limits of extension constitute figure, but figure has no other connexion with extension.

What is Divisibility? 37. DIVISIBILITY.— Divisibility is susceptibility of being divided.

38. To the divisibility of matter there is no known limit, nor can we conceive of anything so small that it is not made up of two halves or four quarters. It is indeed true that our senses are quite limited in their operation, and that we cannot perceive or take cognizance, by means of our senses, of many objects of the existence of which we are convinced without their immediate and direct testimony.

39. Sir Isaac Newton has shown that the thickest part of a soap-bubble does not exceed the two-millionth part of an inch.

40. The microscopic observations of Ehrenberg have proved that there are many species of little creatures, called *Infusoria*, so small that millions of them collected into a single mass would not exceed the bulk of a grain of sand, and thousands of them might swim side by side through the eye of a small needle.

41. In the slate formations in Bohemia these little creatures are found in a fossil state, so small that it would require a hundred and eighty-seven millions of them to weigh a single grain.

42. A single thread of the spider's web has been found to be composed of six thousand filaments.

43. A single grain of gold may be hammered by a gold-beater until it will cover fifty square inches; each square inch may be divided into two hundred strips; and each strip into two hundred parts. One of these parts is only *one two-millionth* part of a grain of gold, and yet it may be seen with the naked eye.

44. The particles which escape from odoriferous objects also afford instances of extreme divisibility.

What is Indestructibility?

45. INDESTRUCTIBILITY. — By the Indestructibility of matter is meant that it cannot be destroyed.

46. A body may be indefinitely divided or altered in its form, color, and other unessential properties, but it can never be destroyed by man. It must continue to exist in some form, with all its essential properties, through all its changes of external appearance. HE alone " who can create can destroy."

47. When water disappears, either by boiling over a fire or by evaporation under the heat of the sun, it is not destroyed, but merely changed from a liquid to a fluid form, and becomes steam or vapor. Some of its *unessential* properties are altered, but its *essential* properties remain the same, under all the changes which it undergoes. In the form of water it has no elasticity * and but a limited degree of compressibility.* But when "it dries up" (as it is called) it rises in the form of steam or vapor, and expands to such a degree as to become invisible. It then assumes other properties, not possessed before (such as elasticity and expansibility); it ascends in the air and forms clouds; these clouds, affected by the temperature of the air and other agents, again fall to the earth in the form of rain, hail, snow or sleet, and form springs, fountains, rivers, &c. The water on or in the earth, therefore, is constantly changing its shape or situation, but no particle of it is ever actually destroyed.

48. Substances used as fuel, whether in the form of wood, coal, or other materials, in like manner undergo many changes by the process of combustion. Parts of them rise in the form of smoke, part ascends in vapor, while the remainder is reduced to the form of ashes; but no part is absolutely destroyed. Combustion merely disunites the simple substances of which the burning materials are composed, forming them into new combinations; but every part still continues in existence, and retains all the *essential* † properties of bodies.

What is Inertia?

49. INERTIA. — Inertia ‡ is the resistance of matter to a change of state, whether of motion or of rest.

---

* Late writers assert that water *has* a slight degree both of elasticity and expansibility.

† The reader will be careful to carry in his mind what is meant by the term *an essential property*. It is explained in the note to No. 31, page 21.

‡ The literal meaning of *inertia* is inactivity, and implies *inability to change a state of rest or of motion*. A clear and distinct understanding of this property of all matter is essential in all the departments of material philosophy. All matter, mechanically considered, must be in a state either

50. A body at rest cannot put itself in motion, nor can a body in motion stop itself. This incapacity to change its state from rest to motion, or from motion to a state of rest, is what is implied by the term *inertia*.

51. It follows, therefore, from what has just been stated, that when a body is in motion its *inertia* will cause it to continue to move until its motion is destroyed by some other force.

52. There are two forces constantly exerted around us which tend to destroy motion, namely, *gravity* and *the resistance of the air*. All motion caused by animal or mechanical power is affected by these two forces. Gravity (which will presently be explained) causes all bodies, *whether in motion or at rest*, to tend towards the centre of the earth, and the air presents a resistance to all bodies moving in it. Could these and all other *direct obstacles* to motion be set aside, a body when once put in motion would always remain in motion, and a body at rest, unaffected by any external force, would always remain at rest.*

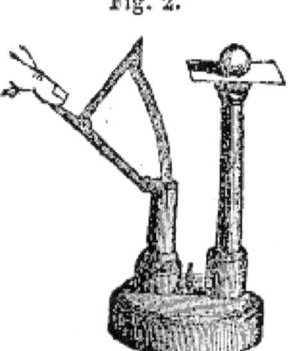

Fig. 2.

53. *Experiment to illustrate Inertia.*—Fig. 2 represents the simple apparatus of Mr. Wightman for illustrating the inertia of a body. A card is placed on the top of a stand, and a ball is balanced on the card. A quick motion is of motion or rest; and, in whatever state it may be, it must remain in that state until a change is effected by some efficient cause, independent of the body itself. A body placed upon another body in motion partakes of the motion of the body on which it is placed. But, if that body be suddenly stopped, the superincumbent body will not stop at the same time, unless it be securely fastened. Thus, if a horse moving at a rapid rate be suddenly stopped, the rider will be thrown forward, on account of this *inertia* of his body, unless by extra exertion he secures himself on the saddle by bracing his feet on the stirrups. On the contrary, if the horse, from a state of rest, start suddenly forward, the rider will be thrown backwards. For the same reason, when a person jumps from a vehicle in motion to the ground, his body, partaking of the motion of the vehicle, cannot be suddenly brought to a state of rest by his feet resting on the ground, but will be thrown forward in the direction of the motion which it has acquired from the vehicle. This is the reason that so many accidents happen from leaping from a vehicle in motion.

* In the absence of all positive proof from the things around us of the statement just made, we may find from the truths which astronomy teaches that inertia is one of the necessary properties of all matter. The heavenly bodies, launched by the hand of their Creator into the fields of infinite space, with no opposing force but gravity alone, have performed their stated revolutions in perfect consistency with the character which this property gives them; and all the calculations which have been made with respect to them, verified as they have repeatedly been by observation, have been predicated on their possession of this necessary property of all matter.

then given to the card by means of a spring, and the card flies off, leaving the ball on the top of the stand.*

54. Nature seems to have engrafted some knowledge of mechanical laws on the instinct of animals. When an animal, and especially a large animal, is in rapid motion, he cannot (on account of the inertia of his body) suddenly stop his motion, or change its direction; and the larger the animal the more difficult does a sudden stoppage become. The hare pursued by the hound often escapes, when the dog is nearly upon him, by a sudden turn, or changing the direction of its flight, thus gaining time upon his pursuer, whose *inertia* is not so readily overcome, and who is thus impelled forward beyond the spot where the hare turned.

55. Children at play are in the same manner enabled "*to dodge*" their elder playmates, and the activity of a boy will often enable him to escape the pursuit of a man.

56. It is the effect of inertia to render us sensible to motion. A person in motion would be quite unconscious of that state, were it not for the obstacles which have a tendency to impede his progress. In a boat on smooth water, motion is perceptible only by the apparent change in the position of surrounding objects; but, if the course of the boat be interrupted by running aground, or striking against a rock, the person in the boat would feel the shock caused by the sudden change from a state of motion to a state of rest, and, unless secured to his seat in the boat, he would be precipitated forward

What is Attraction?
57. ATTRACTION.—Attraction is the tendency which different bodies or portions of matter have to approach or to adhere to each other.

What is the law of Attraction?
58. Every portion of matter is attracted by every other portion of matter, and this attraction is the stronger in proportion to the quantity and the distance. The larger the quantity and the less the distance, the stronger is the attraction.†

---

* The ball remains on the pillar in this case not solely from its *inertia*, but because sufficient motion is not communicated to the ball by the friction of the card to counteract the effect of gravity on the ball. If the ball, therefore, be not accurately balanced on the card, the experiment will not be successful, because the card cannot move without communicating at least a portion of its motion to the ball.

† [N. B. This subject will be more fully treated under the head of *Gravity*. See page 33.]

How many kinds of Attraction are there? 59. There are two kinds of attraction, namely, the Attraction of Gravitation and the Attraction of Cohesion.

The former belongs to all matter, whatever its form; the latter appears to belong principally to solid bodies.

What is the Attraction of Gravitation? 60. The Attraction of Gravitation is the reciprocal attraction of separate portions of matter.

What is the Attraction of Cohesion? 61. The Attraction of Cohesion is that which causes the particles of a body to cohere together. [See No. 31.]

62. The attraction of cohesion appears to exist but in a very slight degree, if at all, in liquids and fluids.

Exemplify the two kinds of Attraction; namely, Gravity and Cohesive Attraction? 63. The attraction of gravitation causes a body, when unsupported, to fall to the ground. The attraction of cohesion holds together the particles of a body, and causes them to unite in masses.*

64. Having described the *essential* properties of bodies, we come now to the consideration of other properties belonging respectively to different kinds of matter; such as Porosity, Density, Rarity, Compressibility, Expansibility, Mobility, Elasticity, Brittleness, Flexibility, Malleability, Ductility, Tenacity.

65. It has already been stated that matter consists of minute particles or atoms, united by different degrees of cohesive attraction. These atoms are probably of different shapes in different bodies, and the different degrees of compactness with which they unite give rise to certain qualities, which differ greatly in different substances. These qualities or properties are described under the names of Porosity, Density and Rarity, which will presently be described.

---

* Besides these two kinds of attraction, there seem to be other kinds of attractive force, active in vegetation and in animal life, known by the names of *Endosmose* and *Exosmose*, terms applied to the transmission of gaseous matter or vapors through membranous substances. See note to Capillary Attraction, under the head of *Hydrostatics*, on page 112.

Other kinds of attraction, called Electrical and Magnetical Attraction, will hereafter be considered under their appropriate head. The subject of Chemical Attraction or Affinity belongs distinctly to the subject of Chemistry and will not, therefore, be considered in this work

66. Besides the property of attraction possessed by the particles or atoms of which a body is composed, there seems to be another property, of a nature directly opposite to attraction, which exerts itself with a *repulsive* force, to prevent a closer approximation of the particles than that which by the law of their nature they assume. This property is called *repulsion*. This repulsion prevents the particles or atoms from coming into perfect contact, so that there must be small spaces between them, where they do not absolutely touch one another. [*See Figure* 1*st.*] These spaces are called *pores*, and where they exist give rise to that property or quality described under the name of *Porosity*.

What is Porosity? 67. POROSITY. — Porosity implies, therefore, that there are spaces, or *pores*, between the particles or atoms which form the mass of a body.

68. DENSITY. — When the pores are few, so that a large number of particles unite in a small mass, the body is called *a dense body*.

What is Density? 69. Density, therefore, implies the closeness or compactness of the particles which compose any substance.

70. RARITY. — When the pores in any substance are numerous so that the particles which form it touch one another in only a few points, the body is called *a rare body*.

What is Rarity? 71. Rarity, therefore, is the reverse of density, and implies extension of bulk without increase in the quantity of matter.

72. From what has now been stated it appears [*See No.* 67] that the particles of a body are connected together by a system of attractions and repulsions which give rise to distinctions which have already been described. It remains to be stated that these *attractions* and *repulsions* differ much in degree in different substances, and this difference gives rise to other properties, which will now be explained, under their appropriate names.

73. COMPRESSIBILITY. — When the repulsion of the particles of any substance can be overcome and the mass can be reduced within narrower limits of extension, it is said to possess the property of Compressibility.*

---

* Compressibility differs from *Contractibility* rather in cause than in effect. Contractibility implies a change of bulk caused by change of temperature, or any other agency *not mechanical*. Compressibility implies that the diminution of bulk is caused by some external mechanical force.

What is Compressibility? 74. Compressibility, therefore, may be defined, the susceptibility of a reduction of the limits of extension.

75. This property is possessed by all known substances, but in very different degrees,—some substances requiring but little force to compress them, others resisting very great forces; but it is not known that there is any substance unsusceptible of compression, if a sufficient force be applied.*

76. Liquids in general are less easily compressed than solid bodies; so much so, indeed, that in practical science they are generally considered as incompressible. Under a very considerable mechanical force, a slight degree of compression has been observed.†

77. EXPANSIBILITY.—The system of attractions and repulsions among the particles of a body are sometimes so equally balanced that they exist, as it were, in an equilibrium. In other cases the repulsive energy is so great as to predominate when the attractive force is unaided. When the repulsive energy is permitted to act without restraint, it forces the particles asunder and increases the limits of extension, giving rise to another property of matter possessed by many bodies, but in an eminent degree by matter in a gaseous form. This property is called *Expansibility*.

What is Expansibility? 78. Expansibility,‡ therefore, may be defined as that property of matter by which it is enabled to increase its limits of extension.

79. ELASTICITY.—When the atoms or particles which constitute a body are so balanced by a system of attractions and repulsions that they resist any force which tends to change the figure of the

---

* Sir Isaac Newton conjectured that if the earth were so compressed as to be absolutely without pores, its dimensions might not exceed a cubic inch.

† Under a pressure of fifteen pounds on a square inch, water has been diminished in bulk only by about forty-nine parts in a million. Under a pressure of fifteen thousand pounds on a square inch, it was compressed by about one-twentieth of its volume. The experiment was tried in a cannon, and the cannon was burst.

‡ Expansibility and Dilatability are but different names for the same property; but *expansion* implies an augmentation of the bulk or volume, dependent on mechanical agency, while dilatation expresses the same condition produced by some physical cause not properly falling under the denomination of mechanical force. Thus heat dilates most substances, while cold contracts them. It is on this principle that the thermometer is constructed. [See page 149, No. 546.]

All gaseous bodies are invested with the property of *dilatability* to an unlimited degree, by means of which, when unrestrained, they will expand spontaneously, and that without the application of any external agency, to a degree to which there is no known limit.

body, they will possess another property, known by the name of *Elasticity*.

What is Elasticity? 80. Elasticity, therefore is the property which causes a body to resume its shape after it has been compressed or expanded.*

81. Thus, when a bow or a steel spring is bent, its elasticity causes it to resume its shape.

82. India rubber (or caoutchouc) possesses the property of elasticity in a remarkable degree, but steam and other bodies in a gaseous form in a still greater.†

83. Ivory is endowed with the property of elasticity in a remarkable degree, but exhibits it not so much by the mere force of pressure, but it requires the force of impact to produce change of form.‡

What is Brittleness? 84. BRITTLENESS.— Brittleness implies aptness to break into irregular fragments.§

---

* This property is possessed, in at least some small degree, by all substances; or, at least, it cannot be said that any substance is wholly destitute of elasticity. Even water and other liquids, which yield with difficulty to compression, recover their volume with a force apparently equal to the compressing force. But, for most practical purposes, many substances, such as putty, wet paste, moist paper, clay, and similar bodies, afford examples of substances possessing the property of elasticity in so slight a degree that they are treated as *non-elastic* bodies.

† The gases or aëriform bodies afford the most remarkable instances of elasticity. When water is converted into steam it occupies a space seventeen hundred times greater than the water from which it is formed, and its elasticity causes it to expand to still larger dimensions on the application of heat. It is this peculiar property of steam, modified, as will be explained in a future part of this work, which is the foundation of its application in the movement of machinery. All gaseous bodies are equally elastic.

‡ The metals which are best adapted to produce sound are those which are most highly elastic. It sometimes happens that two metals, neither of which have any great degree of hardness or elasticity, when combined in certain proportions, will acquire both of these properties. Thus tin and copper, neither of which in a pure state is hard or elastic, when mixed in a certain proportion, produce a compound so hard and elastic that it is eminent for its sonorous property, and is used for making bells, &c

§ Brittleness and hardness are properties which frequently accompany each other, and brittleness is not inconsistent with elasticity. Thus glass, for instance, which is the most brittle of all known substances, is highly elastic. The same body may acquire or be divested of its brittleness according to the treatment which it receives. Thus iron, and some other metals, when heated and suddenly plunged into cold water, become brittle; but if, in a heated state, they are buried in hot sand, and thus be

What is Flexibility? 85. FLEXIBILITY. — Flexibility implies a disposition to yield without breaking when bent.

What is Malleability? 86. MALLEABILITY. — Malleability implies that property by means of which a body may be reduced to the form of thin plates by means of the hammer or the pressure of rollers.

87. This property is possessed in an eminent degree by some of the metals, especially gold, silver, iron and copper, and it is of vast importance in the arts. A knowledge of the uses of iron, and of its malleability, is one of the first steps from a savage to a civilized state of life.

88. The most malleable of the metals is gold, which may be hammered to such a degree of thinness as to require three hundred and sixty thousand leaves to equal an inch in thickness.*

89. DUCTILITY. — Some substances admit of being extended simultaneously both in length and breadth. Others can be extended to a greater degree in length alone; and this property gives rise to another name, called *Ductility*.

What is Ductility? 90. DUCTILITY. — Ductility is that property which renders a substance susceptible of being drawn out into wire.

91. The same metals are not always both *ductile* and malleable to the same degree. Thus iron may be beaten into any form, when heated, but not into very thin plates, but it can be drawn into extremely fine wire. Tin and lead, on the contrary, cannot be drawn out into small wire, but they are susceptible of being beaten into extremely thin leaves.

92. Gold and platinum have a high degree both of ductility and malleability. Gold can be beaten (as has already been stated) into

---

permitted to cool very gradually, they will lose their brittleness and acquire the opposite quality of flexibility. This process in the arts is called *annealing*.

* The malleability of the metals varies according to their temperature. Iron is most malleable at a white heat. Zinc becomes malleable at the temperature of 300° or 400°. Some of the metals, and especially antimony, arsenic, bismuth and cobalt, possess scarcely any degree of this property.

The familiar process of *welding* is dependent on malleability. The two pieces of metal to be welded are first heated to that temperature at which they are most malleable, and, the ends being placed together, the particles are driven into such intimate connexion by the welding-hammer that they cohere. Different metals may in some cases be thus welded together.

leaves so thin that it would require many thousands of them to equal an inch in thickness. It has also been drawn into wire so attenuated that one hundred and eighty yards of it would not weigh more than a single grain. An ounce of such wire would be more than fifty miles in length. But platinum can be drawn even to a finer wire than this.

What is Tenacity? 93. TENACITY.—Tenacity implies the adhesion of the particles of a body.

94. Tenacity is one of the great elements of strength. It is the absence of tenacity which constitutes brittleness. Both imply strength, but in different forms. Thus glass, the most brittle of all substances, has a great degree of tenacity. A slender rod of glass, which cannot resist the slightest lateral pressure, if suspended vertically by one end will sustain a very considerable weight at the other end.*

---

* A knowledge of the tenacity of different substances is of great use in the arts. The tenacity of metals and other substances has been ascertained by suspending weights from wires of the metals, or rods and cords of different materials.

The following table presents very nearly the weights sustained by wires of different metals, each having the diameter of about one-twelfth of an inch.

Lead,	27 pounds.	Silver,	187 pounds.
Tin,	34 "	Platinum,	274 "
Zinc,	109 "	Copper,	302 "
Gold,	150 "	Iron,	549 "

Cords of different materials, but of the same diameter, sustained the following weights:

Common flax,	1175 pounds.	New Zealand flax,	2380 pounds
Hemp,	1633 "	Silk,	3400 "

The following table presents a more extended list of materials. The area of a transverse section of the rods on which the experiment was tried was one square inch.

	Pounds Avoirdupois.		Pounds Avoirdupois.
English Oak,	8,000 to 12,000	Tin,	7,129
Fir,	11,000	Lead,	3,146
Beech,	11,500	Rope, 1 inch in circumference,	1,000
Mahogany,	8,000		
Teak,	15,000	Whale line, 2 inches in circumference, spun by hand,	2,240
Cast Steel,	134,256		
Iron Wire,	93,964		
Swedish Bar-iron,	72,064	Do., by machinery,	3,520
Cast-iron,	18,656	Rope, 3 inches in circumference,	5,628
Wrought Copper,	33,792		
Platinum Wire,	52,987	Do., 4 inches,	9,988
Silver Wire,	38,257	Cable, 14¼ inches,	89,600
Gold,	30,888	Do., 23 inches,	255,360
Zinc,	22,551		

A more particular account of the tenacity of various substances will be

95. The tenacity of metals is much increased by uniting them. A compound consisting of five parts of gold and one of copper has a tenacity of more than double that of the gold or copper alone; and brass, which is composed of copper and zinc, has a tenacity more than double that of the copper, and nearly twenty times as great as that of the zinc alone. A mixture of three parts of tin and one of lead has a tenacity more than double that of the tin; and a mixture of eight parts of lead and one of zinc has a tenacity nearly double that of the zinc, and nearly five times that of the lead alone.*

96. GRAVITY.—It has already been stated that matter in all its forms, whether solid, fluid or gaseous, possesses the property of *attraction*. This property, with its laws, is now to be particularly considered, under the name of Gravity.

What is Gravity?   97. Gravity is the reciprocal attraction of separate portions of matter.

With what force do all bodies attract each other?   All bodies attract each other with a force proportionate to their size, density and distance from each other. [*See No.* 59.]

98. This law explains the reason why a body which is not supported falls to the earth. Two bodies existing in any portion of space mutually attract each other, and would rush together were they not prevented by some superior force. Let us suppose, for instance, that two balls made of the same materials, but one weighing 11 pounds and the other weighing only one pound, were ten feet apart, but both were a hundred feet above the surface of the earth. According to this law, the two balls would rush together, the lighter ball passing over nine feet of the distance, and the heavier ball over one foot; and this they would do, were they not both prevented by a superior force. *That superior force is the earth, which, being a much larger body, attracts them both with a superior force.* This superior force they will both obey, and both will therefore fall to the earth. As the attraction of the earth and of the balls is mutual, the earth will also move towards the balls while the balls are falling to the earth; but the size of the earth is so much greater than that of the balls, that the distance that the earth would move towards the balls would be too small to be appreciated.†

---

found in Barlow's Essay on the Strength of Timber, Rennie's Treatise (in *Phil. Trans.* 1818), Tredgold's Principles of Carpentry, and the 4th vol. of Manchester Memoirs, by Mr. Hodgkinson.

* There are many other specific properties of bodies besides those that have now been enumerated, the consideration of which belongs to the science of Chemistry.

† The earth is one quatrillion, that is, one thousand million millions times larger than the largest body which has ever been known to fall

99. The attraction of the earth is the cause of what we call weight. When we say that a body weighs an ounce, a pound, or a ton, we express by these terms the degree of attraction by which it is drawn towards the earth. Therefore,

What is Weight? 100. Weight is the measure of the earth's attraction.*

101. As this attraction depends upon the quantity of matter which a body contains, it follows that

What bodies have the greatest weight? Those bodies will have the greatest weight which contain the greatest quantity of matter.†

102. TERRESTRIAL GRAVITY. — It has already been stated [see No. 97] that the attraction which one mass of matter has for another is in proportion to the quantity and the distance; and that, the larger the quantity of matter and the less its distance, the stronger will be the attraction. The law of this attraction may be stated as follows:

What is the law of attraction? 103. Every portion of matter attracts every other portion of matter with a force proportioned directly to the quantity, and inversely as the square of the distance.

through our atmosphere. Supposing, then, that such a body should fall through a distance of one thousand feet, the earth would rise no more than the hundred billionth part of an inch, a distance altogether imperceptible to our senses.

The principle of mutual attraction is not confined to the earth. It extends to the sun, the planets, comets and stars. The earth attracts each of them, and each of them attracts the earth, and these mutual attractions are so nicely balanced by the power of God as to cause the regular motions of all the heavenly bodies, the diversity of the seasons, the succession of day and night, summer and winter, and all the grand operations which are described in astronomy.

* When we say that a body weighs an ounce, a pound, or a hundred pounds, we express, by these terms, the degree of attraction by which it is drawn towards the earth.

† The weight of a body is not dependent solely on its size or bulk; its density must also be considered. If we take an equal quantity, by measure, of two substances, — lead and cork, for instance, — we shall find that, although both are of the same size, the lead will weigh much more than the cork. The cork is more porous than the lead, and, consequently, the particles of which it is composed must be further apart, and therefore there must be fewer of them within a given bulk; while, in the lead, the pores are much smaller, and the particles will, therefore, be crowded into a much smaller space

OF GRAVITY.

104. Let us now apply this law to terrestrial gravity — that is, to the earth's attraction; and, for that purpose, let us suppose four balls, of the same size and density, to be placed respectively as follows, namely:

The first at the centre of the earth.

The second on the surface of the earth.

The third above the earth's surface, at twice the distance of the surface from the centre (*that distance being four thousand miles*).

The fourth to be half way between the surface and the centre.

To ascertain the attractive force of the earth on each of these balls, we reason thus:

The first ball (*at the centre*) will be surrounded on all sides by an equal quantity of matter, and it will remain at rest.

The second ball will be attracted downwards to the centre by the whole mass below it.

The third ball, being at twice the distance from the surface (gravity *decreasing* as the *square* of the distance *increases*), will be attracted by a force equal to only one-fourth of that at the surface.

The fourth ball, being attracted *downwards* by that portion of the earth which is below it, and *upwards* by that portion which is above it, will be influenced only by the *difference* between these two opposite attractions; and, as the downward attraction is twice as great as the upward, the downward attraction will prevail with half its original force, the other half being balanced by the upward attraction.

105. As *weight is the measure of the earth's attraction*, we may represent this principle by the weight of the balls, as follows (*supposing the weight of each ball, at the surface of the earth, to be one pound*):

The first ball will weigh nothing.

The second will weigh one pound.

The third will weigh one-quarter of a pound.

The fourth will weigh one-half of a pound.

The law of terrestrial gravity, then, may be stated as follows:

What is the law of Terrestrial Gravity?

106. The force of gravity is greatest at the surface of the earth, and it decreases upwards as the square of the distance from the centre increases, and downwards simply as the distance from the centre decreases.

According to the principles just stated, a body which at the surface of the earth weighs a pound at the centre of the earth will weigh nothing.

1000 miles from the centre it will weigh						¼ of a pound.
2000 " " " " " "						½ of a pound.
3000 " " " " " "						¾ of a pound.
4000 " " " " " "						1 pound.

8000 miles from the centre it will weigh ¼ of a **pound**.
12000 "   "   "   "   "   " $\frac{1}{9}$.
16000 "   "   "   "   "   " $\frac{1}{16}$.
20000 "   "   "   "   "   " $\frac{1}{25}$.
24000 "   "   "   "   "   " $\frac{1}{36}$.
28000 "   "   "   "   "   " $\frac{1}{49}$.
32000 "   "   "   "   "   " $\frac{1}{64}$.

If the principles that have now been stated have been understood, the solution of the following questions will not be difficult.

107. *Questions for Solution.*

[N. B. We use the term *weight* in these questions in its philosophical sense, as "*the measure of the earth's attraction at the surface.*"]

(1.) Suppose that a body weighing 800 pounds could be sunk 500 miles deep into the earth,—what would it weigh?

Solution. 500 miles is ⅛ of 4000 miles; and, as the distance from the centre is decreased by ⅛, its weight would also be decreased in the same proportion, and the body would weigh 700 pounds.

(2.) Suppose a body weighing 2 tons were sunk one mile below the surface of the earth, what would it weigh? *Ans.* 1.9995 T.

(3.) If a load of coal weighs six tons at the surface of the earth, what would it weigh in the mine from which it was taken, supposing the mine were at a perpendicular distance of half a mile from the surface? *Ans.* 5.99925 T.

(4.) If the fossil bones of an animal dug from a depth of 5228 feet from the surface, weigh four tons, what would be their weight at the depth where they were exhumed? *Ans.* 3 T. 19 cwt. 98 lb. +

(5.) If a cubic yard of lead weigh 12 tons at the surface of the earth, what would it weigh at the distance of 1000 miles from the centre? *Ans.* 3 T.

(6.) If a body on the surface of the earth weigh 4 tons, what would be its weight if it were elevated a thousand miles above the surface?

Solution. Square the two distances 4000 and 5000, &c.

	Tons.	cwt.	qrs.	lbs.
Answer.	2	11	0	20.

(7.) Which will weigh the most, a body of 3000 tons at the distance of 4 millions of miles from the earth, or a body of 4000 tons at the distance of 3 millions of miles? *Ans.* .003 T. and .007 T. +

(8.) How far above the surface of the earth must a pound weight be carried to make it weigh one ounce avoirdupois? *Ans.* 12000 mi.

(9.) If a body weigh 2 tons when at the distance of a thousand miles above the surface of the earth, what would it weigh at the surface? *Ans.* 3 T. 2 cwt. 50 lb.

(10.) Suppose two balls ten thousand miles apart were to approach each other under the influence of mutual attraction, the weight of one being represented by 15, that of the other by 30. How far would each move? *Ans.* $6666\frac{2}{3}$ mi. and $3333\frac{1}{3}$ mi.

(11.) Which would have the stronger attraction on the earth, a body at the distance of 95 millions of miles from the earth, with a weight represented by 1000, or a body at the distance represented by 95, and a weight represented by one? *Ans.* As $\frac{1}{9025000000000}$ to $\frac{1}{9025}$.

(12.) Supposing the weight of a body to be represented by 4 and its distance at 6, and the weight of another body to be 6 and its distance at 4, which would exert the stronger power of attraction? *Ans.* The second, as $\frac{3}{8}$ to $\frac{1}{9}$.

108. THE CENTRE OF GRAVITY.—As every part of a body possesses the general property of attraction, it is evident that the attractive force of the mass of a body must be concentrated in some point; and this point is called the centre of gravity of the body.

What is the Centre of Gravity of a body? 109. The Centre of Gravity of a body is the point about which, all the parts balance each other.

110. This point in all spherical bodies of uniform density will be the centre of sphericity.

111. As the earth is a spherical body, its centre of gravity is at the centre of its sphericity.

112. When bodies approach each other under the effect of mutual attraction, they tend mutually to approach the centre of gravity of each other.

113. For this reason, when any body falls towards the earth its motion will be in a straight line towards the centre of the earth. No two bodies from different points can approach the centre of a sphere in a parallel direction, and no two bodies suspended from different points can hang parallel to one another.

Fig. 3.

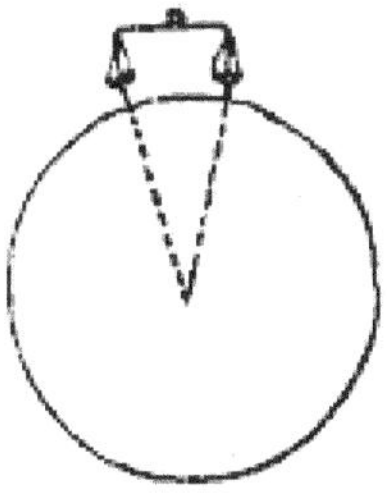

114. Even a pair of scales hanging perpendicularly to the earth, as represented in Fig. 3, cannot be exactly parallel, because they both point to the same spot, namely, the centre of the earth. But their convergency is too small to be perceptible.

What is a Vertical Line? 115. The direction in which a falling body approaches the surface of the earth is called a Vertical Line.

No two vertical lines can be parallel.

116. A weight suspended from any point will always assume a vertical position.*

---

* Carpenters, masons and other artisans, make use of a weight of lead suspended at rest by a string, for the purpose of ascertaining whether their work stands in a vertical position. To this implement they give the name of *plumb-line*, from the Latin word *plumbum*, lead.

117. All bodies under the influence of terrestrial gravity will fall to the surface of the earth in the same space of time, when at an equal distance from the earth, if nothing impede them. But the air presents by its inertia a resistance to be overcome. This resistance can be more easily overcome by dense bodies, and therefore the rapidity of the fall of a body will be in proportion to its density.

To what is the resistance of the air to a falling body proportioned?

118. The resistance of the air to the fall of a body is in direct proportion to the extent of its surface.

119. Heavy bodies can be made to *float* in the air, instead of falling immediately to the ground, by making the extent of their surface counterbalance their weight. Thus gold, which is one of the heaviest of all substances, when spread out into thin leaf is not attracted by gravity with sufficient force to overcome the resistance of the air; it therefore floats in the air, or falls slowly. A sheet of paper also, for the same reason, will fall very slowly if spread open, but, if folded into a small compass, so as to present but a small surface to the air, it will fall much more rapidly.

120. This principle will explain the reason why a person can with impunity leap from a greater height with an expanded umbrella in his hand. The resistance of the air to the broad surface of the umbrella checks the rapidity of the fall.

121. In the same manner the aëronaut safely descends from a balloon at a great height by means of a *parachute*. But, if by any accident the parachute is not *expanded* as he falls, the rapidity of the fall will not be checked. [*See Fig.* 4.]

122. EFFECT OF GRAVITY ON THE DENSITY OF THE AIR.— The air extends to a very considerable distance above the surface of the earth.* That portion which lies near the surface of the earth has to sustain the weight of the portions above; and the pressure of the upper parts

---

* We have no means of ascertaining the exact height to which the air extends. Sir John Herschel says: "Laying out of consideration all nice questions as to the probable existence of a definite limit to the atmosphere, beyond which there is, absolutely and rigorously speaking, *no* air, it is clear that, for all practical purposes, we may speak of those regions which are more distant above the earth's surface than the hundredth part of its diameter as void of air, and, of course, of clouds (which are nothing but visible vapors, diffused and floating in the air, sustained by it, and rendering it turbid, as mud does water). It seems probable, from many indications, that the greatest height at which visible clouds ever exist does not exceed ten miles, at which height the density of the air is about an eighth part of what it is at the level of the sea." Although the exact height to which the atmosphere extends has never been ascertained, it ceases to reflect the sun's rays at a greater height than forty-five miles

of the atmosphere on those beneath renders the air near the surface of the earth much more dense than that in the upper regions.

Fig. 4.

What effect has Gravity upon the air? 123. The air or atmosphere exists in a state of compression, caused by Gravity, which increases its density near the surface of the earth.

124. Gravity causes bodies in a fluid or gaseous form to move in a direction *seemingly* at variance with its own laws.

Thus smoke and steam ascend, and oil poured into a vessel containing a heavier fluid will first sink and then rise to the surface. This seemingly anomalous circumstance, when rightly understood, will be found to be in perfect obedience to the laws of gravitation. Smoke and steam are both substances less dense than air, and are therefore less forcibly attracted by gravitation. The air being more strongly attracted than steam or smoke, on account of its superior density, *falls* into the space occupied by the steam, and forces it upwards. The same reasoning applies in the case of oil; it is forced upwards by the heavier fluid, and both phenomena are thus seen to be the necessary consequences of gravity. The rising of a cork or other similar light substances from the bottom of a vessel of water is explained in the same way. This circumstance leads to the consideration of what is called *specific gravity*.

What is meant by Specific Gravity? 125. SPECIFIC GRAVITY.—Specific Gravity is a term used to express the relative weight of equal bulks of different bodies.*

126. If we take equal bulks of lead, wood, cork and air, we find the lead to be the heaviest, then the wood, then the cork, and lastly the air. Hence we say that the specific gravity of cork is greater than that of air, the specific gravity of wood is greater than that of cork, and the specific gravity of lead greater than that of wood, &c.

127. From what has now been said with respect to the attraction of gravitation and the specific gravity of bodies, it appears that, although the earth attracts all substances, yet this very attraction causes some bodies to rise and others to fall.

128. Those bodies or substances the specific gravity of which is greater than that of air will fall, and those whose specific gravity is less than that of air will rise; or, rather, the air, being more strongly attracted, will get beneath them, and, thus displacing them, will cause them to rise. For the same reason, cork and other light substances will not sink in water, because, the specific gravity of water being greater, the water is more strongly attracted, and will be drawn down beneath them. [For a table of the specific gravity of bodies, see Hydrostatics.]

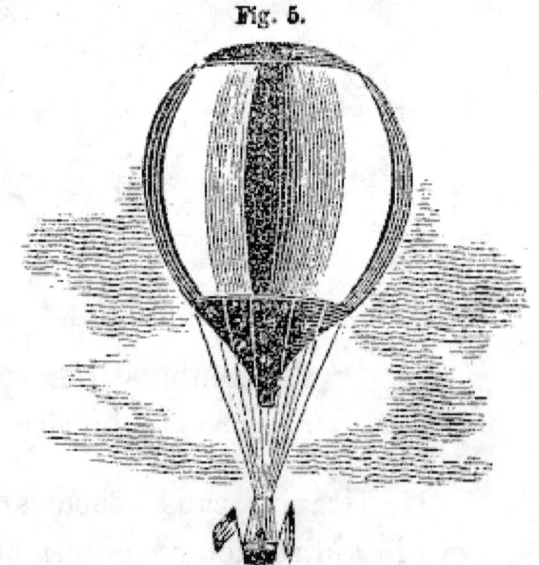

Fig. 5.

129. The principle which causes balloons to rise is the same which occasions the ascent of smoke, steam, &c. The materials of which

---

* The quantity of matter in a body is estimated, not by its apparent size, but by its weight. Some bodies, as cork, feathers, &c., are termed light; others, as lead, gold, mercury, &c., are called heavy. The reason of this is, that the particles which compose the former are not closely packed together, and therefore they occupy considerable space; while in the latter they are joined more closely together, and occupy but little room. A pound of cork and a pound of lead, therefore, will differ very much in apparent size, while they are both equally attracted by the earth,—that is, they weigh the same.

a balloon is made, are heavier than air, but their extension is greatly increased, and they are filled with an elastic fluid of a different nature, specifically lighter than air, so that, on the whole, the balloon when thus filled is much lighter than a portion of air of the same dimensions, and it will rise.

130. Gravity, therefore, causes bodies which are lighter than air to ascend, those which are of equal weight with air to remain stationary, and those which are heavier than air to descend. But the rapidity of their descent is affected by the resistance of the air, which resistance is proportioned to the extent of surface in the falling body.

What is Mechanics?

131. MECHANICS.— Mechanics treats of motion, and the moving powers, their nature and laws, with their effects in machines.

What is Motion?

132. Motion is a continued change of place.

133. On account of the inertia of matter, a body at rest cannot put itself in motion, nor can a body in motion stop itself.

What is meant by a Force?

134. That which causes motion is called *a Force*.

What is meant by Resistance?

135. That which stops or impedes motion is called Resistance.*

What things are to be considered in relation to motion?

136. In relation to motion, we must consider the force, the resistance, the time, the space, the direction, the velocity and the momentum.

What is the velocity, and to what is it proportional?

137. The velocity is the rapidity with which a body moves; and it is always proportional to the force by which the body is put in motion.

138. The velocity of a moving body is determined by the time that it occupies in passing through a given space. The greater the space and the shorter the time, the greater is the velocity. Thus, if one body move at the rate of six miles, and another twelve miles

---

* A force is sometimes a resistance, and a resistance is sometimes a force. The two terms are used merely to denote opposition.

in the same time, the velocity of the latter is double that of the former.

What is the rule for finding the velocity of a moving body?

139. To find the velocity of a body, the space passed over must be divided by the time employed in moving over it.

Thus, if a body move 100 miles in 20 hours, the velocity is found by dividing 100 by 20. The result is five miles an hour.*

140. *Questions for Solution.*

(1.) If a body move 1000 miles in 20 days, what is its velocity? *Ans.* 50 miles a day.

(2.) If a horse travel 15 miles in an hour, what is his velocity? *Ans.* ¼ of a mile in a minute.

(3.) Suppose one man walk 300 miles in 10 days, and another 200 miles in the same time, — what are their respective velocities? *Ans.* 30 & 20.

(4.) If a ball thrown from a cannon strike the ground at the distance of 3 miles in 3 seconds from the time of its discharge, what is its velocity? *A.* 1.

(5.) Suppose a flash of lightning come from a cloud 3 miles distant from the earth, and the thunder be heard in 14 seconds after the flash is seen; how fast does sound travel? *Ans.* $1131\frac{3}{7}$ ft. per sec.

(6.) The sun is 95 millions of miles from the earth, and it takes $8\frac{1}{4}$ minutes for the light from the sun to reach the earth; with what velocity does light move? † *Ans.* 191919 + mi. per sec.

---

* Velocity is sometimes called absolute, and sometimes relative. Velocity is called absolute when the motion of a body in space is considered without reference to that of other bodies. When, for instance, a horse goes a hundred miles in ten hours, his absolute velocity is ten miles an hour. Velocity is called relative when it is compared with that of another body. Thus, if one horse travel only fifty miles in ten hours, and another one hundred in the same time, the absolute velocity of the first horse is five miles an hour, and that of the latter is ten miles; but their relative velocity is as two to one.

† From the table here subjoined, the velocities of the objects enumerated may be ascertained in miles per hour and in feet per second, fractions omitted

TABLE OF VELOCITIES.

	Miles per hour.	Feet per second.
A man walking	3	4
A horse trotting	7	10
Swiftest race-horse	60	88
Railroad train in England	32	47
" " America	18	26
" " Belgium	25	36
" " France	27	40
" " Germany	24	35
English steamboats in channels	14	26
American on the Hudson	18	26
Fast-sailing vessels	10	14

MECHANICS.

How is the time employed by a moving body ascertained?

141. The time employed by a body in motion may be ascertained by dividing the space by the velocity.

Thus, if the space passed over be 100 miles, and the velocity 5 miles in an hour, the time will be 100 divided by 5.   *Ans.* 20 hours.

142.   *Questions for Solution.*

(1.) If a cannon-ball, with a velocity of 3 miles in a minute, strike the ground at the distance of one mile, what is the time employed?   *Ans.* ⅓ of a minute, or 20 seconds.

(2.) Suppose light to move at the rate of 192,000 miles in a second of time, how long will it take to reach the earth from the sun, which is 95 millions of miles distant?   *Ans.* 8 *min.* 14.07 *sec.* +

(3.) If a railroad-car run at the rate of 20 miles an hour, how long will it take to go from Washington to Boston,—distance 432 miles?   *Ans.* 21.6 *hr.*

(4.) Suppose a ship sail at the rate of 6 miles an hour, how long will it take to go from the United States to Europe, across the Atlantic Ocean, a distance of 2800 miles?   *Ans.* 19 *da.* 10 *hr.* 40 *min.*

(5.) If the earth go round the sun in 365 days, and the distance travelled be 540 millions of miles, how fast does it travel?   *Ans.* 1,479,452 $\frac{4}{7}$ *mi.*

(6.) Suppose a carrier-pigeon, let loose at 6 o'clock in the morning from Washington, reach New Orleans at 6 o'clock at night, a distance of 1200 miles, how fast does it fly?   *Ans.* 100 *mi. per hr.*

How may the space passed over by a body in motion be ascertained?

143. The space passed over may be found by multiplying the velocity by the time.

	Miles per hour.	Feet per second.
Slow rivers	3	4
Rapid rivers	7	10
Moderate wind	7	10
A storm	36	52
A hurricane	80	117
Common musket-ball	850	1,240
Rifle-ball	1,000	1,466
24 lb. cannon-ball	1,600	2,346
Air rushing into a vacuum at 32° F.	884	1,296
Air-gun bullet, air compressed to ·01 of its volume	466	683
Sound	743	1,142
A point on the surface of the earth	1,037	1,520
Earth in its orbit	67,374	98,815

The velocity of light is 192,000 miles in a second of time.

The velocity of the electric fluid is said to be still greater, and some authorities state it to be at the rate of 288,000 miles in a second of time.

Thus, if the velocity be 5 miles an hour, and the time 20 hours, the space will be twenty multiplied by 5. *Ans.* 100 miles.

144. (1.) If a vessel sail 125 miles in a day for ten days, how far will it sail in that time ? *Ans.* 1250 *mi.*

(2.) Suppose the average rate of steamers between New York and Albany be about 11 miles an hour, which they traverse in about 14 hours, what is the distance between these two cities by the river ? *Ans.* 154 *mi.*

(3.) Suppose the cars going over the railroad between these two cities travel at the rate of 25 miles an hour and take 8 hours to go over the distance, how far is it from New York to Albany by railroad ? *Ans.* 200 *mi.*

(4.) If a man walking from Boston at the rate of 2½ miles in an hour reach Salem in 6 hours, what is the distance from Boston to Salem ? *Ans.* 15 *mi.*

(5.) The waters of a certain river, moving at the rate of 4 feet in a second, reach the sea in 6 days from the time of starting from the source of the river. What is the length of that river ? *Ans.* $392\frac{3}{11}$ *mi.*

(6.) A cannon-ball, moving at the rate of 2400 feet in a second of time, strikes a target in 4 seconds. What is the distance of the target? *A.* 9600 *ft.*

145. The following formulæ embrace the several ratios of the time, space and velocity :

(1.) The space divided by the time equals the velocity, or $\frac{s}{t} = v$.

(2.) The space divided by the velocity equals the time, or $\frac{s}{v} = t$.

(3.) The velocity multiplied by the time equals the space, or $v \times t = s$.

How many kinds of motion are there? 146. There are three kinds of motion, namely, Uniform, Accelerated and Retarded.

What is Uniform Motion? 147. Uniform Motion is that by which a body moves over equal spaces in equal times.

What is Accelerated Motion? 148. Accelerated Motion is that by which the velocity increases while the body is moving.

What is Retarded Motion? 149. Retarded Motion is that by which the velocity decreases while the body is moving.

How are uniform, accelerated and retarded motion respectively produced? 150. Uniform Motion is produced by the momentary action of a single force. Accelerated Motion is produced by the continued action of one or more forces. Retarded Motion is produced by some resistance.

151. A ball struck by a bat, or a stone thrown from the hand is

in theory an instance of uniform motion; and, if the attraction of gravity and the resistance of the air could be suspended, it would proceed onwards in a straight line, with a uniform motion, forever. But, as the resistance of the air and gravity both tend to deflect it, it in fact becomes first an instance of retarded, and then of accelerated motion.

152. A stone, or any other body, falling from a height, is an instance of accelerated motion. The force of gravity continues to operate upon it during the whole time of its descent, and constantly increases its velocity. It begins its descent with the first impulse of attraction, and, could the force of gravity which gave it the impulse be suspended, it would continue its descent with a uniform velocity. But, while falling it is every moment receiving a new impulse from gravity, and its velocity is constantly increasing during the whole time of its descent.

153. A stone thrown perpendicularly upward is an instance of retarded motion; for, as soon as it begins to ascend, gravity immediately attracts it downwards, and thus its velocity is diminished. The retarding force of gravity acts upon it during every moment of its ascent, decreasing its velocity until its upward motion is entirely destroyed. It then begins to fall with a motion continually accelerated until it reaches the ground.

What time does a body occupy in its ascent and descent?

154. A body projected upwards will occupy the same time in its ascent and descent.

This is a necessary consequence of the effect of gravity, which uniformly retards it in the ascent and accelerates it in its descent.

How can perpetual motion be produced?

155. PERPETUAL MOTION. — Perpetual Motion is deemed an impossibility in mechanics, because action and reäction are always equal and in contrary directions.

What is meant by Action and Reäction?

156. By the *action* of a body is meant the effect which it produces upon another body. By *reäction* is meant the effect which it receives from the body on which it acts.

Thus, when a body in motion strikes another body, it acts upon it, or produces motion; but it also meets with resistance from the body which is struck, and this resistance is the reäction of the body.

NATURAL PHILOSOPHY.

Illustration of Action and Reäction by means of Elastic and Non-elastic Balls.

(1.) Figure 6 represents two ivory* balls, A and B, of equal size, weight, &c., suspended by threads. If the ball A be drawn a little on one side and then let go, it will strike against the other ball B. and drive it off to a distance equal to that through which the first ball fell; but the motion of A will be stopped, because when it strikes B it receives in return a blow equal to that which it gave, but in a contrary direction, and its motion is thereby stopped, or, rather. given to B. Therefore, when a body strikes against another. the quantity of motion communicated to the second body is lost by the first; but this loss proceeds. not from the blow given by the striking body, but from the reäction of the body which it struck.

(2.) Fig. 7 represents six ivory balls of equal weight, suspended by threads. If the ball A be drawn out of the perpendicular and let fall against B, it will communicate its motion to B, and receive a reäction from it which will stop its own motion. But the ball B cannot move without moving C; it will therefore communicate the motion which it received from A to C. and receive from C a reäction, which will stop its motion. In like manner the motion and reäction are received by each of the balls D, E, F; but, as there is no ball beyond F to act upon it, F will fly off.

N. B. This experiment is to be performed with *elastic* balls only.

(3). Fig. 8 represents two balls of clay (which are not elastic), of equal weight, suspended by strings. If the ball D be raised and let fall against E, only part of the motion of D will be destroyed by it (because the bodies are non-elastic), and the two balls will move on together to *d* and *e*, which are less distant from the vertical line than the ball D was before it fell. Still,

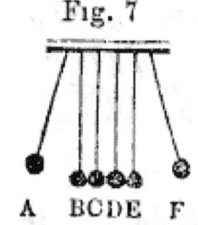

Fig. 8.

* It will be recollected that *ivory* is considered highly *elastic*.

however, action and reäction are equal, for the action on E is only enough to make it move through a smaller space, but so much of D's motion is now also destroyed.*

157. It is upon the principle of action and reäction that birds are enabled to fly. They strike the air with their wings, and the reäction of the air enables them to rise, fall, or remain stationary, at will, by increasing or diminishing the force of the stroke of their wings.†

158. It is likewise upon the same principle of action and reäction that fishes swim, or, rather, make their way through the water, namely, by striking the water with their fins.‡

159. Boats are also propelled by oars on the same principle, and the oars are lifted out of the water, after every stroke, so as completely to prevent any reäction in a backward direction.

How may motion be caused? 160. Motion may be caused either by action or reäction. When caused by action it is called Incident, and when caused by reäction it is called Reflected Motion. §

---

* Figs. 6 and 7, as has been explained, show the effect of action and reäction in elastic bodies, and Fig. 8 shows the same effect in non-elastic bodies. When the elasticity of a body is imperfect, an intermediate effect will be produced; that is, the ball which is struck will rise higher than in case of non-elastic bodies, and less so than in that of perfectly elastic bodies; and the striking ball will be retarded more than in the former case, but not stopped completely, as in the latter. They will, therefore, both move onwards after the blow, but not together, or to the same distance; but this, as in the preceding cases, the whole quantity of motion destroyed in the striking ball will be equal to that produced in the ball struck. Connected with "the Boston school apparatus" is a stand with ivory balls, to give a visible illustration of the effects of collision.

† The muscular power of birds is much greater in proportion to their weight than that of man. If a man were furnished with wings sufficiently large to enable him to fly, he would not have sufficient strength or muscular power to put them in motion.

‡ The power possessed by fishes, of sinking or rising in the water, is greatly assisted by a peculiar apparatus furnished them by nature, called an air-bladder, by the expansion or contraction of which they rise or fall, on the principle of specific gravity.

§ The word *incident* implies *falling upon*, or *directed towards*. The word *reflected* implies *turned back*. Incident motion is motion directed towards any particular object, against which a moving body strikes. Reflected motion is that which is caused by the reäction of the body which is struck. Thus, when a ball is thrown against a surface, it rebounds or is turned back. This return of the ball is called reflected motion. As reflected motion is caused by reäction, and reäction is increased by elasticity, it follows that reflected motion is always greatest in those bodies which are most elastic. For this reason, a ball filled with air rebounds better than one stuffed with bran or wool, because its elasticity is greater. For the same reason, balls made of caoutchouc, or India-rubber, will rebound more than those which are made of most other substances.

48 NATURAL PHILOSOPHY.

What is an angle of Incidence?

161. The angle * of incidence is the angle formed by the line which the incident body makes in its passage towards any object, with a line perpendicular to the surface of the object.

* As this book may fall into the hands of some who are unacquainted with geometrical figures, a few explanations are here subjoined:

1. An angle is the opening made by two lines which meet each other in a point. *The size of the angle depends upon the opening, and not upon the length of the lines.*

2. A circle is a perfectly round figure, every part of the outer edge of which, called the circumference, is equally distant from a point within, called the centre. [*See Fig.* 9.]

3. The straight lines drawn from the centre to the circumference are called *radii*. [The singular number of this word is *radius*.] Thus, in Fig. 9, the lines C D, C O, C R, and C A, are radii.

Fig. 9.

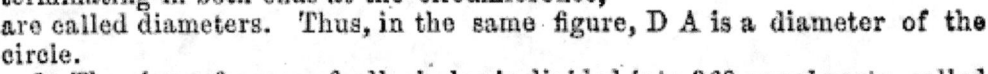

4. The lines drawn through the centre, and terminating in both ends at the circumference, are called diameters. Thus, in the same figure, D A is a diameter of the circle.

5. The circumference of all circles is divided into 360 equal parts, called degrees. The diameter of a circle divides the circumference into two equal parts, of 180 degrees each.

6. All angles are measured by the number of degrees which they contain Thus, in Fig. 9, the angle R C A, as it includes one-quarter of the circle, is an angle of 90 degrees, which is a quarter of 360. And the angles R C O and O C D are angles of 45 degrees.

7. Angles of 90 degrees are right angles; angles of less than 90 degrees, acute angles; and angles of more than 90 degrees are called obtuse angles. Thus, in Fig. 9, R C A is a right angle, O C R an acute, and O C A an obtuse angle.

8. A perpendicular line is a line which makes an angle of 90 degrees on each side of any other line or surface; therefore, it will incline neither to the one side nor to the other. Thus, in Fig. 9, R C is perpendicular to D A.

9. The tangent of a circle is a line which touches the circumference, without cutting it when lengthened at either end. Thus, in Fig. 9, the line R T is a tangent.

10. A square is a figure having four equal sides, and four equal angles. These will always be right angles. [*See Fig.* 11.]

11. A parallelogram is a figure whose opposite sides are equal and parallel [*See Figs.* 12 *and* 13.] A square is also a parallelogram.

12 A rectangle is a parallelogram whose angles are right angles.

[N. B. It will be seen by these definitions that both a square and a rectangle are parallelograms, but all parallelograms are not rectangles nor squares. A square is both a parallelogram and a rectangle. *Three things are essential to a square;* namely, the four sides must all be equal, they must also be parallel, and the angles must all be right angles. Two things only are essential to a rectangle; namely, the angles must all be right angles, and the opposite sides must be equal and parallel. One thing only is essential to a parallelogram; namely, the opposite sides must be equal and parallel.]

13. The diagonal of a square, of a parallelogram, or a rectangle, is a line

MECHANICS. 49

Explain Fig. 10 162. Thus, in Fig. 10, the line A B C represents a wall, and P B a line perpendicular to its surface. O is a ball moving in the direction of the dotted line, O B. The angle O B P is the angle of incidence.

Fig. 10.

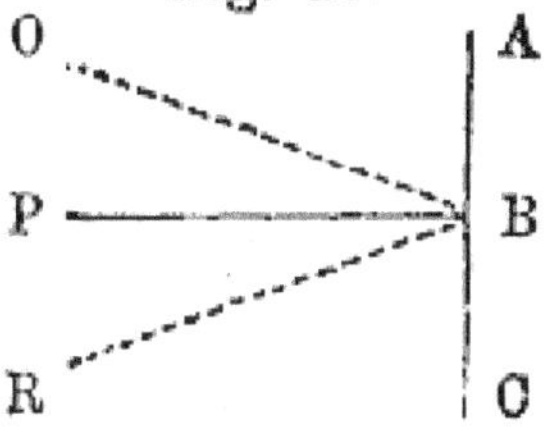

What is the angle of reflection? 163. The angle of reflection is the angle formed by the perpendicular with the line made by the reflected body as it leaves the surface against which it struck.

Thus, in Fig. 10, the angle P B R is the angle of reflection.

What is the proportion of the angle of incidence to the angle of reflection? 164. The angles of incidence and reflection are always equal to one another.*

(1.) Thus, in Fig. 10, the angle of incidence, O B P, and the angle of reflection, P B R, are equal to one another; that is, they contain an equal number of degrees.

What will be the course of a body in motion which strikes against another fixed body? 165. From what has now been stated with regard to the angles of incidence and reflection, it follows, that *when a ball is thrown perpendicularly against an object which it cannot penetrate, it will return in the same direction; but, if it be thrown obliquely, it will return obliquely on the opposite side of the perpendicular. The more obliquely the ball is thrown, the more obliquely it will rebound.* †

drawn through either of them, and terminating at the opposite angles. Thus, in Figs. 11, 12, and 13, the line A C is the diagonal of the square, parallelogram, or rectangle.

* An understanding of this law of reflected motion is very important, because it is a fundamental law, not only in Mechanics, but also in Pyronomics, Acoustics and Optics.

† It is from a knowledge of these facts that skill is acquired in many different sorts of games, as Billiards, Bagatelle, &c. A ball may also, on the same principle, be thrown from a gun against a fortification so as to reach an object out of the range of a direct shot.

What is the Momentum of a body?

166. MOMENTUM. — The Momentum* of a body is its quantity of motion,† and it expresses the force with which it would strike against another body.

How is the Momentum of a body calculated?

The Momentum of a body is ascertained by multiplying its weight by its velocity.

167. Thus, if the velocity of a body be represented by 5 and its weight by 6, its momentum will be 30.

How can a small or a light body be made to do as much damage as a large one?

168. A small or a light body may be made to strike against another body with a greater force than a heavier body simply by giving it sufficient velocity,—that is, by making it have greater momentum.

Thus, a cork weighing $\frac{1}{4}$ of an ounce, shot from a pistol with the velocity of 100 feet in a second, will do more damage than a leaden shot weighing $\frac{1}{8}$ of an ounce, thrown from the hand with a velocity of 40 feet in a second, because the momentum of the cork will be the greater.

The momentum of the cork is $\frac{1}{4} \times 100 = 25$.
That of the leaden shot is $\frac{1}{8} \times 40 = 5$.

169. *Questions for Solution.*

(1.) What is the momentum of a body weighing 5 pounds, moving with the velocity of 50 feet in a second? *Ans.* 250.

(2.) What is the momentum of a steam-engine, weighing 3 tons, moving with the velocity of 60 miles in an hour? *Ans.* 180.

[N. B. It must be recollected that, in comparing the momenta of bodies, the velocities and the time of the bodies compared must be respectively of the same denomination. If the time of one be minutes and of the other be hours, they must both be considered in minutes, or both in hours. So, with regard to the spaces and the weights, if one be feet all must be expressed in feet; if one be in pounds, all must be in pounds. It is better, however, to express the weight, velocities and spaces, by abstract numbers, as follows :]

(3.) If a body whose weight is expressed by 9 and velocity by 6 is in motion, what is its momentum? *Ans.* 54.

(4.) A body whose momentum is 63 has a velocity of 9; what is its weight? *Ans.* 7.

* The plural of this word is *momenta*.
† The quantity of motion communicated to a body does not affect the duration of the motion. If but little motion be communicated, the body will move slowly. If a great degree be imparted, it will move rapidly. But in both cases the motion will continue until it is destroyed by some external force.

MECHANICS. 51

[N. B. The momentum being the product of the weight and velocity, the weight is found by dividing the momentum by the velocity, and the velocity is found by dividing the momentum by the weight.]

(5.) The momentum is expressed by 12, the weight by 2 ; what is the velocity ? *Ans.* 6.

(6.) The momentum 9, velocity 9, what is the weight ? *Ans.* 1.

(7.) Momentum 36, weight 6, required the velocity. *Ans.* 6.

(8.) A body with a momentum of 12 strikes another with a momentum of 6 ; what will be the consequence ? *Ans.* Both have mom. of 6.

[N. B. *When two bodies, in opposite directions, come into collision, they each lose an equal quantity of their momenta.*]

(9.) A body weighing 15, with a velocity of 12, meets another coming in the opposite direction, with a velocity of 20, and a weight of 10 ; what will be the effect ? *Ans.* Both move with mom. of 20.

(10.) Two bodies meet together in opposite directions. A has a velocity of 12 and a weight of 7, B has a momentum expressed by 84. What will be the consequence ? *Ans.* Both mom. destroyed.

(11.) Suppose the weight of a comet be represented by 1 and its velocity by 12, and the weight of the earth be expressed by 100 and its velocity by 10, what would be the consequence of a collision, supposing them to be moving in opposite directions ? *Ans.* Both have mom. of 988.

(12.) If a body with a weight of 75 and a velocity of 4 run against a man whose weight is 150, and who is standing still, what will be the consequence, if the man uses no effort but his weight ? *Ans.* Man has vel. of $1\tfrac{1}{3}$.

(13.) With what velocity must a 64 pound cannon-ball fly to be equally effective with a battering-ram of 12,000 pounds propelled with a velocity of 16 feet in a second ? *Ans.* 3000 *ft.*

170. ATTRACTION — LAW OF FALLING BODIES. — When one body strikes another it will cause an effect proportional to its own weight and velocity (or, in other words, its momentum) ; and the body which receives the blow will move on with a uniform velocity (if the blow be sufficient to overcome its inertia) in the direction of the motion of the blow. But, when a body moves by the force of a constant attraction, it will move with a constantly accelerated motion.

171. This is especially the case with falling bodies. The earth attracts them with a force sufficient to bring them down through a certain number of feet during the first second of time. While the body is thus in motion with a velocity, say of sixteen feet, the earth still attracts it, and during the second second it communicates an additional velocity, and every successive second of time the attraction of the earth adds to the velocity in a similar proportion, so that during any given time, a falling body will acquire a velocity which, in the same time, would carry it over twice the space through which it has already fallen. Hence we deduce the following law:

What is the law of falling bodies? 172. A body falling from a height will fall sixteen feet in the first second of time,* three

---

* This is only an approximation to the truth ; it actually falls sixteen feet and one inch during the first second, three times that distance in the second, &c The questions proposed to be solved assume sixteen feet only.

times that distance in the second, five times in the third, seven in the fourth, its velocity increasing during every successive second, as the odd numbers 1, 3, 5, 7, 9, 11, 13, &c.*

The laws of falling bodies are clearly demonstrated by a mechanical arrangement known by the name of "*Attwood's Machine,*" in which a small weight is made to communicate motion to two others attached to a cord passing over friction-rollers (causing one to ascend and the other to descend), and marking the progress of the descending weight by the oscillations of a pendulum on a graduated scale, attached to one of the columns of the machine. It has not been deemed expedient to present a cut of the machine, because without the machine itself the explanation of its operation would be unsatisfactory, with the machine itself in view the simplicity of its construction would render an explanation unnecessary.

* The entire spaces through which a body will have fallen in any given number of seconds *increase as the squares* of the times. This law was discovered by Galileo, and may thus be explained. If a body fall sixteen feet in one second, in two seconds it will have fallen four times as far, in three seconds nine times as far, in four seconds sixteen times as far, in the fifth second twenty-five times, &c., in the sixth thirty-six times, &c.

ANALYSIS OF THE MOTION OF A FALLING BODY.

Number of Seconds.	Spaces.	Velocities.	Total Space.
1	1	2	1
2	3	4	4
3	5	6	9
4	7	8	16
5	9	10	25
6	11	12	36
7	13	14	49
8	15	16	64
9	17	18	81
10	19	20	100

From this statement it appears that the spaces passed through by a falling body, in any number of seconds, increase as the odd numbers 1, 3, 5, 7, 9, 11, &c.; the velocity increases as the even numbers 2, 4, 6, 8, 10, 12, &c.; and the total spaces passed through in any given number of seconds increase as the squares of the numbers indicating the seconds, — thus, 1, 4, 9, 16, 25, 36, &c.

Aristotle maintained that the velocity of any falling body is in direct proportion to its weight; and that, if two bodies of unequal weight were let fall from any height at the same moment, the heavier body would reach the ground in a shorter time, in exact proportion as its weight exceeded that of the lighter one. Hence, according to his doctrine, a body weighing two pounds would fall in half the time required for the fall of a body weighing only one pound. This doctrine was embraced by all the followers of that distinguished philosopher, until the time of Galileo, of Florence, who flourished about the middle of the sixteenth century. He maintained that the velocity of a falling body is not affected by its weight, and challenged the adherents of the Aristotelian doctrine to the test of experiment. The leaning tower of Pisa was selected for the trial, and there the experiment was tried which proved the truth of Galileo's theory. A distinguished writer thus describes the scene. "On the appointed day the disputants

MECHANICS.

173. The height of a building, or the depth of a well, may thus be estimated very nearly by observing the length of time which a stone takes in falling from the top to the bottom.

174. *Exercises for Solution.*

(1.) If a ball, dropped from the top of a steeple, reaches the ground in 5 seconds, how high is that steeple?

$16+48+80+112+144=400$ feet; or, $5\times 5=25$, square of the number of seconds, multiplied by the number of feet it falls through in one second, namely, 16 feet; that is, $25\times 16=400$ feet.

(2.) Suppose a ball, dropped from the spire of a cathedral, reach the ground in 9 seconds, how high is that spire?

$16+48+80+112+144+176+208+240+272=1296$ feet.

Or, squaring the time in seconds, $9^2=81$, multiplied by $16=1296$. *Ans.*

[It will hereafter be shown that this law of falling bodies applies to *all* bodies, whether falling perpendicularly or obliquely. Thus, whether a stone be thrown from the top of a building horizontally or dropped perpendicularly downwards, in both cases the stone will reach the ground in the same time; and this rule applies equally to a ball projected from a cannon, and to a stone thrown from the hand.]

(3.) If a ball, projected from a cannon from the top of a pyramid, reach the ground in 4 seconds, how high is the pyramid?   *Ans.* 256 *ft.*

(4.) How deep is a well, into which a stone being dropped, it reaches the water 6 feet from the bottom of the well in 2 seconds?   *Ans.* 70 *ft.*

(5.) The light of a meteor bursting in the air is seen, and in 45 seconds a meteoric stone falls to the ground. Supposing the stone to have proceeded from the explosion of the meteor perpendicularly, how far from the earth, in feet, was the meteor? $45^2\times 16=32,400$ feet.

(6.) What is the difference in the depth of two wells, into one of which a stone being dropped, is heard to strike the water in 5 seconds, and into the other in 9 seconds, supposing that the water be of equal depth in both, and making no allowance for the progressive motion of sound?   *A.* 896 *ft.*

repaired to the tower of Pisa, each party, perhaps, with equal confidence. It was a crisis in the history of human knowledge. On the one side stood the assembled wisdom of the universities, revered for age and science, venerable, dignified, united and commanding. Around them thronged the multitude, and about them clustered the associations of centuries. On the other there stood an obscure young man (Galileo), with no retinue of followers, without reputation, or influence, or station. But his courage was equal to the occasion; confident in the power of truth, his form is erect and his eye sparkles with excitement. But the hour of trial arrives. The balls to be employed in the experiments are carefully weighed and scrutinized, to detect deception. The parties are satisfied. The one ball is exactly twice the weight of the other. The followers of Aristotle maintain that, when the balls are dropped from the tower, the heavy one will reach the ground in exactly half the time employed by the lighter ball. Galileo asserts that the weights of the balls do not affect their velocities, and that the times of descent will be equal; and here the disputants join issue. The balls are conveyed to the summit of the lofty tower. The crowd assemble round the base; the signal is given; the balls are dropped at the same instant; and, swift descending, at the same moment they strike the earth. Again and again the experiment is repeated, with uniform results Galileo's triumph was complete; not a shadow of a doubt remained." ["*The Orbs of Heaven.*"]

(7.) A boy raised his kite in the night, with a lantern attached to it. Unfortunately, the string which attached the lantern broke, and the lantern fell to the ground in 6 seconds. How high was the kite ? *Ans.* 576 *ft.*

175. RETARDED MOTION OF BODIES PROJECTED UPWARDS. — All the circumstances attending the accelerated descent of falling bodies are exhibited when a body is projected upwards, but in a reversed order.

How can we determine the height to which a body, projected upwards with a given velocity, will ascend?

176. To determine the height to which a body, projected upwards, will rise, with a given velocity, it is only necessary to determine the height from which a body would fall to acquire the same velocity.

177. Thus, if it be required to ascertain how high a body would rise when projected upwards with a force sufficient to carry it 144 feet in the first second of time, we reverse the series of numbers $16+48+80+112+144$ [*see table on page* 52], and, reading them backward, $144+112+80+48+16$, we find their sum to be 400 feet, and the time employed would be 5 seconds.

How does the time of the ascent of a body compare with the time of its descent?

178. The time employed in the ascent and descent of a body projected upwards will, therefore, always be equal.

Questions for Solution.

(1.) Suppose a cannon-ball, projected perpendicularly upwards, returned to the ground in 18 seconds ; how high did it ascend, and what is the velocity of projection ? *Ans.* 1296 *ft.*; 272 *ft.* 1*st sec.*

(2.) How high will a stone rise which a man throws upward with a force sufficient to carry it 48 feet during the first second of time ? *Ans.* 64 *ft.*

(3.) Suppose a rocket to ascend with a velocity sufficient to carry it 176 feet during the first second of time ; how high will it ascend, and what time will it occupy in its ascent and descent ? *Ans.* 576 *ft.*; 12 *sec.*

(4.) A musket-ball is thrown upwards until it reaches the height of 400 feet. How long a time, in seconds, will it occupy in its ascent and descent, and what space does it ascend in the first second? *Ans.* 10 *sec.*; 144 *ft.*

(5.) A sportsman shoots a bird flying in the air, and the bird is 3 seconds in falling to the ground. How high up was the bird when he was shot ? *Ans.* 144 *ft.*

(6.) How long time, in seconds, would it take a ball to reach an object 5000 feet above the surface of the earth, provided that the ball be projected with a force sufficient only to reach the object ? *Ans.* 17.67 *sec.* +

179. COMPOUND MOTION. — Motion may be produced either by a single force or by the operation of two or more forces.

MECHANICS. 55

In what direction is the motion of a body impelled by a single force?
180. Simple Motion is the motion of a body impelled by a single force, and is always in a straight line in the same direction with the force that acts.

What is Compound Motion?
181. Compound Motion is caused by the operation of two or more forces at the same time.

When a body is struck by two equal forces, in opposite directions, how will it move?
182. When a body is struck by two equal forces, in opposite directions, it will remain at rest.

183. If the forces be *unequal*, the body will move with diminished force in the direction of the greater force. Thus, if a body with a momentum of 9 be opposed by another body with a momentum of 6, both will move with a momentum of 3 in the direction of the greater force.

How will a body move when struck by two forces in different directions?
184. A body, struck by two forces in different directions, will move in a line between them, in the direction of the diagonal of a parallelogram, having for its sides the lines through which the body would pass if urged by each of the forces separately.

How will the body move, if the forces are equal and at right angles to each other?
185. When the forces are equal and at right angles to each other, the body will move in the diagonal of *a square*.

186. Let Fig. 11 represent a ball struck by the two equal forces X and Y. In this figure the forces are inclined to each other at an angle of 90°, or a right angle. Suppose that the force X would send it from C to B, and the force Y from C to D. As it cannot obey both, it will go between them to A, and the line C A,

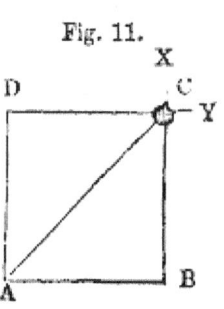
Fig. 11.

through which it passes, is the diagonal of the square, A B C D. This line also represents the *resultant* of the two forces.

The time occupied in its passage from C to A will be the same as the force X would require to send it to B, or the force Y to send it to D.

How will a body move under the influence of two unequal forces at right angles to each other?

187. If two *unequal* forces act at right angles to each other on a body, the body will move in the direction of the diagonal of a rectangle.

Explain Fig. 12.

188. *Illustration.* — In Fig. 12 the ball C is represented as acted upon by two unequal forces, X and Y. The force X would send it to B, and the force Y to D. As it cannot obey both, it will move in the direction C A, the diagonal of the rectangle A B C D.

Fig. 12.

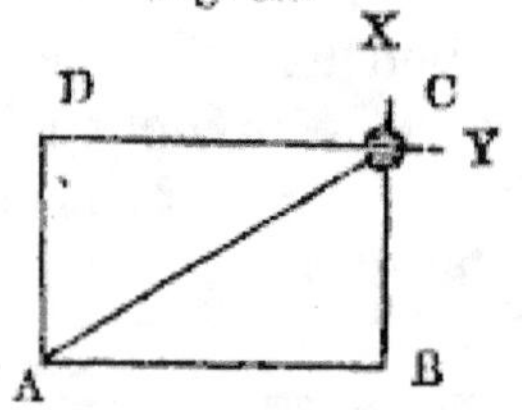

How will the body move if the forces act in the direction of any other than a right angle? How will a body move if the forces act in the direction of an acute or obtuse angle?

189. When two forces act in the direction of an acute or an obtuse angle, the body will move in the direction of the diagonal of a parallelogram.

Explain Fig. 13.

190. *Illustration.* — In figure 13 the ball C is supposed to be influenced by two forces, one of which would send it to B and the other to D, the forces acting in the direction of an acute angle. The ball will, therefore, move between them in the line C A, the *longer* diagonal of the parallelogram A B C D.

Fig. 13.

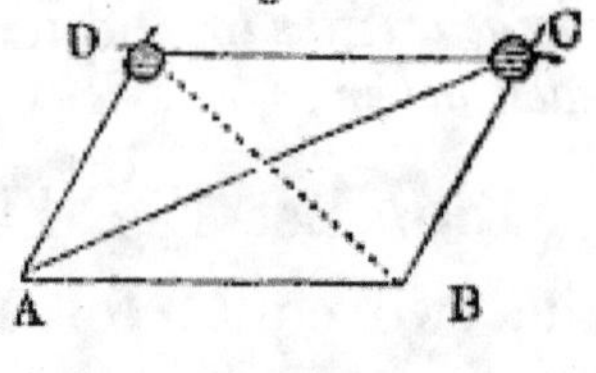

191. The same figure explains the motion of a ball when the two forces act in the direction of an obtuse angle.

192. *Illustration.* — The ball D, under the influence of two

forces, one of which would send it to C, and the other to A, which, it will be observed, is in the direction of an obtuse angle, will proceed in this case to B, the *shorter* diagonal of the parallelogram A B C D.

[N. B. A parallelogram containing acute and obtuse angles has two diagonals, the one which joins the acute angles being the longer.]

What is Resultant Motion? 193. Resultant Motion is the effect or result of two motions compounded into one.

194. If two men be sailing in separate boats, in the same direction, and at the same rate, and one toss an apple to the other, the apple would appear to pass directly across from one to the other, in a line of direction perpendicular to the side of each boat. But its real course is through the air in the diagonal of a parallelogram, formed by the lines representing the course of each boat, and perpendiculars drawn to those lines from the spot where each man stands as the one tosses and the other catches the apple. In Fig. 14 *Explain Fig. 14.* the lines A B and C D represent the course of each boat. E the spot where the man stands who tosses the apple; while the apple is in its passage, the boats have passed from E and G to H and F respectively. But the apple, having a motion, with the man, that would carry it from E to H, and likewise a projectile force which would carry it from E to G, cannot obey them both, but will pass through the dotted line E F, which is the diagonal of the parallelogram E G F H.*

How can we ascertain the direction of the 195. When a body is acted upon by *three* or more forces at the same time, we may take any

---

* On the principle of resultant motion, if two ships in an engagement be sailing before the wind, at equal rates, the aim of the gunners will be exactly as though they both stood still. But, if the gunner fire from a ship standing still at another under sail, or a sportsman fire at a bird on the wing, each should take his aim a little forward of the mark, because the ship and the bird will pass a little forward while the shot is passing to them.

motion when the body is influenced by three or more forces?

two of them alone, and ascertain the resultant of those two, and then employ the resultant as a new force, in conjunction with the third,* &c.

What is Circular Motion?

196. CIRCULAR MOTION.— Circular Motion is motion around a central point.

What causes Circular Motion?

197. Circular motion is caused by the continued operation of two forces, by one of which the body is projected forward in a straight line, while the other is constantly deflecting it towards a fixed point. [*See No.* 184.]

198. The whirling of a ball, fastened to a string held by the hand, is an instance of circular motion. The ball is urged by two forces, of which one is the force of projection, and the other the string which confines it to the hand. The two forces act at right angles to each other, and (according to No. 184) the ball will move in the diagonal of a parallelogram. But, as the force which confines it to the hand only keeps it within a certain distance, without drawing it nearer to the hand, the motion of the ball will be through the diagonals of an indefinite number of minute parallelograms, formed by every part of the circumference of the circle.

How many centres require to be noticed in Mechanics?

199. There are three different centres which require to be distinctly noticed; namely, the Centre of Magnitude, the Centre of Gravity, and the Centre of Motion.

---

* The resultant of two forces is always described by the third side of a triangle, of which the two forces may be represented, in quantity and direction, by the other two sides. When three forces act in the direction of the three sides of the same triangle, the body will remain *at rest*.

When two forces act at right angles, the resultant will form the hypothenuse of a right-angled triangle, either of the sides of which may be found, when the two others are given, by the common principles of arithmetic or geometry.

From what has now been stated, it will easily be seen, that if any number of forces whatever act upon a body, and in any directions whatever, the resultant of them all may easily be found, and this resultant will be their mechanical equivalent. Thus, suppose a body be acted upon at the same time by six forces, represented by the letters A, B, C, D, E, F. First find the resultant of A and B by the law stated in No. 184, and call this resultant G. In the same manner, find the resultant of G and C, calling it H. Then find the resultant of H and D, and thus continue until each of the forces be found, and the last resultant will be the mechanical equivalent of the whole

MECHANICS.

What is the Centre of Magnitude?
200. The Centre of Magnitude is the central point of the bulk of a body.

What is the Centre of Gravity?
201. The Centre of Gravity is the point about which all the parts balance each other.

What is the Centre of Motion?
202. The Centre of Motion is the point around which all the parts of a body move.

What is the Axis of Motion?
203. When the body is not of a size nor shape to allow every point to revolve in the same plane, the line around which it revolves is called the Axis of Motion.*

Does the centre or the axis of motion revolve?
204. The centre or the axis of motion is generally supposed to be at rest.

205. Thus the axis of a spinning-top is stationary, while every other part is in motion around it. The axis of motion and the centre of motion are terms which relate only to circular motion.

What are Central Forces?
206. The two forces by which circular motion is produced are called *Central* Forces. Their names are, the Centripetal Force and the Centrifugal Force.†

What is the Centripetal Force?
207. The Centripetal Force is that which confines a body to the centre around which it revolves.

What is the Centrifugal Force?
208. The Centrifugal Force is that which impels the body to fly off from the centre.

---

* Circles may have a centre of motion; spheres or globes have an axis of motion. Bodies that have only length and breadth may revolve around their own centre, or around axes; those that have the three dimensions of length, breadth and thickness, must revolve around axes.

† The word *centripetal* means seeking the centre, and *centrifugal* means flying from the centre. In circular motion these two forces constantly balance each other; otherwise the revolving body will either approach the centre, or recede from it, according as the centripetal or centrifugal force is the stronger.

What follows if the centripetal or centrifugal force be destroyed?

209. If the centrifugal force of a revolving body be destroyed, the body will immediately approach the centre which attracts it; but if the centripetal force be destroyed, the body will fly off in the direction of a tangent to the curve which it describes in its motion.*

210. Thus, when a mop filled with water is turned swiftly round by the handle, the threads which compose the head will fly off from the centre; but, being confined to it at one end, they cannot part from it; while the water they contain, being unconfined, is thrown off in straight lines.

When a body is revolving around its centre or its axis, what parts move with the greatest velocity?

211. The parts of a body which are furthest from the centre of motion move with the greatest velocity; and the velocity of all the parts diminishes as their distance from the axis of motion diminishes.

Explain Fig. 15.

212. Fig. 15 represents the vanes of a windmill. The circles denote the paths in which the different parts of the vanes move. M is the centre or axis of motion around which all the parts revolve. The outer part revolves in the circle D E F G, another part revolves in the circle H I J K, and the inner part in the circle L N O P. Consequently, as they all revolve around M in the same time, the velocity of the parts which revolve in the outer circle is as much greater than the velocity of the parts which revolve in the inner circle, L N O P, as the diameter of the outer circle is greater than the diameter of the inner.

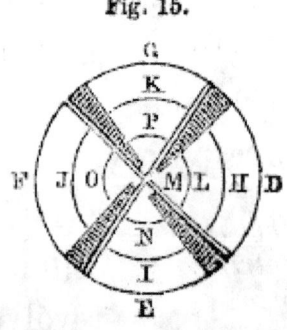

Fig. 15.

* The centrifugal force is proportioned to the square of the velocity of a moving body. Hence, a cord sufficiently strong to hold a heavy body revolving around a fixed centre at the rate of fifty feet in a second, would require to have its strength increased four-fold, to hold the same ball, if its velocity should be doubled.

In the daily revolution of the earth around its own axis, what parts of the earth move most slowly, and what parts most rapidly?

213. As the earth revolves round its axis, it follows, from the preceding illustration, that the portions of the earth which move most rapidly are nearest to the equator, and that the nearer any portion of the earth is to the poles the slower will be its motion.

What is required in order to produce curvilinear motion? and why?

214. Curvilinear motion requires the action of two forces; for the impulse of one single force always produces motion in a straight line.

What effect has the centrifugal force on a body revolving around its longer axis?

215. A body revolving rapidly around its longer axis, if suspended freely, will gradually change the direction of its motion, and revolve around its shorter axis.

This is due to the centrifugal force, which, impelling the parts from the centre of motion, causes the most distant parts to revolve in a larger circle.*

---

* This law is beautifully illustrated by a simple apparatus, in which a hook is made to revolve rapidly by means of multiplying wheels. Let an oblate spheroid, a double cone, or any other solid having unequal axes, be suspended from the hook by means of a flexible cord attached to the extremity of the longer axis. If, now, it be caused rapidly to revolve, it will immediately change its axis of motion, and revolve around the shorter axis.

The experiment will be doubly interesting if an endless chain be suspended from the hook, instead of a spheroid. So soon as the hook with the chain suspended is caused to revolve, the sides of the chain are thrown outward by the centrifugal force, until a complete ring is formed, and then the circular chain will commence revolving horizontally. This is a beautiful illustration of the effects of the centrifugal force. An apparatus, with a chain and six bodies of different form, prepared to be attached to the multiplying wheels in the manner described, accompanies most sets of philosophical apparatus.

Attached to the same apparatus is a thin hoop of brass, prepared for connexion with the multiplying wheels. The hoop is made rapidly to revolve around a vertical axis, loose at the top and secured below. So soon as the hoop begins to revolve rapidly, the horizontal diameter of the ring begins to increase and the vertical diameter to diminish, thus exhibiting the manner in which the equatorial diameter of a revolving body is lengthened, and the polar diameter is shortened, by reason of the centrifugal force. The daily revolution of the earth around its axis has produced this effect, so that the equatorial diameter is at least twenty-six miles longer than the polar. In those planets that revolve faster than the earth the effect is still

62 NATURAL PHILOSOPHY.

What is Projectiles?   **216. Projectiles.** — Projectiles is a branch of Mechanics which treats of the motion of bodies thrown or driven by an impelling force above the surface of the earth.

What is a Projectile?   217. A Projectile is a body thrown into the air, — as a rocket, a ball from a gun, or a stone from the hand.

How are projectiles affected in their motion?   The force of gravity and the resistance of the air cause projectiles to form a curve both in their ascent and descent; and, in descending, their motion is gradually changed from an oblique towards a perpendicular direction.

Explain Fig. 16.   218. In Fig. 16 the force of projection would carry a ball from A to D, while gravity would bring it to C. If these two forces alone prevailed, the ball would proceed in the dotted line to B. But, as the resistance of the air operates in direct opposition to the force of projection, instead of reaching the ground at B, the ball will fall somewhere about E.*

What is the course of a body thrown obliquely in a horizontal direction?   219. When a body is thrown in a horizontal direction, or upwards or downwards, *obliquely*, its course will be in the direction of a curve-line, called a *parabola*†

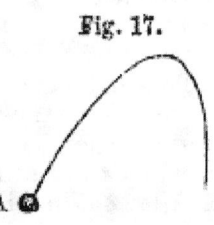

Fig. 17.

more striking, as is the case with the planet Jupiter, whose figure is nearly that of an oblate spheroid.

The developments of Geology have led some writers to the theory that the earth, during one period of its history, must have had a different axis of motion; but it will be exceedingly difficult to reconcile such a theory to the law of rotations which has now been explained, especially as a much more rational explanation can be given to the phenomena on which the theory was built.

* It is calculated that the resistance of the air to a cannon-ball of two pounds' weight, with the velocity of two thousand feet in a second, is more than equivalent to sixty times the weight of the ball.

† The science of *gunnery* is founded upon the laws relating to projectiles.

(see Fig. 17); but when it is thrown *perpendicularly* upwards or downwards, it will move perpendicularly, because the force of projection and that of gravity are in the same line of direction.

The force of gunpowder is accurately ascertained, and calculations are predicated upon these principles, which enable the engineer to direct his guns in such a manner as to cause the fall of the shot or shells in the very spot where he intends. The knowledge of this science saves an immense expenditure of ammunition, which would otherwise be idly wasted, without producing any effect. In attacks upon towns and fortifications, the skilful engineer knows the means he has in his power, and can calculate, with great precision, their effects. It is in this way that the art of war has been elevated into a science, and much is made to depend upon skill which, previous to the knowledge of these principles, depended entirely upon physical power.

The force with which balls are thrown by gunpowder is measured by an instrument called the *Ballistic pendulum*. It consists of a large block of wood, suspended by a rod in the manner of a pendulum. Into this block the balls are fired, and to it they communicate their own motion. Now, the weight of the block and that of the ball being known, and the motion or velocity of the block being determined by machinery or by observation, the elements are obtained by which the velocity of the ball may be found; for *the weight of the ball is to the weight of the block as the velocity of the block is to the velocity of the ball*. By this simple apparatus many facts relative to the art of gunnery may be ascertained. If the ball be fired from the same gun, at different distances, it will be seen how much resistance the atmosphere opposes to its force at such distances. Rifles and guns of smooth bores may be tested, as well as the various charges of powder best adapted to different distances and different guns. These, and a great variety of other experiments, useful to the practical gunner or sportsman, may be made by this simple means.

The velocity of balls impelled by gunpowder from a musket with a common charge has been estimated at about 1650 feet in a second of time, when first discharged. The utmost velocity that can be given to a cannon-ball is 2000 feet per second, and this only at the moment of its leaving the gun.

In order to increase the velocity from 1650 to 2000 feet, one-half more powder is required; and even then, at a long shot, no advantage is gained, since, at the distance of 500 yards, the greatest velocity that can be obtained is only 1200 or 1300 feet per second. Great charges of powder are, therefore, not only useless, but dangerous; for, though they give little additional force to the ball, they hazard the lives of many by their liability to burst the gun.

Experiment has also shown that, although long guns give a greater velocity to the shot than short ones, still that, on the whole, short ones are preferable; and, accordingly, armed ships are now almost invariably furnished with short guns, called carronades.

The length of sporting guns has also been greatly reduced of late years. Formerly, the barrels were from four to six feet in length; but the best fowling-pieces of the present day have barrels of two feet or two and a half only in length. Guns of about this length are now universally employed for such game as woodcocks, partridges, grouse, and such birds as are taken on the wing, with the exceptions of ducks and wild geese, which require longer and heavier guns

64 NATURAL PHILOSOPHY.

What forces affect a horizontal projectile, and what effect do they produce?

220. A ball thrown in a horizontal direction is influenced by three forces; namely, first, the force of projection (which gives it a horizontal direction); second, the resistance of the air through which it passes, which diminishes its velocity, without changing its direction; and third, the force of gravity, which finally brings it to the ground.

How is the force of gravity affected by the force of projection?

221. The force of gravity is neither increased nor diminished by the force of projection.*

Explain Fig. 18.

222. Fig. 18 represents a cannon, loaded with a ball, and placed on the top of a tower, at such a height as to require just three seconds for another ball to descend perpendicularly. Now, suppose the cannon to be fired in a horizontal direction, and at the same instant the other ball to be dropped towards the ground. They will both reach the horizontal line at the base of the tower at the same instant. In this figure C a represents the perpendicular line of the falling ball. C b is the curvilinear path of the projected ball, 3 the horizontal line at the base of the tower. During the first second of time, the falling ball reaches 1, the next second 2, and at the end of the

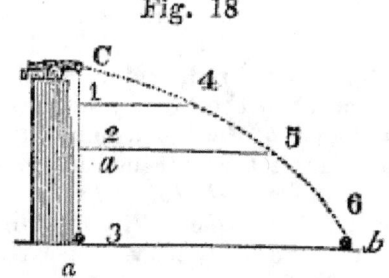

Fig. 18

---

* The action of gravity being always the same, the shape of the curve of every projectile depends on the velocity of its motion; but, whatever this velocity be, the moving body, if thrown horizontally from the same elevation, will reach the ground at the same instant. Thus, a ball from a cannon, with a charge sufficient to throw it half a mile, will reach the ground at the same instant of time that it would had the charge been sufficient to throw it one, two, or six miles, from the same elevation. The distance to which a ball will be projected will depend entirely on the force with which it is thrown, or on the velocity of its motion. If it moves slowly, the distance will be short; if more rapidly, the space passed over in the same time will be greater; but in both cases the descent of the ball towards the earth, in the same time, will be the same number of feet, whether it moves fast or slow, or even whether it move forward at all, or not.

third second it strikes the ground. Meantime, that projected from the cannon moves forward with such velocity as to reach 4 at the same time that the falling ball reaches 1. But the projected ball falls downwards exactly as fast as the other, since it meets the line 1 4, which is parallel to the horizon, at the same instant. During the next second the ball from the cannon reaches 5, while the other falls to 2, both having an equal descent. During the third second the projected ball will have spent nearly its whole force, and therefore its downward motion will be greater, while the motion forward will be less than before.

What effect has the projectile force on gravity? 223. *Hence it appears that the horizontal motion does not interfere with the action of gravity, but that a projectile descends with the same rapidity while moving forward that it would if it were acted on by gravity alone.* This is the necessary result of the action of two forces.

What is the Random of a projectile? 224. The Random of a projectile is the horizontal distance from the place whence it is thrown to the place where it strikes.

At what angle does the greatest random take place? 225. The greatest random takes place at an angle of 45 degrees; that is, when a gun is pointed at this angle with the horizon, the ball is thrown to the greatest distance.

What will be the effect if a ball be thrown at any angle above 45 degrees? Let Fig. 19 represent a gun or a carronade, from which a ball is thrown at an angle of 45 degrees with the horizon. If the ball be thrown at any angle above 45 degrees, the random will be the same as it would be at the same number of degrees below 45 degrees.*

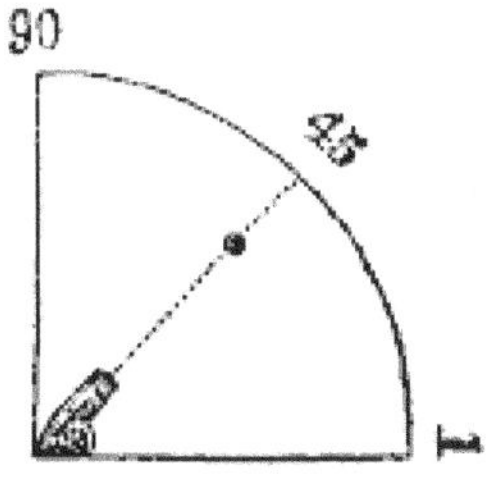

Fig. 19.

---

* A knowledge of this fact, and calculations predicated on it, enables the engineer so to direct his guns as to reach the object of attack when within the range of shot.

66 NATURAL PHILOSOPHY.

What is the Centre of Gravity of a body? 226. CENTRE OF GRAVITY. — It has already been stated [*see Nos.* 109 & 110] that *the Centre of Gravity of a body is the point around which all the parts balance each other.* It is, in other words, the centre of the weight of a body.

What is the Centre of Magnitude? 227. The Centre of Magnitude is the central point of the bulk of a body.

Where is the centre of gravity of a body? 228. When a body is of uniform density, the centre of gravity is in the same point with the centre of magnitude. But when one part of the body is composed of heavier materials than another part, the centre of gravity (being the centre of the weight of the body) no longer corresponds with the centre of magnitude.

Thus the centre of gravity of a cylinder plugged with lead is not in the same point as the centre of magnitude.

If a body be composed of different materials, not united in chemical combination, the centre of gravity will not correspond with the centre of magnitude, unless all the materials have the same specific gravity.

When will a body stand, and when will it fall? 229. When the centre of gravity of a body is supported, the body itself will be supported; but when the centre of gravity is unsupported, the body will fall.*

What is the Line of Direction? 230. A line drawn from the centre of gravity, perpendicularly to the horizon, is called the Line of Direction.

231. The line of direction is merely a line indicating the path which the centre of gravity would describe, if the body were permitted to fall freely.

---

* The Boston School Apparatus contains a set of eight Illustrations for the purpose of giving a clear idea of the centre of gravity, and showing the difference between the centre of gravity and the centre of magnitude.

MECHANICS.

When will a body stand, and when will it fall? 232. When the line of direction falls within the base * of any body, the body will stand; but when that line falls outside of the base, the body will fall, or be overset.

Explain Fig. 21. 233. (1.) Fig. 21 represents a loaded wagon on the declivity of a hill. The line C F represents a horizontal line, D E the base of the wagon. 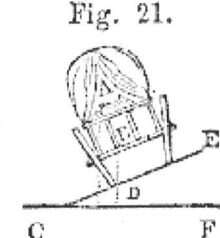 If the wagon be loaded in such a manner that the centre of gravity be at B, the perpendicular B D will fall within the base, and the wagon will stand. But if the load be altered so that the centre of gravity be raised to A, the perpendicular A C will fall outside of the base, and the wagon will be overset. From this it follows that a wagon, or any carriage, will be most firmly supported when the line of direction of the centre of gravity falls exactly between the wheels; and that is the case on a level road. The centre of gravity in the human body is between the hips, and the base is the feet.

234. So long as we stand uprightly, the line of direction falls within this base. When we lean on one side, the centre of gravity not being supported, we no longer stand firmly.

How does a rope-dancer perform his feats of agility? 235. A rope-dancer performs all his feats of agility by dexterously supporting the centre of gravity. For this purpose, he carries a heavy pole in his hands, which he shifts from side to side as he alters his position, in order to throw the weight to the side which is deficient; and thus, in changing the situation of the centre of gravity, he keeps the line of direction within the base, and he will not fall.†

---

* The base of a body is its lowest side. The base of a body standing on wheels or legs is represented by lines drawn from the lowest part of one wheel or leg to the lowest part of the other wheel or leg.
Thus, in Figs. 20 and 21, D E represents the base of the wagon and of the table.

† The shepherds in the south of France afford an interesting instance of the application of the art of balancing to the common business of life. These men walk on stilts from three to four feet high, and their children,

236. A spherical body will roll down a slope, because the centre of gravity is not supported.*

237. Bodies, consisting of but one kind of substance, as wood, stone or lead, and whose densities are consequently uniform, will stand more firmly than bodies composed of a variety of substances, of different densities, because the centre of gravity in such cases more nearly corresponds with the centre of magnitude.

238. When a body is composed of different materials, it will stand most firmly when the parts whose specific gravity is the greatest are placed nearest to the base.

When will a body stand most firmly? 239. The broader the base and the nearer the centre of gravity to the ground, the more firmly a body will stand.

240. For this reason, high carriages are more dangerous than low ones.

241. A pyramid also, for the same reason, is the firmest of all

Fig. 22.

structures, because it has a broad base, and but little elevation.

when quite young, are taught to practise the same art. By means of these odd additions to the length of the leg, their feet are kept out of the water, or the heated sand, and they are also enabled to see their sheep at a greater distance. They use these stilts with great skill and care, and run, jump, and even dance on them, with great ease.

* A cylinder can be made to roll up a slope, by plugging one side of it with lead; the body being no longer of a uniform density, the centre of gravity is removed from the middle of the body to some point in the lead as that substance is much heavier than wood. Now, in order that the cylinder may roll down the plane, as it is here situated, the centre of gravity must rise, which is impossible; the centre of gravity must always descend in moving, and will descend by the nearest and readiest means, which will be by forcing the cylinder up the slope, until the centre of gravity is supported, and then it stops.

A body also in the shape of two cones united at their bases can be made to roll up an inclined plane formed by two bars with their lower ends inclined towards each other. This is illustrated by a simple contrivance in the "Boston School Set," and the fact illustrated is called "*the mechanical paradox.*"

MECHANICS. 69

242. A cone has also the same stability; but, mathematically considered, a cone is a pyramid with an infinite number of sides.

243. Bodies that have a narrow base are easily overset, because, if they are but slightly inclined, the line of direction will fall outside of the base, and consequently their centre of gravity will not be supported.

Why can a person carry two pails of water more easily than one?
244. A person can carry two pails of water more easily than one, because the pails balance each other, and the centre of gravity remains supported by the feet. But a single pail throws the centre of gravity on one side, and renders it more difficult to support the body.

Where is the centre of gravity of two bodies connected together?
245. COMMON CENTRE OF GRAVITY OF TWO BODIES. — When two bodies are connected, they are to be considered as forming but one body, and have but one centre of gravity. If the two bodies be of equal weight, the centre of gravity will be in the middle of the line which unites them. But, if one be heavier than the other, the centre of gravity will be as much nearer to the heavier one as the heavier exceeds the light one in weight.

Explain Figures 23, 24, and 25.
246. Fig. 23 represents a bar with an equal weight fastened at each end; the centre of gravity is at A, the middle of the bar, and whatever supports this centre will support both the bodies and the pole.

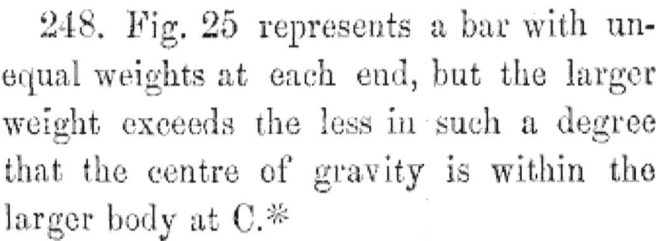

Fig. 23.

247. Fig. 24 represents a bar with an unequal weight at each end. The centre of gravity is at C, nearer to the larger body.

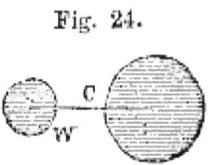

Fig. 24.

248. Fig. 25 represents a bar with unequal weights at each end, but the larger weight exceeds the less in such a degree that the centre of gravity is within the larger body at C.*

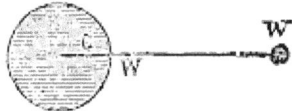

Fig. 25.

---

* There are no laws connected with the subject of Natural Science so grand and stupendous as the laws of attraction. Long before the sublime fiat, "*Let there be light,*" was uttered, the Creator's voice was heard amid

What things in Mechanics require distinct consideration?

249. THE MECHANICAL POWERS. — There are five things in mechanics which require a distinct consideration, namely:

First, the power that acts.

Secondly, the resistance which is to be overcome by the power.

Thirdly, the centre of motion, or, as it is sometimes called, the fulcrum.*

Fourthly, the respective velocities of the power and the resistance; and,

the expanse of universal emptiness, calling matter into existence, and subjecting it to these laws. Obedient to the voice of its Creator, matter sprang from "primeval nothingness," and, in atomic embryos, prepared to cluster into social unions. Spread abroad in the unbounded fields of space, each particle felt that it was "*not good to be alone.*" Invested with the social power, it sought companionship. The attractive power, thus doubled by the union, compelled the surrounding particles to join in close embrace, and thus were worlds created. Launched into regions of unbound space, the new-created worlds found that their union was but a part of a great social system of law and order. Their bounds were set. A central point controls the Universe, and in harmonious revolution around this central point for ages have they rolled. Nor can one lawless particle escape. The sleepless eye of Nature's law, vicegerent of its God, securely binds them all.

"Could but one small, rebellious atom stray,
Nature itself would hasten to decay."

With this sublime view of Creation, how can we escape the conclusion that the very existence of *a law* necessarily implies a Law-giver, and that Law-giver must be the Creator? Shall we not then say, with the Psalmist, "*It is the* FOOL *who hath said in his heart that there is no God*"?

Who, then, will not see and admire the beautiful language of Mr. Alison, while his heart burns with the rapture and gratitude which the sentiments are so well fitted to kindle:

"When, in the youth of Moses, 'the Lord appeared to him in Horeb,' a voice was heard, saying, 'Draw nigh hither, and put off thy shoes from off thy feet, for the place where thou standest is holy ground.' It is with such a reverential awe that every great or elevated mind will approach to the study of nature, and with such feelings of adoration and gratitude that he will receive the illumination that gradually opens upon his soul."

"It is not the lifeless mass of matter, he will then feel, that he is examining; it is the mighty machine of Eternal Wisdom, — the workmanship of Him 'in whom everything lives, and moves, and has its being.' Under an aspect of this kind, it is impossible to pursue knowledge without mingling with it the most elevated sentiments of devotion; — it is impossible to perceive the laws of nature without perceiving, at the same time, the presence and the providence of the Law-giver: — and thus it is that, in every age, the evidences of religion have advanced with the progress of true philosophy; and that SCIENCE, IN ERECTING A MONUMENT TO HERSELF, HAS, AT THE SAME, ERECTED AN ALTAR TO THE DEITY."

* The word *fulcrum* means a prop, or support.

THE MECHANICAL POWERS. 71

Fifthly, the instruments employed in the construction of the machine.

250. (1.) The power that acts is the muscular strength of men or animals, the weight and momentum of solid bodies, the elastic force of steam, springs, the pressure of the air, the weight of water and its force when in motion, &c.

(2.) The resistance to be overcome is the attraction of gravity or of cohesion, the inertness of matter, friction, &c.

(3.) The centre of motion, or the fulcrum, is the point about which all the parts of the body move.

(4.) The velocity is the rapidity with which an effect is produced.

(5.) The instruments are the mechanical powers which enter into the construction of the machine.

What are the Mechanical Powers? 251. The powers which enter into the construction of a machine are called the Mechanical Powers. They are contrivances designed to increase or to diminish force, or to alter its direction.

What is the fundamental principle of Mechanics? 252. All the Mechanical Powers are constructed on the principle that *what is gained in power is lost in time*. This is the fundamental law of Mechanics.

253. If 1 lb. is required to overcome the resistance of 2 lbs., the 1 lb. must move over two feet in the same time that the resistance takes to move over one. Hence the resistance will move only half as fast as the power; or, in other words, the resistance requires double the time required by the power to move over a given space.

Explain Fig. 26. 254. Fig. 26 illustrates the principle as applied to the lever. W represents the weight, F the fulcrum, P the power, and the bar W F P the lever. To raise the weight W to *w*, the power P must descend to *p*. But, as the radius of the circle in which the power P moves is double that of the radius of the circle in which the weight W moves,

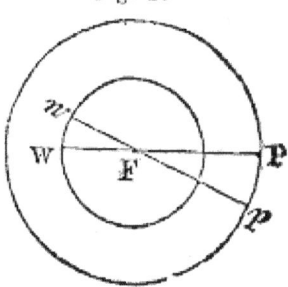

Fig. 26.

the arc P p is double the arc W w; or, in other words, the distance P p is double the distance of W w. Now, as these distances are traversed in the same time by the power and the weight respectively, it follows that the velocity of the power must be double the velocity of the weight; that is, the power must move at the rate of two feet in a second, in order to move the weight one foot in the same time.

This principle applies not only to the lever, but to all the Mechanical Powers, and to all machines constructed on mechanical principles.

How many Mechanical Powers are there, and their names? 255. There are six Mechanical Powers: the Lever, the Wheel and Axle, the Pulley, the Inclined Plane, the Wedge and the Screw.

All instruments and machines are constructed on the principle of one or more of the Mechanical Powers.

All the Mechanical Powers may be reduced to three classes, namely: 1st, a body revolving on an axis; 2d, a flexible cord; and, 3d, an inclined surface, smooth and hard. To the first belongs the lever, and the wheel and axle; to the second, the pulley; to the third, the inclined plane, the wedge and the screw.

What is the Lever, and how is it used? 256. The Lever is an inflexible bar, movable on a fulcrum or prop.

It is used by making one part to rest on a fulcrum, applying the power to bear on another part, while a third part of the lever opposes its motion to the resistance which is to be overcome.

257. In every lever, therefore, whatever be its form, there are three things to be distinctly considered, namely: the position of the fulcrum, of the power, and of the weight, respectively. It is the position of these which makes the distinction between the different kinds of levers.

How many kinds of levers are there? 258. There are three kinds of levers, called the first, second and third, according to the respective position of the fulcrum, the power, and the weight.

These may be represented thus:

Power,	Fulcrum,	Weight.
Power,	Weight,	Fulcrum.
Weight,	Power,	Fulcrum.

THE MECHANICAL POWERS. 73

What is the position of the power, the weight, and the fulcrum, respectively, in the three kinds of lever?

That is, (1.) The power * is at one end, the weight at the other, and the fulcrum between them.

(2.) Power at one end, the fulcrum at the other, and the weight between them.

(3) The weight is at one end, the fulcrum at the other, and the power between them.

Describe a lever of the first kind by figure 27, and tell the advantage gained by it.

259. In a lever of the first kind the fulcrum is placed between the power and the weight.

Fig. 27 represents a lever of the first kind, resting on the fulcrum F, and movable upon it. W is the weight to be moved, and P is the power which moves it. *The advantage gained in raising a weight, by the use of this kind of lever, is in proportion as the distance of the power from the fulcrum exceeds that of the weight from the fulcrum.* Thus, in this figure, if the distance between P and F be double that between W and F, then a man, by the exertion of a force of 100 pounds with the lever, can move a weight of 200 pounds. From this it follows that *the nearer the power is applied to the end of the lever, the greater is the advantage gained.* Thus, a greater weight can be moved by the same power when applied at B than when it is exerted at P.

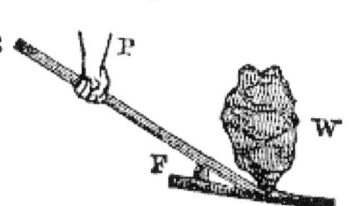
Fig. 27.

On what principle is the common steelyard constructed?

Describe the steelyard.

260. The common steelyard, an instrument for weighing articles, is constructed on the principle of the lever of the first kind. It consists of a rod or bar, marked with notches to designate the pounds and ounces, and a weight, which is mova-

---

* It is to be understood, in the consideration of all instruments and machines, that some effect is to be produced by some power. The names *power* and *weight* are not always to be taken literally. They are terms used to express the cause and the effect. Thus, in the movement of a clock, the weight is the cause, the movement of the hands is the effect. The cause of motion, whether it be a weight or a resistance, *is technically called the power;* the effect, whether it be the raising of a weight, the overcoming of resistance or of cohesion, the separation of the parts of a body, compression or expansion, is technically called the *weight.*

ble along the notches. The bar is furnished with three hooks, *on the longest of which the article to be weighed is always to be hung.* The other two hooks serve for the handle of the instru

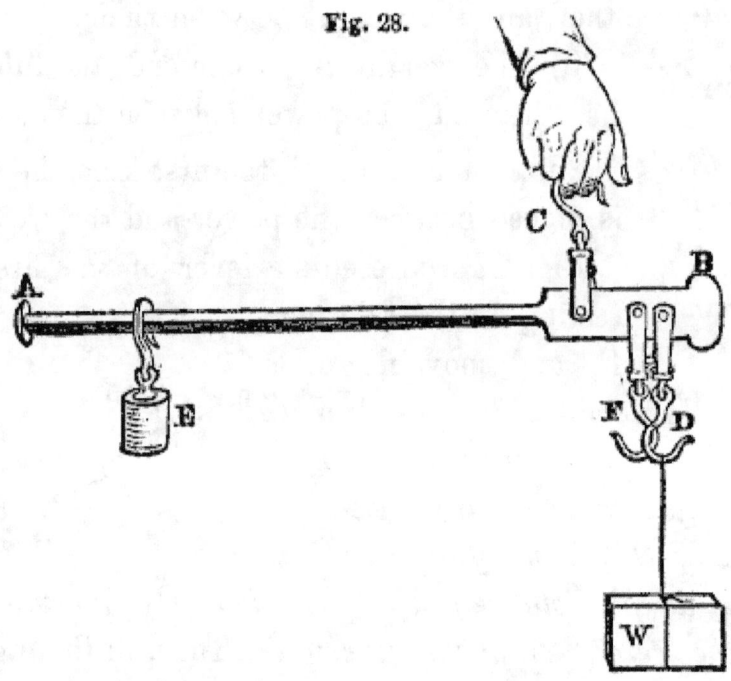

Fig. 28.

ment when in use. The pivot of each of these two hooks serves for the fulcrum.

Of what use are the three hooks in the steelyard?

261. When suspended by the hook C, as in Fig. 28, it is manifest that a pound weight at E will balance as many pounds at W as the distance between the pivot of D and the pivot of C is contained in the space between the pivot of C and the ring from which E is suspended.

The same instrument may be used to weigh heavy articles, by using the middle hook for a handle, where, as will be seen in Fig. 29, the space between the pivot of F (which in this case is the fulcrum) and the pivot of D (from which the weight is suspended) being lessened, is contained a greater number of times in the distance between the fulcrum and the notches on the bar. The steelyard is furnished with two sets of notches on opposite sides of the bar. An equilibrium* will always be

---

* *Of Equilibrium.* — In the calculations of the powers of all machines it is

produced when the product of the weights on the opposite sides of the fulcrum into their respective distances from it are equal to one another.

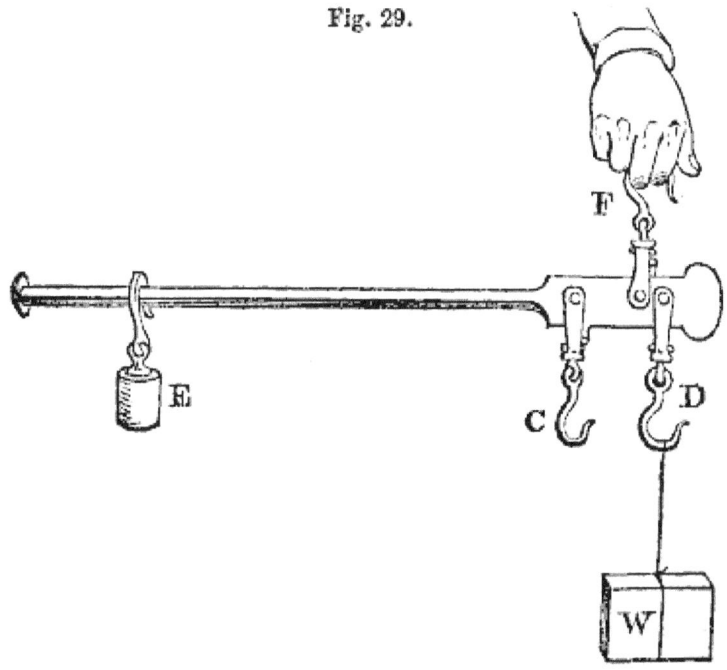

Fig. 29.

A balance, or pair of scales, is a lever of the first kind, with equal arms. Steelyards, scissors, pincers, snuffers, and a poker used for stirring the fire, are all levers of the first kind. The longer the handles of scissors, pincers, &c., and the shorter the points, the more easily are they used.

262. The lever is made in a great variety of forms and of many different materials, and is much used in almost every kind of mechanical operation. Sometimes it is detached from the fulcrum,

---

necessary to have clearly in mind the difference between action and equilibrium.

By equilibrium is meant an equality of forces; as, when one force is opposed by another force, if their respective momenta are equal, an equilibrium is produced, and the forces merely counterbalance each other. To produce any action, there must be inequality in the condition of one of the forces. Thus, a power of one pound on the longer arm of a lever will balance a weight of two pounds on the shorter arm, if the distance of the power from the fulcrum be exactly double the distance of the weight from the fulcrum; and the reason why they exactly balance is, because their momenta are equal. No motion can be produced or destroyed without a difference between the force and the resistance. In calculating the mechanical advantage of any machine, therefore, the condition of equilibrium must first be duly considered. After an equilibrium is produced, whatever is added upon the one side or taken away on the other destroys the equilibrium, and causes the machine to move.

but most generally the fulcrum is a pin or rivet by which the lever is permanently connected with the frame-work of other parts of the machinery.

263. When two weights are equal, and the fulcrum is placed exactly in the centre of the lever between them, they will mutually balance each other; or, in other words, the centre of gravity being supported, neither of the weights will sink. This is the principle of the common scale for weighing.

How is power gained by the use of the lever? 264. To gain power by the use of the lever, the fulcrum must be placed near the weight to be moved, and the power at the greater distance from it. *The force of the lever, therefore, depends on its length, together with the power applied, and the distance of the weight from the fulcrum.**

What is a Compound Lever? 265. A Compound Lever, represented in Fig. 30, consists of several levers, so arranged that the shorter arm of one may act on the longer arm of the other. Great power is obtained in this way, but its exercise is limited to a very small space.

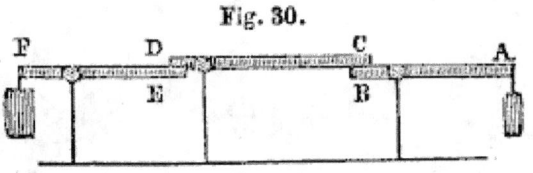
Fig. 30.

Describe the lever of the second kind, with Fig. 31. 266. In a lever of the second kind, the fulcrum is at one end, the power at the other, and the weight between them.

(1.) Let Fig. 31 represent a lever of the second kind. F is the fulcrum, P the power, and W the weight. The advantage gained by a lever of this kind is in proportion as the distance of the power from the fulcrum exceeds that of the weight from the fulcrum. Thus, in this figure, if the distance

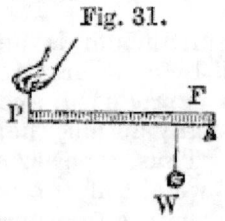
Fig. 31.

---

* This being the case, it is evident that the shape of the lever will not influence its power, whether it be *straight* or *bent*. The *direct distance* between the fulcrum and the weight, compared with the same distance between the fulcrum and the power, being the only measure of the mechanical advantage which it affords.

from P to F is four times the distance from W to F, then a power of one pound at P will balance a weight of four pounds at W.

(2.) On the principle of this kind of lever, two persons, carrying a heavy burden suspended on a bar, may be made to bear unequal portions of it, by placing it nearer to the one than the other.

267. Two horses, also, may be made to draw unequal portions of a load, by dividing the bar attached to the carriage in such a manner that the weaker horse may draw upon the longer end of it.

268. Oars, rudders of ships, doors turning on hinges, and cutting-knives which are fixed at one end, are constructed upon the principle of levers of the second kind.*

Fig. 32.

Describe the lever of the third kind by Fig. 33.

269. In a lever of the third kind the fulcrum is at one end, the weight at the other, and the power is applied between them.

In levers of this kind *the power must always exceed the weight in the same proportion as the distance of the weight from the fulcrum exceeds that of the power from the fulcrum.*

In Fig. 33 F is the fulcrum, W the weight, and P the power between the fulcrum and the weight; and the power must exceed the weight in the same proportion that the distance between W and F exceeds the distance between P and F.

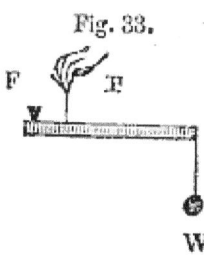

Fig. 33.

270. A ladder, which is to be raised by the strength of a man's arms, represents a lever of this kind, where the fulcrum is that end which is fixed against the wall; the weight may be considered as at the top part of the ladder, and the power is the strength applied in raising it.

271. The bones of a man's arm, and most of the movable bones of animals, are levers of the third kind. But the loss of power in limbs of animals is compensated by the beauty and compactness of

---

* It is on the same principle that, in raising a window, the hand should be applied to the middle of the sash, as it will then be easily raised, whereas, if the hand be applied nearer to one side than the other, the centre of gravity being unsupported, will cause the further side to bear against the frame, and obstruct its free motion.

the limbs, as well as the increased velocity of their motion. The wheels in clock and watch work, and in various kinds of machinery, may be considered as levers of this kind, when the power that moves them acts on the pinion, near the centre of motion, and the resistance to be overcome acts on the teeth at the circumference. But here the advantage gained is the change of slow into rapid motion.

272. PRACTICAL EXAMPLES OF LEVERAGE.

Questions for Solution

(1.) Suppose a lever, 6 feet in length, to be applied to raise a weight of 50 pounds, with a power of only 1 pound, where must the fulcrum be placed? *Ans.* 1.41 *in.* +

(2.) If a man wishes to move a stone weighing a ton with a crow-bar 6 feet in length, he himself being able, with his natural strength, to move a weight of 100 pounds only, what must be the greatest distance of the fulcrum from the stone? *Ans.* 3.42 *in.* +

(3.) If the distance of the power from the fulcrum be eighteen times greater than the distance of the weight from the fulcrum, what power would be required to lift a weight of 1000 pounds? *Ans.* 55.55 *lb.* +

(4.) If the distance of the weight from the fulcrum be only a tenth of the distance of the power from the fulcrum, what weight can be raised by a power of 170 pounds? *Ans.* 1700 *lb.*

(5.) In a pair of steelyards the distance between the hook on which the weight is hung and the hook by which the instrument is suspended is 2 inches; the length of the steelyards is 30 inches. How great a weight may be suspended on the hook to balance a weight of 2 pounds at the extremity of the longer arm? *Ans.* 28 *lb.*

(6.) Archimedes boasted that, if he could have a place to stand upon, he could move the whole earth. Now, suppose that he had a fulcrum with a lever, and that his weight, compared with that of the earth, was as 1 to 270 millions. Suppose, also, that the fulcrum were a thousand miles from the earth; what must be his distance from the fulcrum?
Ans. 270,000,000,000 *mi.*

(7.) Which will cut the more easily, a pair of scissors 9 inches long, with the rivet 5 inches from the points, or a pair of scissors 6 inches long, with the rivet 4 inches from the points? *Ans.* The first.

(8.) Two persons, of unequal strength, carry a weight of 200 pounds suspended from a pole 10 feet long. One of them can carry only 75 pounds, the other must carry the rest of the weight. How far from the end of the pole must the weight be suspended? *Ans.* 3.75 *ft.*

(9.) How must the whiffle-tree * of a carriage be attached, that one horse may draw but 3 cwt. of the load, while the other draws 5 cwt.? *Ans.* At ⅜.

(10.) On the end of a steelyard, 3 feet long, hangs a weight of 4 pounds. Suppose the hook, to which articles to be weighed are attached, to be at the extremity of the other end, at the distance of 4 inches from the hook by which the steelyards are held up. How great a weight can be estimated by the steelyard? *Ans.* 32 *lb.*

What is the Wheel and Axle?

273. THE WHEEL AND AXLE. — The Wheel and Axle consists of a cylinder with a wheel attached, both revolving around the same axis of motion.

---

* The whiffle-tree is generally attached to a carriage by a hook or leather band *in the centre,* so that the draft shall be equal on both sides. The hook or leather band thus becomes a fulcrum.

THE MECHANICAL POWERS.   79

How are the power and the weight applied to the wheel and axle?

274. The weight is supported by a rope or chain wound around the cylinder; the power is applied to another rope or chain wound around the circumference of the cylinder. Sometimes projecting spokes from the wheel supply the place of the chain.*

275. The place of the cylinder is sometimes supplied by a small wheel.

Explain the construction of the wheel and axle by Fig. 34.

276. The wheel and axle, though made in many forms, will easily be understood by inspecting Figs. 34 and 35. In Fig. 34 P represents the larger wheel, where the power is applied; C the smaller wheel, or cylinder, which is the axle; and W the weight to be raised.

Fig. 34.

What is the advantage gained by the use of the wheel and axle?

The advantage gained is in proportion as the circumference of the wheel is greater than that of the axle. That is, if the circumference of the wheel be six times the circumference of the axle, then a power of one pound applied at the wheel will balance a power of six pounds on the axle.

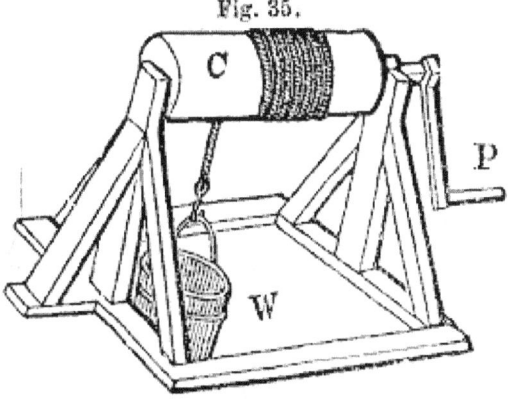
Fig. 35.

How does the wheel and axle described in Fig. 35 differ from that described in Fig. 34?

277. Sometimes the axle is constructed with a winch or handle, as in Fig. 35, and sometimes the wheel has projecting spokes, as in Fig. 34.

* A cylinder is a long circular body of uniform diameter, with extremities forming equal and parallel circles.

On what principle is the wheel and axle constructed?

278. The principle upon which the wheel and axle is constructed is the same with that of the other Mechanical Powers, the want of power being compensated by velocity. It is evident (from the Figs. 34 and 35) that the velocity of the circumference of the wheel is as much greater than that of the axle as it is further from the centre of motion; for the wheel describes a great circle in the same time that the axle describes a small one; therefore the power is increased in the same proportion as the circumference of the wheel is greater than that of the axle. If the velocity of the wheel be twelve times greater than that of the axle, a power of one pound on the wheel will support a weight of twelve pounds on the axle.

279. The wheel and axle are sometimes called "*the perpetual lever,*" the diameter of the wheel representing the longer arm, the diameter of the axle representing the shorter arm, the fulcrum being at the common centre.

280. The capstan,* on board of ships and other vessels, is constructed on the principle of the wheel and axle. It consists of an axle placed uprightly, with a head or drum, pierced with holes for the lever, or levers, which supply the place of the wheel.

281. Windmills, lathes, the common windlass, used for drawing water from wells, and the large wheels in mills, are all constructed on the principle of the wheel and axle.

282. Wheels are a very essential part to most machines. They are applied in different ways, but, when affixed to the axle, their mechanical power is always in the same proportion; that is, as the circumference of the wheel exceeds that of the axle, so much will the power be increased. Therefore, the larger the wheel, and the smaller the axle, the greater will be the power obtained.

What are Cranks, and how are they made?

283. CRANKS.—Cranks are sometimes connected with the axle of a wheel, either to give or to receive its motion. They are made by bending the axle in such a manner as to form four right angles facing in different directions, as is represented in Fig. 36. *They are, in fact, nothing more than a double winch.*

Fig. 36.

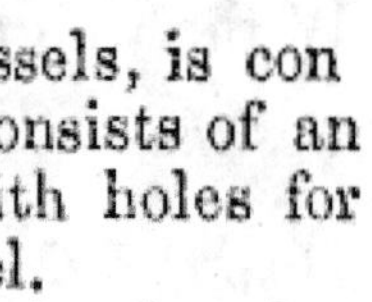

* The difference between a capstan and a windlass lies only in the position of the wheel. If the wheel turn horizontally, it is called a capstan; if vertically, a windlass.

284. A rod connects the crank with other parts of the machinery, either to communicate motion to or from a wheel. When the rod which communicates the motion stands perpendicular to the crank, which is the case twice during each revolution, it is at what is commonly called the *dead point*, and the crank loses all its power. But, when the rod stands obliquely to the crank, the crank is then effective, and turns or is turned by the wheel.

285. Cranks are used in the common foot-lathe to turn the wheel. They are also common in other machinery, and are very convenient for changing rectilinear to circular motion, or circular to rectilinear.

286. When they communicate motion *to* the wheel they operate like the shorter arm of a lever; and, on the contrary, when they communicate the motion *from* the wheel they act like the longer arm.

What are Fly-wheels, and what is their use?
287. FLY-WHEELS are heavy rims of metal secured by light spokes to an axle. They are used to accumulate power, and distribute it equally among all the parts of a machine. They are caused to revolve by a force applied to the axle, and, when once set in motion, continue by their inertia to move for a long time. As their motion is steady, and without sudden jerks, they serve to steady the power, and cause a machine to work with regularity.

288. Fly-wheels are particularly useful in connexion with cranks, especially when at the *dead points*, as the momentum of the fly-wheel, received from the cranks when they acted with most advantage, immediately carries the crank out of the neighborhood of the dead points, and enables it to again act with advantage.

289. There are two ways in which the wheel and axle is supported, namely, first on pointed pivots, projecting into the extremities of the axle,* and, secondly, with the extremities of the axle resting on gudgeons. As by the former mode a less extensive area is subjected to friction, it is in many cases to be preferred.

How many kinds of
290. WATER-WHEELS. — There are three kinds of Water-wheels, called, respectively,

---

* The terms *axle, axis, arbor* and *shaft,* are synonymously used by mechanics to express the bar or rod which passes through the centre of a wheel. The terminations of a horizontal arbor are called gudgeons, and of an upright one frequently pivots; but gudgeons more frequently denote the *beds* on which the extremities of the axle revolve, and pivots are either the pointed extremities of an axle, or short pins in the frame of a machine which receive the extremities of the axle. The term axis, in a more exact sense, may mean merely the longest central diameter, or a **diameter** about which motion takes place.

82 NATURAL PHILOSOPHY.

Water-wheels are there? the Overshot, the Undershot and the Breast Wheel.

291. The Overshot Wheel receives its motion from the weight of the water flowing in at the top.

Describe the Overshot Wheel. Fig. 37 represents the Overshot Wheel. It consists of a wheel turning on an axis (not represented in the figure), with compartments called buckets, *a b c d*, &c., at the circumference, which are successively filled with water from the stream S. The weight of the water in the buckets causes the wheel to turn, and the buckets, being gradually inverted, are emptied as they descend. It will be seen, from an inspection of the figure, that the buckets in the descending side of the wheel are always filled, or partly filled, while those in the opposite or ascending part are always empty until they are again presented to the stream. This kind of wheel is the most powerful of all the water-wheels.

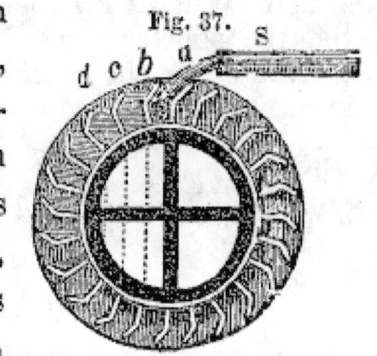

Fig. 37.

292. The Undershot Wheel is a wheel which is moved by the motion of the water. It receives its impulse at the bottom.

Describe the Undershot Wheel. Fig. 38 represents the Undershot Wheel. Instead of buckets at the circumference, it is furnished with plane surfaces, called float-boards, *a b c d*, &c., which receive the impulse of the water, and cause the wheel to revolve.

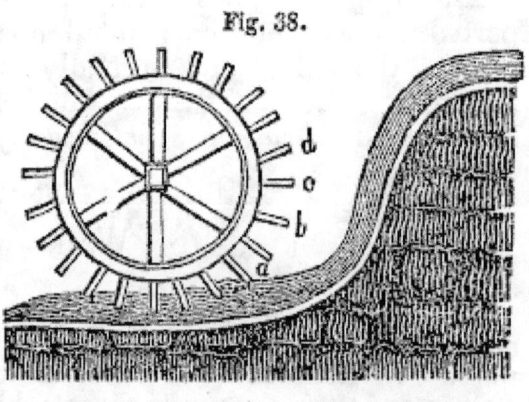

Fig. 38.

Describe the Breast Wheel. 293. The Breast Wheel is a wheel which receives the water at about half its own height, or at the level of its own axis. It is moved both by the weight and the motion of the water.

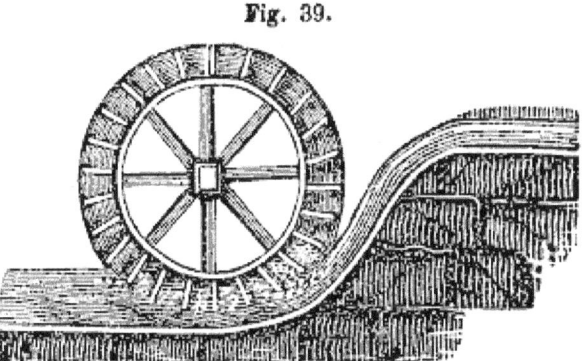

Fig. 39.

Fig. 39 represents a Breast Wheel. It is furnished either with buckets, or with float-boards, fitting the water-course, receiving the weight of the water with its force, while in motion it turns with the stream.

294. In the water-wheels which have now been described, the motion is given to the circumference of the larger wheel, either by the weight of the water or by its force when in motion.

295. All wheels used in machinery are connected with the different parts of the machine by other parts, called *gearing*. Sometimes they are turned by the friction of endless bands or cords, and sometimes by cogs, teeth, or pinions. When turned by bands, the motion may be direct or reversed by attaching the band with one or two centres of motion respectively.

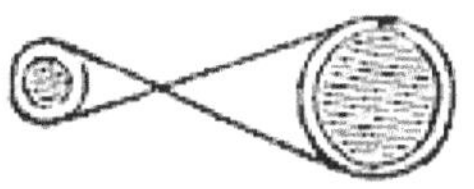

296. When the wheel is intended to revolve in the same direction with the one from which it receives its motion, the band is attached as in Fig. 40; but when it is to revolve in a contrary direction, it is crossed as in Fig. 41. In Fig. 40 the band has but one centre of motion; in Fig. 41 it has two.

Fig. 40.

Fig. 41.

297. Instead of the friction of bands, the rough surfaces of the wheels themselves are made to communicate their motion. The wheels and axles thus rubbing together are sometimes coated with rough leather, which, by increasing the friction, prevents their slipping over one another without communicating motion.

298. Figure 42 represents such a combination of wheels. As the wheel a is turned by the weight S, its axle presses against the circumference of the wheel b, causing it to turn; and, as it turns, its axle rubs against the circumference of the wheel c, which in like manner communicates its motion to d. Now, as the circumference of the wheel a is equal to six times the circumference of its axle, it is evident that when the wheel a has made one revolution b will have performed only one-sixth of a revolution.

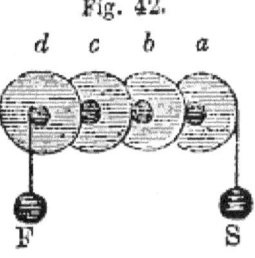

Fig. 42.

The wheel a must therefore turn round six times to cause b to turn once. In like manner b must perform six revolutions to cause c to turn once, and c must turn as many times to cause d to revolve once. Hence it follows that while *d* revolves once on its axis *c* must revolve six times, *b* thirty-six times, and *a* two hundred and sixteen times.

299. If, on the contrary, the power be applied at F, the conditions will all be reversed, and *c* will revolve six times, *b* thirty-six, and *a* two hundred and sixteen times. Thus it appears that we may obtain rapid or slow motion by the same combination of wheels.

How may rapid or slow motion be obtained at pleasure by a combination of wheels with their axles?

300. To obtain rapid motion, the power must be applied to the axle; to obtain slow motion, the power must be applied to the circumference of the wheel.

301. Wheels are sometimes moved by means of cogs or teeth articulating one with another, on the circumference of the wheel and the axle. The cogs on the surface of the wheels are generally called teeth, and those on the surface of the axle are called leaves. The axle itself, when furnished with leaves, is called a *pinion*.

302. Fig. 43 represents a connexion of cogged wheels. The wheel B, being moved by a string around its circumference, is a simple wheel, without teeth. Its axle, being furnished with cogs or *leaves*, to which the teeth of the wheel D are fitted, communicates its motion to D, which, in like manner, moves the wheel C. The power P and the weight W must be attached to the circumference of the wheel or of the axle, according as a slow or a rapid motion is desired.

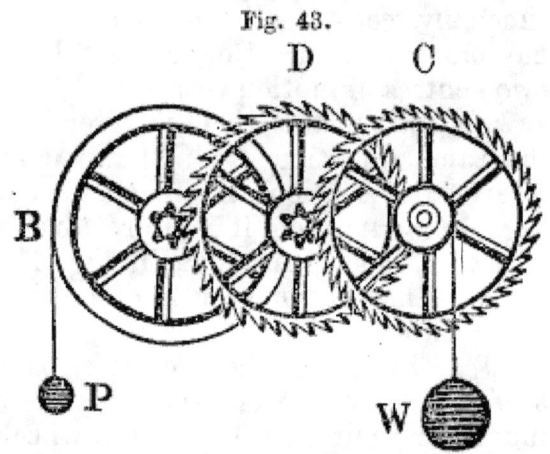

Fig. 43.

303. Wheels with teeth or cogs are of three kinds, according to the position of the teeth. When the teeth are raised perpendicular to the axis, they are called *spur wheels*, or spur gear. When the teeth are parallel with the axis, they are called *crown wheels*. When

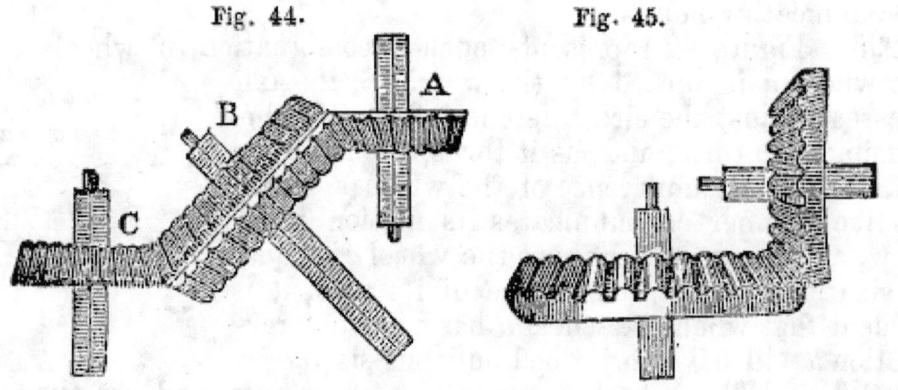

Fig. 44.   Fig. 45.

they are raised on a surface inclined to the axis, they are called *bevelled wheels.* In Fig. 43 the wheels are spur wheels. In Figs. 44 and 45 the wheels are bevelled wheels.

304. Different directions may be given to the motion produced by wheels, by varying the position of their axles, and causing them to revolve in different planes, as in Fig. 44; or by altering the shape and position of the cogs, as in Fig. 45.

How may the power of toothed wheels be estimated? 305. The power of toothed wheels may be estimated by substituting the number of teeth in the wheel and the number of leaves in the pinion for the diameter or the circumference of the wheel and axle respectively.

306. SUSPENSION OF ACTION.—In the arrangement of machinery, it is often necessary to cut off the action of the moving power from some parts, while the rest continues in motion. This is done by causing a toothed wheel to slide aside in the direction of its axis to and from the cogs or leaves into which it articulates, or, when the motion is communicated by a band, by causing the band to slip aside from the wheel to another wheel, which revolves freely around the axle, without communicating its motion.

307. Wheels are used on vehicles to diminish the friction of the road. The larger the circumference of the wheel, the more readily it will overcome obstacles, such as stones or inequalities in the surface of the road.

308. A large wheel is also attended with two additional advantages; namely, first, in passing over holes, ruts and excavations, a large wheel sinks less than a small one, and consequently causes less jolting and expenditure of power; and, secondly, the wear of large wheels is less than that of small ones, for, if we suppose a wheel six feet in diameter, it will turn round but once while a wheel three feet in diameter will turn round twice, its tire will come twice as often to the ground, and its spokes will twice as often have to bear the weight of the load.

309. But wheels must be limited in size by two considerations: first, the strength of the materials; and secondly, the centre of the wheel should never be higher than the breast of the horse, or other animal by which the vehicle is drawn; for otherwise the animal would have to draw obliquely downward, as well as forward, and thus expend part of his strength in drawing against the ground.*

---

* In descending a steep hill, the wheels of a carriage are often *locked* (as it is called), that is, fastened in such a manner as to prevent their turning; and thus the rolling is converted into the sliding friction, and the vehicle descends more safely.

Castors are put on the legs of tables and other articles of furniture, to facilitate the moving of them; and thus the sliding is converted into the rolling friction.

310. Practical Examples of Power applied to the Wheel and Axle.

Questions for Solution.

(1.) With a wheel 5 feet in diameter and a power of 6 pounds, what must be the diameter of the axle to support 3 cwt. ? *Ans.* 1.2 *in.*

(2.) How large must be the diameter of the wheel to support with 10 lbs. a weight of 5 cwt. on an axle 9 inches in diameter ? *Ans.* 37.5 *ft.*

(3.) A wheel has a diameter of 4 feet, an axle of 6 inches. What power must be applied to the wheel to balance 2 cwt. on the axle ? *Ans.* 25 *lb.*

(4.) There is a connexion of cogged wheels having 6 leaves on the pinion and 36 cogs on the wheel. What is the proportion of the power to the weight in equilibrium ? *Ans.* As 1 to 6.

(5.) Suppose a lever of six feet inserted in a capstan 2 feet in diameter, and six men whose united strength is represented by ½ of a ton at the capstan, how heavy an anchor can they draw up, allowing the loss of ⅓ of their power from friction ? *Ans.* 2 *T.*

(6.) What must be the proportion of the axle to the wheel, to sustain a weight 30 cwt. with a power of 3 cwt. ? *Ans.* As 1 to 10.

(7.) The weight is to the power in the proportion of six to one. What must be the proportion of the wheel to the axle ? *Ans.* 6 to 1.

(8.) The power is represented by 10, the axle by 2. How can you represent the wheel and axle ? *Ans.* 10 : *weight* : : 2 : *wheel.*

(9.) The weight is expressed by 15, the power by 3. What will represent the wheel and axle ? *Ans.* 5 and 1.

(10.) The axle is represented by 16, the power by 4. Required the proportion of the wheel and axle. *Ans.* 4 : *weight* : : 16 : *wheel.*

(11.) What is the weight of an anchor requiring 6 men to weigh it, by means of a capstan 2 feet in diameter, with a lever 8 feet long, 2 feet of its length being inserted in the capstan ; supposing the power of each man to be represented by 2 cwt., and a loss of ⅓ the power by friction ? *Ans.* 56 *cwt.*

(12.) A stone weighing 2 tons is to be raised by a windlass with spokes 2 feet in length, projecting from an axle 9 inches in diameter. How many men must be employed, supposing each man's power equal to 2 cwt., including the loss by friction ? *Ans.* 2.5 *men.*

What is a Pulley? 311. THE PULLEY.—The Pulley is a small wheel turning on an axis, with a string or rope in a groove running around it.

How many kinds of pulleys are there? There are two kinds of pulleys—the fixed and the movable. The fixed pulley is a pulley that has no other motion than a revolution on its axis, and it is used only for changing the direction of motion.

Explain Fig. 46. 312. Fig. 46 represents a fixed pulley. P is a small wheel turning on its axis, with a string running round it in a groove. W is a weight to be raised, F is the force or power applied. It is evident that, by pulling the string at F, the weight must rise just as much as the string is drawn down. As, therefore, the velocity of the weight and the power is precisely the same, it is manifest that they balance each other, and that no mechanical advantage is gained.* But this pulley is very useful for changing the direction of motion. If, for instance, we wish to raise a weight to the top of a high building, it can be done with the assistance of a fixed pulley, by a man standing below. A curtain, or a sail, also, can be raised by means of a fixed pulley, without ascending with it, by drawing down a string running over the pulley.

Fig. 46.

On what principle does the fixed pulley act? 313. The fixed pulley operates on the same principle as a lever of the first kind with equal arms, where the fulcrum being in the centre of gravity, the power and the weight are equally distant from it, and no mechanical advantage is gained.

How does the movable pulley differ from the fixed? 314. The movable pulley differs from the fixed pulley by being attached to the weight; it therefore rises and falls with the weight.

Fig. 47.

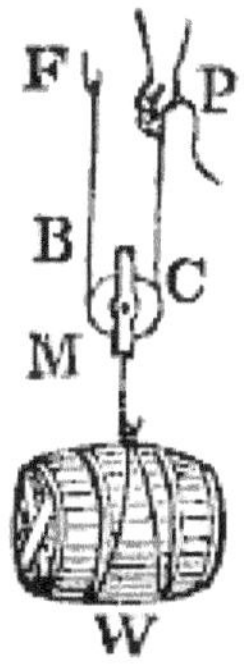

Explain Fig. 47. 315. Fig. 47 represents a movable pulley, with the weight W attached to it by a hook below. One end of the rope is fastened at F; and, as the power P draws the weight upwards, the pulley rises with the weight. Now, in order to raise the weight one inch, it is evident that both sides of the string

---

* Although the fixed pulley gives no direct mechanical advantage, a man may advantageously use his own strength by the use of it. Thus, if he seat himself on a chair suspended from one end of a rope passing over a fixed pulley, he may draw himself up by the other end of the rope by exerting a force equal only to one-half of his own weight. One half of his weight is supported by the chair and the other half by his hands, and the effect is the same as if he drew *only one half of himself at a time*; for, the rope being doubled across the pulley, two feet of the rope must pass through his hands before he can raise himself one foot. In this manner laborers and others frequently descend into wells, and from the upper floors of stores, by means of a rope passing over a fixed wheel or pulley.

must be shortened; in order to do which, the power P must pass over two inches. As the velocity of the power is double that of the weight, it follows that a power of one pound will balance a weight on the movable pulley of two pounds.*

What is the advantage gained in the use of the movable pulley?

316. The power gained by the use of pulleys is ascertained by multiplying the number of movable pulleys by 2.†

317. A weight of 72 pounds may be balanced by a power of 9 pounds with four pulleys, by a power of 18 pounds with two pulleys, or by a power of 36 pounds with one pulley. But in each case the space passed over by the power must be double the space passed over by the weight, multiplied by the number of movable pulleys. That is, to raise the weight one foot, with one pulley, the power must pass over two feet, with two pulleys four feet, with four pulleys eight feet.

Explain Fig. 48.

318. Fig. 48 represents a system of fixed and movable pulleys. In the block F there are four fixed pulleys, and in the block M there are four movable pulleys, all turning on their common axis, and rising and falling with the weight W. The movable pulleys are connected with the fixed ones by a string attached to the hook H, passing over the alternate grooves of the pulleys in each block, forming eight cords, and terminating at the power P. Now, to raise the weight one foot, it is evident that each of the eight cords must be

Fig. 48.

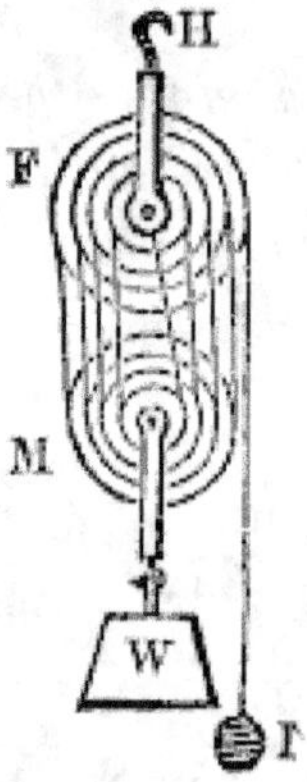

---

* Thus, it is seen that pulleys act on the same principle with the lever and the wheel and axle, the deficiency of the strength of the power being compensated by superior velocity. Now, as we cannot increase our natural strength, but can increase the velocity of motion, it is evident that we are enabled, by pulleys, and other mechanical powers, to reduce the resistance or weight of any body to the level of our strength.

† This rule applies only to the movable pulleys in the same block, or when the parts of the rope which sustains the weight are parallel to each other. The mechanical advantage, however, which the pulley seems to possess in theory, is considerably diminished in practice by the stiffness of the ropes and the friction of the wheels and blocks. When the parts of the cord, also, are not parallel, the pulley becomes less efficacious; and when the parts of the cord which supports the weight very widely depart from parallelism, the pulley becomes wholly useless. There are certain arrangements of the cord and the pulley by which the effective power of the

shortened one foot, and, consequently, that the power P must descend eight times that distance. The power, therefore, must pass over eight times the distance that the weight moves.

319. The movable pulley, as well as the fixed, acts on the same principle with the lever, the deficiency of the strength of the power with the movable pulley being compensated by its superior velocity.

On what principle is the movable pulley constructed? 320. The fixed pulley acts on the principle of a lever with *equal* arms. [*See No.* 313.] The movable pulley, on the contrary, by giving a superior velocity to the power, operates like a lever with unequal arms.

321. *Practical use of Pulleys.* — Pulleys are used to raise goods into warehouses, and in ships, &c., to draw up the sails. Both kinds of pulleys are in these cases advantageously applied: for the sails are raised up to the masts by the sailors on deck by means of the fixed pulleys, while the labor is facilitated by the mechanical power of the movable ones.

322. Both fixed and movable pulleys are constructed in a great variety of forms, but the principle on which all kinds are constructed is the same. What is generally called a *tackle and fall*, or a *block and tackle*, is nothing more than a pulley. Pulleys have likewise lately been attached to the harness of a horse, to enable the driver to govern the animal with less exertion of strength.

What law applies to all the Mechanical Powers? 323. It may be observed, in relation to the Mechanical Powers in general, that *power is always gained at the expense of time and velocity; that is, the same power which will raise one pound in one minute will raise two pounds in two minutes, six pounds in six minutes, sixty pounds in sixty minutes, &c.: and that the same quantity of force used to raise two pounds one foot will raise one pound two feet, &c.* And, further, it may be stated that the product of the weight multiplied by the velocity of the weight will always be equal to the product of the power multiplied by the velocity of the power.

pulley may be augmented in a three-fold instead of a two-fold proportion But, when such an advantage is secured, it must be by contriving to make the power pass over three times the space of the weight.

In what proportion is the power to the weight when the movable pulley is used?

Hence we have the following rule: *The power is in the same proportion to the weight as the velocity of the weight is to the velocity of the power.**

324. PRACTICAL EXAMPLES OF APPLICATION OF THE PULLEY.

Questions for Solution.

(1.) Suppose a power of 9 lbs. applied to a set of 3 movable pulleys. Allowing ½ loss for friction, what weight can be sustained by them? *A.* 36 *lb.*

(2.) Six movable pulleys are attached to a weight of 1800 lbs.; what power will support them, allowing a loss of two-thirds of the power from friction? *Ans.* 450 *lb.*

(3.) Six men, with a block and tackle containing nine movable pulleys, are required to raise a sail. Suppose each man's strength to be represented by two cwt. and two-thirds of the power lost by friction, what is the weight of the sail, with its appendages? *Ans.* 72 *cwt.*

(4.) If a stone weighing 3 tons is to be raised by horse power to the wall of a building in process of erection, by means of a derrick from which are suspended 3 movable pulleys, how many horses must be employed, supposing each horse capable of drawing as much as eight men, each of whom can lift 2 cwt., making an allowance of two-thirds for friction? *Ans.* 1½.

(5.) A block contains 5 movable pulleys, connected with a beam containing 5 fixed pulleys. A weight of half a ton is to be raised. Allowing a loss of two-thirds for friction, what power must be applied to raise it? *A.* 3 *cwt.*

(7.) The power is 3, the weight is 27; how many pulleys must be used, if friction requires an allowance of two-thirds? *Ans.* 27.

(8.) Friction one-third of the power, power 6, weight 72,— how many pulleys? *Ans.* 18.

(9.) Weight 84, friction nothing, pulleys, 3 fixed, 3 movable; required the power. *Ans.* 14.

(10.) Power 12, friction 8, four pulleys, two of them fixed; required the weight. *Ans.* 16.

(11.) Six movable and six fixed pulleys. The weight is raised 3 feet. How far has the power moved? *Ans.* 36 *ft.*

(12.) The power has moved 12 feet; how far has the weight moved under two pulleys, one fixed, the other movable? *Ans.* 6 *ft.*

(13.) The weight, suspended from a fixed pulley, has moved 6 feet. How far has the power moved? *Ans.* 6 *ft.*

(14.) The power has moved 20 feet under a fixed pulley; how far has the weight moved? *Ans.* 20 *ft.*

What is the Inclined Plane?

325. THE INCLINED PLANE.— The Inclined Plane consists of a hard plain surface, inclined to the horizon.

326. The principle on which the inclined plane acts as a mechanical power is simply the fact that *it supports part of the weight.* If a body be placed on a horizontal plane, its whole weight will be

---

* The stiffness of the cords and the friction of the blocks frequently require large deduction to be made from the effective power of pulleys. The loss thus occasioned will sometimes amount to two-thirds of the power.

supported; but, if the plane be elevated at one end, by degrees, it will support less of the weight in proportion to the elevation, until the plane becomes at right angles to the horizon, when it will support no part of the weight, and the body will fall perpendicularly.

327. A body, in ascending or descending an inclined plane, will have a greater space to traverse than if it should rise or fall perpendicularly. The time, therefore, of its ascent or descent will be longer, and thus it will oppose less resistance, and thus, also, a less force will be required to cause its ascent. Hence, we see that the fundamental principle of Mechanics, "*What is gained in power is lost in time*," applies to the Inclined Plane as well as to the Mechanical Powers that have already been described.

What is the advantage gained by the use of the inclined plane?

328. The advantage gained by the use of the inclined plane is in proportion as the length of the plane exceeds its perpendicular height.

Fig. 49 represents an inclined plane. C A its height, C B its length, and W a weight which is to be moved on it. If the length C B be four times the height C A, then a power of one pound at C will balance a weight of four pounds on the inclined plane C B.

Fig. 49.

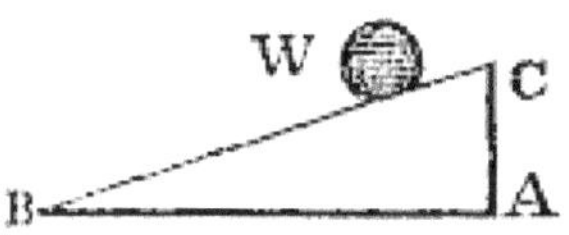

329. The greater the inclination of the plane, the greater must be its perpendicular height, compared with its length; and, of course, the greater must be the power to elevate a weight along its surface.

330. Instances of the application of the inclined plane are very common. Sloping planks or pieces of timber leading into a cellar, and on which casks are rolled up and down; a plank or board with one end elevated on a step, for the convenience of trundling wheelbarrows, or rolling barrels into a store, &c., are inclined planes.

331. Chisels and other cutting instruments, which are *chamfered*, or sloped only on one side, are constructed on the principle of the inclined plane.*

332. Roads which are not level may be considered as inclined planes, and the inclination of the road is estimated by the height corresponding to some proposed length. To raise a load up an inclined plane requires a power sufficient to carry it along the whole distance of the length of the base, and then to lift it up to

---

* Chisels for cutting wood should have their edges at an angle of about 30°; for cutting iron from 50° to 60°, and for cutting brass at about 80° or 90°. Tools urged by pressure may be sharper than those which, like the wedge, are driven by percussion.

the elevation; but in the inclined plane a feebler force will accomplish the desired object, because the resistance is spread equally over the whole distance.*

What is the Wedge?

333. THE WEDGE.— The Wedge consists of two inclined planes united at their bases.

What is the advantage gained by the use of the wedge?

334. The advantage gained by the wedge is in proportion as its length exceeds the thickness between the converging sides.

In what proportion is the power of the wedge?

It follows that the power of the wedge is in proportion to its sharpness.

335. Fig. 50 represents a wedge. The line ab represents the base of each of the inclined planes of which it is composed, and at which they are united.

Fig. 50.

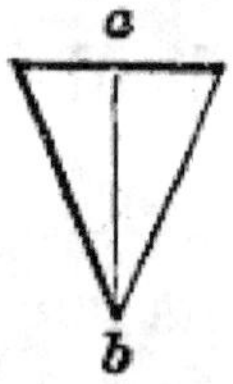

336. The wedge is a very important mechanical power, used to split rocks, timber, &c., which could not be effected by any other power.†

337. Axes, hatchets, knives, and all other cutting instruments, chamfered, or sloped on both sides, are constructed on the principle of the wedge; also pins, needles, nails, and all piercing instruments.

On what does the effective power of the wedge depend?

338. The effective power of the wedge depends on friction; for, if there were no friction, the wedge would fly back after every stroke.

---

* Mention has already been made of the sagacity of animals in a former page [see No. 54], and a sort of intuitive knowledge which they appear to possess of philosophical principles. In ascending a steep hill, a common dray-horse will drag his load from side to side, as if he were conscious that he thus made the plane longer in proportion to its height, and thereby made his load the lighter.

† The wedge is an instrument of exceedingly effective power, and is frequently used in presses for extracting the juice of seeds, fruits, &c. It is used especially in the *oil mill*, by which the oil is extracted from seeds. The seeds are placed in hair bags, between planes of hard wood, which are pressed together by wedges. The pressure thus exerted is so intense that the seeds, after the extraction of the oil, are converted into masses as hard and compact as the most dense woods.

Wedges are used also in the launching of vessels, and also for restoring buildings to the perpendicular which have been inclined by the sinking of the foundation.

339. The wedge derives much of its power from the force of percussion, which in its nature is so different from continued force, such as the pressure of weights, the force of springs, &c., that it would be difficult to submit it to numerical calculation; and, therefore, we cannot properly represent the proportion which a blow bears to the weight.

What is the Screw? 340. THE SCREW.—The Screw is an inclined plane wound around a cylinder, thus producing a circular inclined plane, forming what is called the threads of the screw.

341. Cut a piece of paper in the shape of an inclined plane, as represented by Fig. 49, and, beginning with the end represented by the height C A, in that Figure, wind it around a pencil, or a round ruler. The edge of the paper will be a circular inclined plane, and will represent the threads of the screw. The distance between any two threads on the same side of the rule will represent the perpendicular height of the inclined plane that extends once around the cylinder, and the advantage gained in the use of the screw (when used without a lever) will be the same as in the inclined plane; namely, as the length of the plane exceeds the perpendicular height. But the screw is seldom used alone. A lever is generally attached to the screw, and it is with this attachment the screw will now be considered.

What appendage generally attends the Screw? 342. The Screw is generally accompanied by an appendage called the *nut*, which consists of a concave cylinder or block, with a hollow spiral cavity cut so as to correspond exactly with the threads of the screw. When thus fitted together, the screw and the nut form two inclined planes, the one resting on the other.

Is the screw, or the nut movable? 343. Sometimes the screw is movable and the nut is stationary, and sometimes the screw is stationary and the nut is movable.

344. At every revolution the screw or the nut advances or retreats through a space equal to the distance between the threads of the screw.

In what manner does the power applied to the screw move? 345. The power applied to a screw generally describes a circle around the screw, perpendicular to the plane in which the screw or nut moves.

94 NATURAL PHILOSOPHY.

What is the advantage gained by the screw?

346. The advantage gained by the screw is in proportion as the circumference described by the power exceeds the distance between the threads of the screw.

What is meant by the Convex and Concave Screw?

347. The cylinder with its threads is called the Convex Screw, and the nut is called the Concave Screw. The lever is sometimes attached to the screw, and sometimes to the nut.

Explain Fig. 51.

348. Fig. 51 represents a fixed screw S, with a movable nut N, to which is attached the lever L. By turning the lever in one direction the nut descends, and by turning it in the opposite direction the nut ascends, at every revolution of the lever, through a space equal to the distance between the threads of the screw; to accomplish which, the hand or power applied to the end of the lever L will describe a circle around the screw S, of which the radius is L S. The power thus passes over a space represented by the circumference of this circle, and the advantage gained is in the same proportion as the space exceeds the distance between each thread of the screw

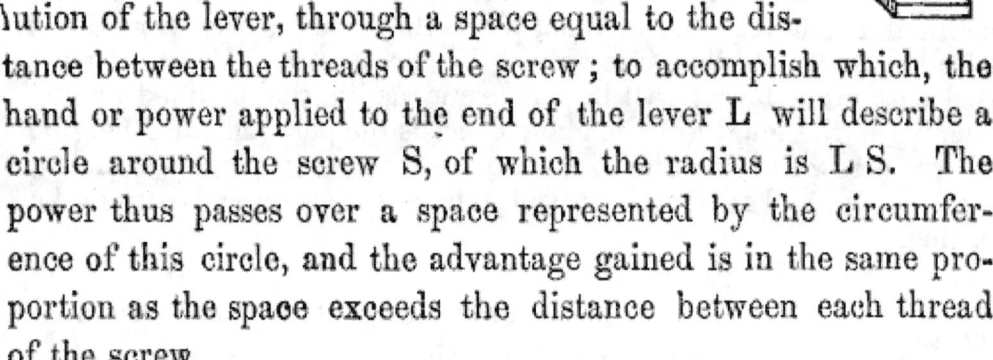

Fig. 51.

Explain Fig. 52.

349. Fig. 52 represents a movable screw, with a nut fixed in a frame, and consequently immovable. As the lever L is turned, the screw ascends or descends at every revolution of the lever through a space equal to the distance between the threads of the screw, and the advantage gained is in the same proportion as in the case of the movable nut in Fig. 51.

Fig. 52.

350. It will thus be seen that, although the screw is usually considered distinctly as a mechanical power, it is in fact a compound power, consisting of two circular inclined planes, moved by a lever.

351. The power of the screw being estimated by the distance between the threads, it follows that the closer the threads are together, the greater will be the power, but the slower will be the motion produced; for, every revolution of the lever advances the

screw or the nut only through a space as great as the distance of the threads from each other.

352. The screw is applied to presses and engines of all kinds where great power is to be applied, without percussion, through small distances. It is used in bookbinders' presses, in cider and

Fig. 53.

wine presses, in raising buildings. It is also used for coining, and for punching square or circular holes through thick plates of metal. When used for this purpose, the lever passes through the head of the screw and terminates at both ends with heavy balls or weights, the momentum of which adds to the force of the screw, and invests it with immense power.

Fig. 54.

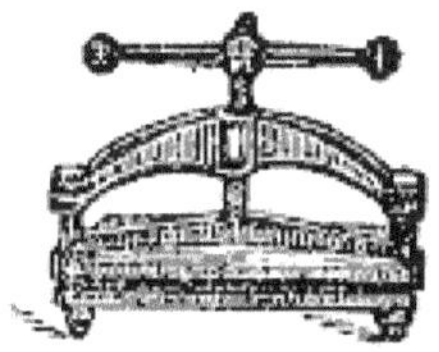

353. HUNTER'S SCREW.—The ingenious contrivance known by the name of Hunter's Screw consists of two screws of different threads playing one within the other; and such will be the effect, that while one is advancing forward the other will retreat, and the resistance will be urged forward through a distance equal only to the difference between the threads of the two screws. An indefinite increase in the power is thus obtained, without diminishing the thread of the screw.*

---

* From what has been stated with regard to the Mechanical Powers, it appears that by their aid a man is enabled to perform works to which his unassisted natural strength is wholly inadequate. But the power of all machines is limited by the strength of the materials of which they are composed. Iron, which is the strongest of all substances, will not resist a strain beyond a certain limit. Its cohesive attraction may be destroyed, and it can withstand no resistance which is stronger than its cohesive attraction. Besides the strength of the materials, it is necessary, also, to consider the *time* which is expended in the application of mechanical assistance. Archimedes is said to have boasted to Hiero, King of Syracuse, that, if he would give him a place to stand upon, he would move the whole world. In order to do this, Archimedes must himself have moved over as much more space than he moved the world as the weight of the world exceeded his own weight; and it has been computed that he must have moved with the velocity of a cannon-ball for a million of years, in order to move the earth the twenty-seven millionth part of an inch.

354. Practical Examples of the Application of the Inclined Plane and the Screw.

Questions for Solution.

(1.) With an inclined plane the power moves 16 feet, the power is to the weight as 6 to 24. How far does the weight move? *Ans.* 4 *ft.*

(2.) The length of an inclined plane is 5 feet, the proportion of the power to the weight is as 2 to 10. What is the height of the plane? *A.* 1 *ft.*

(3.) An inclined plane is 4 feet high, a power of 6 lbs. draws up 30 lbs. What is the length of the plane? *Ans.* 20 *ft.*

(4.) The length of a plane is 12 feet, the height is 3 feet. What is the proportion of the power to the weight to be raised? *Ans.* As 1 to 4.

(5.) The distance between the threads of a screw is 1 inch, the length of the lever is 2 feet. What is the proportion *Ans.* 1 to 150.79 +

(6.) Which will exert the greater force, a lever 3 feet long with the fulcrum 6 inches from one end, or a screw with a distance of 1 inch between the threads and a lever one foot long? *Ans. The screw.*

(7.) A screw with the threads 2 inches apart, and a lever 6 feet long, draws a ship of 200 tons up an inclined plane whose length is to the height in the proportion of 1 to 16. What power must be applied to the lever of the screw? *Ans.* 11.05 *lb.* +

(8.) If a man can lift a weight of 150 lbs., how much can he draw up an inclined plane whose length is to its height as 24 to 3? *Ans.* 1200 *lb.*

(9.) A Hunter's screw has a lever four feet long. The distance between the threads of the larger screw is 1 inch, between those of the smaller ¾ of an inch. How much weight can a man whose power is represented by 175 lbs. move with such a screw? *Ans.* 211115.52 *lb.*

(10.) A screw with a lever of 2 feet in length, and a distance of ½ of an inch between its threads, acts on the teeth or cogs of a wheel whose diameter is to that of the axle as 4 to 1. Fastened to the axle is a rope, one end of which is attached to a weight at the bottom of an inclined plane, the length of which is to the height as 12 to 3. Suppose this weight to require the strength of a man who can lift 200 lbs. to be applied to the lever of the screw to move it. What is the weight? *Ans.* 9650995200 *lb.*

What is the Toggle Joint? 355. THE KNEE JOINT, OR TOGGLE JOINT. — The Toggle Joint, or Knee Joint, consists of two bars united by a hinge or ball and socket, which, being urged by a power perpendicular to the resistance, acts with rapidly-increasing force, until the bars form a straight line.

The toggle (or knee) joint affords a very useful mode of converting velocity into power, the motion produced being very nearly at right angles with the direction of the force. It is a combination of levers, and the same law applies to it as to all machinery, namely, that the power is to the resistance inversely as the space of the power is to the space of the resistance.

Explain Fig. 55. 356. Fig 55 represents a toggle joint. A C and B C are the two rods connected by a joint at C. A moving force applied at C, in the direction C D, acts with great and constantly increasing power to separate the parts A and B.

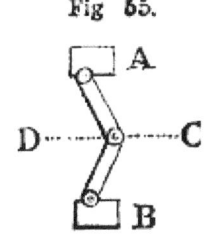

357. The operation of the toggle joint is seen in the iron joints which are used to uphold the tops of chaises. It is also used in various kinds of printing-presses to obtain the greatest power at the moment of impression.*

358. MEDIA. — The motion of all bodies is affected by the substance or element in which they move, and by which they are on all sides surrounded. Thus the bird flies in the air, the fish swims in the water. Air therefore is the medium in which the former moves, while water is the medium in which the motion of the latter is made.

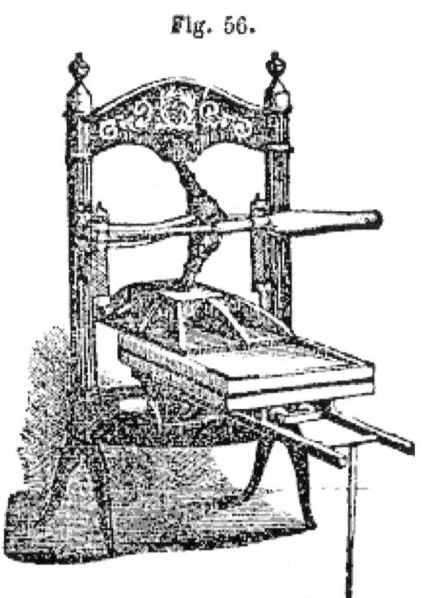

What is a Medium? 359. A Medium is the substance, solid or fluid, which surrounds a body, and which the body must displace as it moves.

360. When the fish swims or the bird flies, each must force its way through the air or the water; and the element thus displaced must rush into the spot vacated by the body in its progress. It has already been stated that the body of the fish or of the bird is propelled in its motion in the one case by the reäction of the air on the wings of the bird, and in the other of the water on the fins of a fish. The fish moves in the denser medium and needs therefore to present a less surface for the reäction of the water; while the bird, living in a comparatively rare medium, presents in his wings a much larger extent of surface to receive the reäction of the air. In making the fins of a fish, therefore, so much smaller, in proportion to its size, than the wings of a bird, nature herself has taught us that,

In what proportion is the resistance of a medium? 361. The resistance of a medium is in exact proportion to its density.

---

* A similar effect, but with a reversed action, is produced when a long rope tightly strained between two points, is forcibly pulled in the middle.

362. A body falling through water will move more slowly than one falling in the air, because it meets with more resistance from the inertia of the water, on account of the greater density of the water.

What is a Vacuum?

363. A VACUUM.—A Vacuum is unoccupied space; that is, a space which contains absolutely nothing.

364. From this definition of a vacuum, it appears that it does not mean a space which *to our eyes appears empty*. What we call an empty bottle is, in fact, full of air, or some other invisible fluid. If we sink an empty bottle in water or any other liquid, neither the water nor any other liquid can enter until some portion of the air is expelled. A small portion of water enters the bottle immersed, and the air issues in bubbles from the mouth of the bottle. Other portions of water then enter the bottle, expelling the air in similar manner, until the water entirely fills the bottle, and then the air-bubbles cease to rise.

365. From this statement of the meaning of the term "*a vacuum*," it will be seen that if a machine be worked in a vacuum (or, as it is more commonly expressed in Latin, "*in vacuo*") its motion will be rendered easier, because the parts receive no resistance from a surrounding medium.

What is Friction, and how many kinds of friction are there? Describe each.

366. FRICTION.—Friction is the resistance which bodies meet with in rubbing against each other.

There are two kinds of friction, namely, the rolling and the sliding friction. The rolling friction is caused by the rolling of a circular body.

367. The sliding friction is produced by the sliding or dragging of one surface over another.

368. Friction is caused by the unevenness of the surfaces which come into contact.* It is diminished in proportion as the surfaces are smoothed and well polished. The sliding friction is overcome with more difficulty than the rolling.

---

* All bodies, how well soever they may be polished, have inequalities in their surfaces, which may be perceived by a microscope. When, therefore, the surfaces of two bodies come into contact, the prominent parts of the one will often fall into the hollow parts of the other, and cause more or less resistance to motion.

THE MECHANICAL POWERS.

What portion of the power of a machine is lost by friction?

369. Friction destroys, but never can generate, motion. It is usually computed that friction destroys one-third of the power of a machine. In calculating the power of a machine, therefore, an allowance of one-third must be made for loss by friction.*

What is used to lessen friction? and why?

370. Oil, grease, black-lead or powdered soapstone, is used to lessen friction, because they act as a polish by filling up the cavities of the rubbing surfaces, and thus make them slide more easily over each other.

How does friction increase?

371. Friction increases:

(1.) As the weight or pressure is increased.

(2.) As the extent of the surfaces in contact is increased.

(3.) As the roughness of the surface is increased.

How may friction be diminished?

372. Friction may be diminished:

(1.) By lessening the weight of the body in motion.

(2.) By mechanically reducing the asperities of the sliding surfaces.

(3.) By lessening the amount of surface of homogeneous bodies in contact with each other.

(4.) By converting a sliding into a rolling motion.

(5.) By applying some suitable unguent.†

---

* When finely-polished iron is made to rub on bell-metal, the friction is said to be reduced to about one-eighth. Mr. Babbit, of Boston, has prepared a composition for the wheel-boxes of locomotive engines and other machinery, which, it is said, has still further reduced the amount of friction. This composition is now much in use. As the friction between rolling bodies is much less than in those that drag, the axle of large wheels is sometimes made to move on small wheels or rollers. These are called friction wheels, or friction rollers. They turn round their own centre as the wheel continues its motion.

† From the experiments made by Coulomb, it appears that the friction of heterogeneous bodies is generally less than that of homogenous that is, that if a body rub against another composed of the same kind of wood or metal, the friction is greater than that of different kinds of metal, or of wood.

Ferguson's experiments go to prove that the friction of polished steel against polished steel is greater than that of polished steel on copper or on

What are the uses of friction?

373. Friction, although it retards the motion of machines, and causes a great loss of power, performs important benefits in full compensation. Were there no friction, all bodies on the surface of the earth would be clashing against each other. Rivers would dash with unbounded velocity, and we should see little but motion and collision. But, whenever a body acquires a great velocity, it soon loses it by friction against the surface of the earth.

374. The friction of water against the surfaces it runs over soon reduces the rapid torrent to a gentle stream; the fury of the tempest is lessened by the friction of the air on the face of the earth; and the violence of the ocean is soon subdued by the attrition of its own waters. Our garments, also, owe their strength to friction; and the strength of ropes, cords, sails and various other things, depends on the same cause, for they are all made of short fibres pressed together by twisting, and this pressure causes a sufficient degree of friction to prevent the fibres sliding one upon another. Without friction it would be impossible to make a rope of the fibres of hemp, or a sheet of the fibres of flax; neither could the short fibres of cotton have ever been made into such an infinite variety of forms as they have received from the hands of ingenious workmen. Wool, also, has been converted into a thousand textures of comfort and luxury, and all these are constituted of fibres united by friction.

What is the Pendulum?

375. REGULATORS OF MOTION. — THE PENDULUM. — The Pendulum* consists of a

brass. In a combination where gun-metal rubs against steel, the same weight may be moved with a force of fifteen and a half pounds that it would require twenty-two pounds to move when cast-iron moves against steel.

* The pendulum was invented by Galileo, a great astronomer of Florence, in the beginning of the seventeenth century. Perceiving that the chandeliers suspended from the ceiling of a lofty church vibrated long and with great uniformity, as they were moved by the wind or by any accidental disturbance, he was led to inquire into the cause of their motion, and this inquiry led to the invention of the pendulum. From a like apparently insignificant circumstance arose the great discovery of the principle of gravitation. During the prevalence of the plague, in the year 1665, Sir Isaac Newton retired into the country to avoid the contagion. Sitting in his orchard, one day, he observed an apple fall from a tree. His inquisitive mind was immediately led to consider the cause which brought the apple to the ground, and the result of his inquiry was the discovery of that grand principle of gravitation which may be considered as the first and most important law of material nature. Thus, out of what had been before the eyes of men, in one shape or another, from the creation of the world, did these philosophers bring the most important results.

REGULATORS OF MOTION. 101

weight or ball suspended by a rod, and made to swing backwards and forwards.

What are the motions of a pendulum called, and how are they caused?

376. The motions of a pendulum are called its vibrations or oscillations, and they are caused by gravity.*

What is the arc of a pendulum?

The part of a circle through which it moves is called its *arc*.

What difference is there in the time of the vibrations of pendulums of equal length?

377. The vibrations of pendulums of equal length are very nearly equal, whether they move through a greater or less part of their arcs.†

378. In Fig. 57 A B represents a pendulum, D F E C the arc in which it vibrates. If the pendulum be raised to E it will return to F, if it be raised to C it will return to D, in nearly the same length of time, because that, in proportion as the *arc* is more extended, the steeper will be its beginnings and endings, and, therefore, the more rapidly will it fall.‡

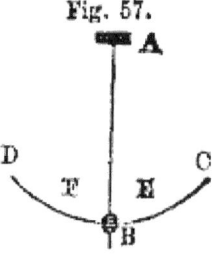

Fig. 57.

---

* When a pendulum is raised from a perpendicular position, its weight will cause it to fall, and, in the act of falling, it acquires a degree of motion which impels it to a height beyond the perpendicular almost as great as that to which it was raised. Its motion being thus spent, gravity again acts upon it to bring it to its original perpendicular position, and it again acquires an impetus in falling which carries it nearly as high on the opposite side. It thus continues to swing backwards and forwards, until the resistance of the air wholly arrests its motion.

It will be understood that gravity affects every part of the length of the pendulum. A ball or flattened weight is attached to the lower end of the pendulum to concentrate the effects of gravity in a single point.

In the construction of clocks, an apparatus connected with the weight or the spring is made to act on the pendulum with such a force as to enable it to overcome the resistance of the air, and keep up a continued motion.

† It has already been stated that a body takes the same time in rising and falling when projected upwards. Gravity brings the pendulum down, and inertia causes it to continue its motion upwards.

‡ The length of the arc in which a pendulum oscillates is called its **amplitude**.

On what does the time of the oscillations of a pendulum depend?

379. The time occupied in the vibration of a pendulum depends upon its length. The longer the pendulum, the slower are its vibrations.*

What is the length of a pendulum that vibrates once every second of time?

380. The length of a pendulum which vibrates sixty times in a minute (or, in other words, which vibrates seconds) is about thirty-nine inches. But in different parts of the earth this length must be varied.

Which must be the longer, to vibrate seconds, a pendulum at the equator or one at the poles?

A pendulum, to vibrate seconds at the equator, must be shorter than one which vibrates seconds at the poles.†

How is a clock regulated?

381. A clock is regulated by lengthening or shortening the pendulum. By lengthening the pendulum, the clock is made to go slower; by shortening it, it will go faster.‡

---

* The weight of the ball at the end of a pendulum does not affect the duration of its oscillations.

† The equatorial diameter of the earth exceeds the polar diameter by about twenty-six miles; consequently the poles must be nearer to the centre of the earth's attraction than the equator, and gravity must also operate with greater force at the poles than at the equator. Hence, also, the length of a pendulum, to vibrate in any given time, must vary with the latitude of the place.

‡ The pendulum of a clock is made longer or shorter by means of a screw beneath the weight or ball of the pendulum. The clock itself is nothing more than a pendulum connected with wheel-work, so as to record the number of vibrations. A weight is attached in order to counteract the retarding effect of friction and the resistance of the air. The wheels show how many swings or beats of the pendulum have taken place in a given time, because at every beat the tooth of a wheel is allowed to pass. Now, if this wheel have sixty teeth, it will turn round once in sixty vibrations of the pendulum, or in sixty seconds; and a hand, fixed on the axis of the wheel projecting through the dial-plate, will be the second-hand of the clock. Other wheels are so connected with the first, and the number of teeth in them is so proportioned, that the second wheel turns sixty times slower than the first, and to this is attached the minute-hand; and the third wheel, moving twelve times slower than the second, carries the hour-hand. On account of the expansion of the pendulum by heat, and its contraction by cold, clocks will go slower in summer than in winter, because the pendulum is thereby lengthened at that season.

In what proportion are the lengths of pendulums?

382. The lengths of pendulums are to each other as the square of the time of their vibration.

383. According to this law, a pendulum, to vibrate once in two seconds, must be four times as long as one that vibrates once in one second; to vibrate once in three seconds, it must be nine times as long; to vibrate once in four seconds, it must be sixteen times as long; once in five seconds, twenty-five times as long, &c.

The seconds employed in the vibrations being

1, 2, 3, 4, 5, 6, 7, 8, 9,

the length of the pendulums would be as

1, 4, 9, 16, 25, 36, 49, 64, 81.

A pendulum, therefore, to vibrate *once in five seconds*, must be *over eighty feet in length*.

384. As the oscillations of a pendulum are dependent upon gravitation, the instrument becomes useful in ascertaining the force of gravity at different distances from the centre of the earth.

385. It has already been stated that the centrifugal force at the equator is greater than in those parts of the earth which are near the poles. As the centrifugal force operates in opposition to that of gravity, it follows that the pendulum must also be affected by it; and this affords additional reason why a pendulum, to vibrate seconds at the equator, must be shorter than one at the poles. It has been estimated that, if the revolution of the earth around its axis were seventeen times faster than it is, the centrifugal force at the equator would be equal to the force of gravity, and, consequently, neither could a pendulum vibrate, nor would bodies there have any weight.

386. As every part of a pendulum-rod tends to vibrate in a different time, it is necessary that all pendulums should have a weight attached to them, which, by its inertia, shall concentrate the attractive force of gravity.

387. Pendulums are subject to variation in warm and cold weather, on account of the dilatation and contraction of the materials of which the rod is composed, by heat and cold. For this reason, the same pendulum is always longer in summer than it is in winter; and a clock will, therefore, always be slower in summer than in winter, unless some means are employed by which the effects of heat and cold on the length of the pendulum can be counteracted. This is sometimes effected in what is called the gridiron pendulum by combining bars or rods of steel and brass, and in the mercurial pendulum by enclosing a quantity of quicksilver in a tube near the bottom of the pendulum.

388. In order to secure a continuous motion to the pendulum (or, in other words, to keep a clock in motion), it is necessary that the pendulum should hang in a proper position. A practised ear can easily detect any error in this respect by the irregularity in the

ticking, or (as it is called) by its being "*out of beat.*" To remedy this fault, it is necessary either to incline the clock to the one side or the other, until the tickings are *synchronous*; or, in other words, are made at equal intervals of time. It can sometimes be done without moving the clock, by slightly bending the upper appendage of the pendulum in such a manner that the two teeth, or projections, shall properly articulate with the escapement-wheel. [See No. 303.]

389. *Table of the Lengths of Pendulums to vibrate Seconds in different latitudes.*

	Inches.		Inches.
At the equator,	39.	At the equator,	39.
Lat. 10° North,	39.01	Lat. 10° South,	39.02
20 "	39.04	20 "	39.04
30 "	39.07	30 "	39.07
40 "	39.10	40 "	39.10
50 "	39.13	50 "	39.13
60 "	39.16	60 "	

390. The observations have been extended but little further, north or south of the equator. Different observers have arrived at different results; probably on account of their different positions in relation to the level of the sea in which the observations were made. In such a work as this, a table of this kind, without pretending to extreme accuracy, is useful, as showing that theory has been confirmed by observation.

391. The moving power of a clock is a weight, which, being wound up, makes a constant effort to descend, and is prevented by a small appendage of the pendulum, furnished with two teeth, or projections, which the vibrations of the pendulum cause alternately to fall between the teeth of a wheel called the escapement-wheel. The escapement-wheel is thus permitted to turn slowly, one tooth at a time, as the pendulum vibrates. If the pendulum with its appendage be removed from the clock, the weight will descend very rapidly, causing all the wheels to revolve with great velocity, and the clock becomes useless as a time-piece.

392. The moving power of a watch * is a spring, called the main-spring, which being tightly wound around a central pin, or axis, its elasticity makes a constant effort to loosen. This power is communicated to a balance-wheel, acted upon by a hair-spring, and having an escapement similar to that of the clock. If the hair-spring, with the escapement, be removed, the main-spring, being unrestrained,

---

* A watch differs from a clock in having a vibrating wheel, instead of a pendulum. This wheel is moved by a spring, called the *hair-spring*. The place of the weight is supplied by another larger spring, called the *main-spring*.

will cause the wheels to revolve with great rapidity, and the watch, also, becomes useless as a time-piece.*

What is a Battering Ram?

393. THE BATTERING RAM.—The Battering Ram was a military engine of great power, used to beat down the walls of besieged places.

Explain Fig. 58.

394. Its construction, and the principle on which it was worked, may be understood by inspection of Fig 58, in which A B represents a large beam, heavily loaded, with

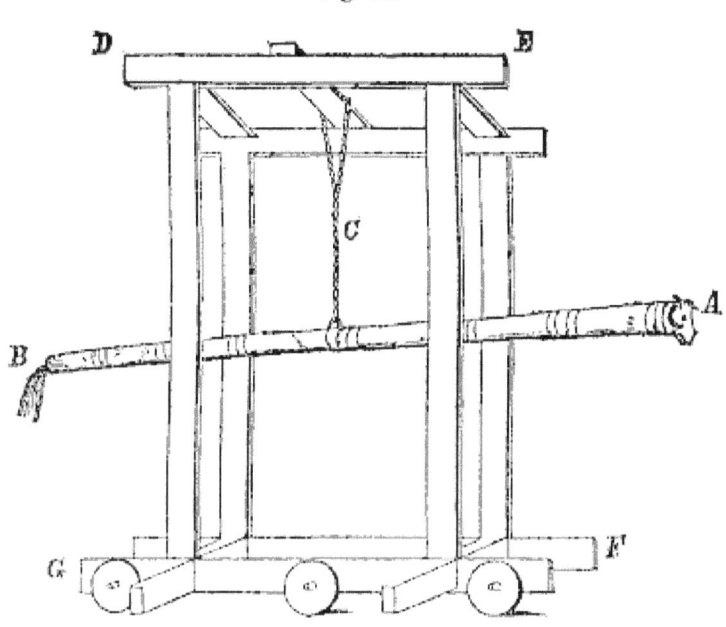

Fig. 58.

a head of iron, A, resembling the head of a ram, from which it takes its name. The beam is accurately balanced, and suspended by a rope or chain C, hanging from another beam, supported by the frame D E F G. At the extreme end B, ropes or chains were attached, by which it could be drawn upwards through the arc of a circle, like a pendulum. The frame was sometimes mounted on wheels.

395. Battering rams were frequently from fifty to a hundred feet in length, and, moving with a force compounded of their weight and velocity, were almost irresistible.†

---

* As a regulator of motion, the pendulum of the clock is to be lengthened or shortened, and the hair-spring of a watch is to be tightened or loosened. This is to be done in the former case in the manner already explained in the text; in the latter, by turning what is called the regulator, which tightens or loosens the hair-spring.

† The ram used by Demetrius Poliorcetes at the siege of Rhodes was

396. The force of a battering ram is estimated by its momentum; that is its weight multiplied by its velocity.

397. *Questions for Solution.*

(1.) Suppose a battering ram weighing 5760 lbs., with a velocity of 11 feet in a second, could penetrate a wall, with what velocity must a cannon-ball weighing 24 lbs. move to do the same execution?

$5760 \times 11 = 63360 \div 24 = 2640$ feet, or one half of a mile in a second.

(2.) If a battering ram have a momentum of 58,000 and a velocity of 8, what is its weight? *Ans.* 7250

(3.) If a ram have a weight of 90,000 and a momentum 81,000, what is its velocity? *Ans.* .9

(4.) What is the weight of a ram with a velocity of 12 and a momentum 60,000? *Ans.* 5000.

(5.) Will a cannon-ball of 9 lbs. and a velocity of 3,000, or a ram with a weight of 15,000 and a velocity of 2, move with the greater force? *Ans.* The ram.

What is the Governor? 398. THE GOVERNOR.—The Governor is an ingenious piece of mechanism, constructed on the principle of the centrifugal force, by means of which the supply of power in machinery is regulated.*

Explain Fig. 59. 399. Fig. 59 represents a governor. A B and A C are two levers, or arms, loaded with heavy

one hundred and six feet long. At the siege of Jerusalem Vespasian employed a ram fifty feet long, armed with an iron butt, with twenty-five projecting points, two feet apart, each as thick as the body of a man. The counter weight at the hindmost end amounted to 1075 cwt., and 1500 men were required to work the machine.

* This very useful appendage to machinery, though long used in mills and other mechanical arrangements, owes its happy adaptation to the steam-engine to the ingenuity of Mr. James Watt.

In manufactures, there is one certain and determinate velocity with which the machinery should be moved, and which, if increased or diminished, would render the machine unfit to perform the work it is designed to execute. Now, it frequently happens that the resistance is increased or diminished by some of the machines which are worked being stopped, or others put on. The moving power, having this alteration in the resistance, would impart a greater or less velocity to the machinery, were it not for the regulating power of the governor, which increases or diminishes the supply of water or of steam, which is the moving power.

But, besides the alteration in the resistance just noticed, there is, also, frequently, greater changes in the power. The heat by which steam is generated cannot always be perfectly regulated. At times it may afford an excess, and at other times too little expansive power to the steam. Water, also, is subject to change of level, and to consequent alteration as a moving power. The wind, too, which impels the sails of a wind-mill, is subject to great increase and diminution. To remedy all these inconveniences is the duty assigned to the governor.

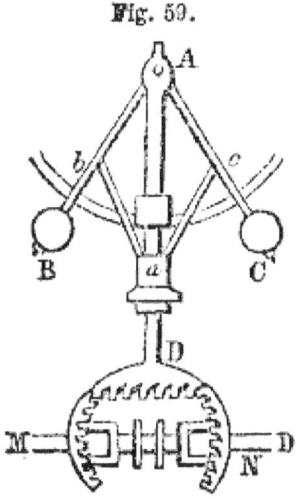

Fig. 59.

balls at their extremities B and C, and suspended by a joint at A upon the extremity of a revolving shaft A D. A *a* is a collar, or sliding box, connected with the levers by the rods *b a* and *c a*, with joints at their extremities. When the shaft A D revolves rapidly, the centrifugal force of the balls B and C will cause them to diverge in their attempt to fly off, and thus raise the collar *a*, by means of the rods *b a* and *c a*. On the contrary, when the shaft A D revolves slowly, the weights B and C will fall by their own weight, and the rods *b a* and *c a* will cause the collar *a* to descend. The steam-valve in a steam-engine, or the sluice-gate of a water-wheel, being connected with the collar *a*, the supply of steam or water, which puts the works in motion, is thus regulated.

What is the Main-spring of a watch? 400. The Main-spring of a watch consists of a long ribbon of steel, closely coiled, and contained in a round box. It is employed instead of a weight, to keep up the motion.

401. As the spring, when closely coiled, exerts a stronger force than when it is partly loosened, in order to correct this inequality the chain through which it acts is wound upon an axis surrounded by a spiral groove (called a *fusee*), gradually increasing in diameter from the top to the bottom; so that, in proportion as the strength of the spring is diminished, it may act on a larger lever, or a larger wheel and axle.

Explain Fig. 60. 402. Fig. 60 represents a spring coiled in a round box. A B is the fusee, surrounded by a spiral groove, on which the chain C is wound. When the watch is recently wound, the spring is in the greatest state of tension, and will, therefore, turn the fusee

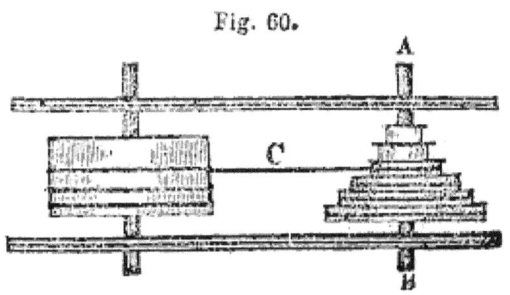

Fig. 60.

by the smallest groove, on the principle of the wheel and axle. As the spring loses its force by being partly unwound, it acts upon the larger circles of the fusee; and the want of strength in the spring is compensated by the mechanical aid of a larger wheel and axle in the larger grooves. By this means the spring is made at all times to exert an equal power upon the fusee. The motion is communicated from the fusee by a cogged wheel, which turns with the fusee.

Of what does Hydrostatics treat?

403. HYDROSTATICS.*—Hydrostatics treats of the nature, gravity and pressure of fluids.

What is the difference between Hydraulics and Hydrostatics?

404. Hydrostatics is generally confined to the consideration of fluids at rest, and Hydraulics to fluids in motion.

What is a Fluid?

405. A Fluid is a substance which yields to the slightest pressure, and the particles of which, having but a slight degree of cohesion, move easily among themselves.†

---

* The subjects of Hydraulics and Hydrostatics are sometimes described under the general name of *Hydrodynamics*. The three terms are from the Greek language, compounded of υδωρ (*hūdor*), signifying *water*, and δυναμις (*dunamis*), *force* or *power*; στατικος (*staticos*), *standing*, and αυλος (*aulos*), *a tube or pipe*. Hence Hydrodynamics would imply, the science which treats of the properties and relations of water and other fluids, whether in a state of motion or rest; while the term *Hydrostatics* would be confined to the consideration of fluids in a state of rest, and Hydraulics to fluids in motion through *tubes* or channels, natural or artificial.

† There is this remarkable difference between bodies in a fluid and bodies in a solid form, namely, that every particle of a fluid is perfectly independent of every other particle. They do not cohere in masses, like the particles of a solid, nor do they repel one another, as is the case with the particles composing a gas. They can move among one another *with the least degree of friction*, and, when they press down upon one another in virtue of their own weight, the downward pressure is communicated *in all directions*, causing a pressure upwards, sideways, and in every possible manner. Herein the particles of a fluid differ from the particles of a solid, even when reduced to the most impalpable powder; and *this it is which constitutes fluidity, namely, the power of transmitting pressure in every direction, and that, too, with the least degree of friction*. The particles which compose a fluid must be very much smaller than the finest grain of an impalpable powder.

How does a liquid differ from a fluid? 406. A liquid differs from a fluid in its degree of compressibility and elasticity. Fluids are highly compressible and elastic. Liquids, on the contrary, have but a slight degree either of compressibility or of elasticity.*

407. Another difference between a liquid and a fluid arises from the propensity which fluids have to expand whenever all external pressure is removed. Thus, whenever a portion of air or gas is removed from a closed vessel, the remaining portion will expand, and, in a rarer state, will fill the whole vessel. Liquids, on the contrary, will not expand without a change of temperature. Liquids, also, have a slight degree of *cohesion*, in virtue of which the particles will form themselves into drops; but the particles of fluids seem to possess the opposite quality of *repulsion*, which causes them to expand without limit, unless confined within the bounds of some vessel, or restricted within a certain bulk by external pressure.

408. The fluid form of bodies seems to be in great measure, if not wholly, attributed to heat. This subtle agent insinuates itself between the particles of bodies, and forces them asunder. Thus, for instance, water divested of its heat becomes ice, which is a solid. In the form of water it is a liquid, having but in a very slight degree the properties either of compressibility or elasticity. An additional supply of heat converts it into steam, endowed with a very great degree both of elasticity and compressibility. But, so soon as steam loses its heat, it is again converted into water. Again, the metals become liquid when raised to certain temperatures, and it is known that many, and supposed that all, of them would be volatilized if the required supply of heat were applied.

---

* The celebrated experiment made at Florence, many years ago, to test the compressibility of water, led to the conclusion that water is wholly incompressible. Later experiments have proved that it may be compressed, and that it also has a slight degree of elasticity. In a voyage to the West Indies, in the year 1839, an experiment was made, at the suggestion of the author, with a bottle filled with fresh water from the tanks on the deck of the Sea Eagle. It was hermetically sealed, and let down to the depth of about seven hundred feet. On drawing it up, the bottle was still full, but the water was brackish, proving that the pressure at that great depth had forced a portion of the deep salt water into the bottle, previously compressing the water in the bottle to make room for it. As it rose to the surface, its elasticity restored it to its normal state of density.

At great depths in the sea the pressure of the superincumbent mass increases the density by compression, and it has been calculated that, at a depth of about ninety miles, water would be compressed into one-half of its volume, and at a depth of 360 miles its density would be nearly equal to that of mercury. Under a pressure of 15,000 lbs. to a square inch, Mr Perkins, of Newburyport, subsequently of London, has shown that water is reduced in bulk one part in twenty-four.

The science of Geology furnishes sufficient reasons for believing that all known substances were once not only in the liquid form, but also previously existed in the form of gas.*

How do fluids gravitate?

409. GRAVITATION OF FLUIDS.—Fluids gravitate in a more perfect manner than solids, on account of their want of cohesive attraction. The particles of a solid body cohere so strongly that, when the centre of gravity is supported, the whole mass will be supported. *But every particle of a fluid gravitates independently of every other particle.*

Why cannot fluids be moulded into figures?

410. On account of the independent gravitation and want of cohesion of the particles of a fluid, they cannot be formed into figures, nor preserved in heaps. Every particle makes an effort to descend, and to preserve what is called the level or equilibrium.

What is the equilibrium of fluids?

411. The level or equilibrium of fluids is the tendency of the particles so to arrange themselves that every part of the surface shall be equally distant from the centre of the earth; that is, from the point towards which gravity tends.

What is the form of the surface of all fluids?

412. Hence the surface of all fluids, when in a state of rest, partakes the spherical form of the earth.

413. For the same reason, a fluid immediately conforms itself to the shape of the vessel in which it is contained. The particles of a solid body being united by cohesive attraction, if any one of them be supported it will uphold those also with which it is united. But, when any particle of a fluid is unsupported, it is attracted down to the level of the surface of the fluid; and the readiness with which fluids yield to the slightest pressure will enable the particle, by its own weight, to penetrate the surface of the fluid, and mix with it.

* The science of Chemistry unfolds the fact that all the great changes in the constitution of bodies are accompanied by the exhibition of heat either in a free or latent condition.

HYDROSTATICS. 111

What is Capillary Attraction?
What are Capillary Tubes?

414. CAPILLARY ATTRACTION.—Capillary Attraction is that attraction which causes fluids to ascend above their level in capillary tubes. Capillary * tubes are tubes with very fine bore.

415. This kind of attraction exhibits itself not only in tubes, but also between surfaces which are very near together. This may be beautifully illustrated by the following experiment. Take two pieces of flat glass, and, having previously wet them, separate their edges *on one side* by a thin strip of wood, card or other material; tie them together, and partly immerse them perpendicularly in colored water. The water will then rise the highest on that side where the edges of the glass meet, forming a beautiful curve downwards towards the edges which are separated by the card.

416. Immerse a number of tubes with fine bores in a glass of colored water, and the water will rise above its equilibrium in all, but highest in the tube with the finest bore.

417. The cause of this seems to be nothing more than the ordinary attraction of the particles of matter for each other. The sides of a small orifice are so near to each other as to attract the particles of the fluid on their opposite sides, and, as all attraction is strongest in the direction of the greatest quantity of matter, the water is raised upwards, or in the direction of the length of the tube. On the outside of the tube, the opposite surfaces cannot act on the same column of water, and, therefore, the influence of attraction is here imperceptible in raising the fluid.

418. All porous substances, such as sponge, bread, linen, sugar, &c., may be considered as collections of capillary tubes; and, for this reason, water and other liquids will rise in them when they are partly immersed.

419. It is on the same principle that the wick of a lamp will carry up the oil to supply the flame, although the flame is several inches above the level of the oil.† If the end of a towel happen to

---

* The word *capillary* is derived from the Latin word *capilla* (hair), and it is applied to this kind of attraction because it is exhibited most prominently in tubes *the bores of which are as fine as a hair*, and hence called capillary tubes.

† The reason why well-filled lamps will sometimes fail to give light is, that the wick is too large for its tube, and, being thus compressed, the capillary attraction is impeded by the compression. The remedy is to reduce the size of the wick. Another cause, also, that prevents a clear light, is that the flame is too far from the surface of the oil. As capillary attraction acts only at short distances, the surface of the oil should always be within a short distance of the flame. But another reason, which requires particular attention, is, that all kinds of oil usually employed for lamps contain a glutinous matter, of which no treatment can wholly divest them. This matter fills the pores or capillary tubes of the wick, and prevents the

be left in a basin of water, it will empty the basin of its contents. On the same principle, when a dry wedge of wood is driven into the crevice of a rock, as the rain falls upon it, it will absorb the water, swell, and sometimes split the rock. In this manner mill-stone quarries are worked in Germany.

420. ENDOSMOSE AND EXOSMOSE.—In addition to the capillary attraction just noticed as peculiar to fluids, another may be mentioned, as yet but imperfectly understood, which seems to be due partly to capillary and partly to chemical attraction, known under the names *endosmose* and *exosmose*.* These phenomena are manifested in the transmission of thin fluids, vapor and gaseous matter, through membranes and porous substances. The ascent of the sap in vegetable, and the absorption of nutritive matter by the organs of animal life, are to be ascribed to these causes.

421. When two liquids of different densities are separated by a membranous substance or by porcelain unglazed, *endosmose* will carry a current *inwards*, and *exosmose* will force one *outwards*, thus causing a partial mixture of the fluids.

422. *Experiment.*—Take a glass tube, and, tying a piece of bladder or clean leather over one end for a bottom, put some sugar into it, and having poured a little water on the sugar, let it stand a few hours in a tumbler of water. It will then be found that the water has risen in the tube through the membranous substance. This is due to *endosmose*. If allowed to stand several days, the liquid will rise several feet.

If the experiment be reversed, and pure water be put into the tube, and the moistened sugar into the tumbler, the tube will be emptied by *exosmose*.

423. The liquid that has the less density will generally pass to the denser liquid and dilute it.

What peculiarity is there in the gravitation of fluids of different densities?

424. GRAVITATION OF FLUIDS OF DIFFERENT DENSITIES.—When solid bodies are placed one above another, they will remain in the position in which they are placed so long as their respective centres of gravity are supported, without regard to their specific gravity. With fluids the case is different.

ascent of the oil to feed the flame. For this reason, the wicks of lamps should be often renewed. A wick that has been long standing in a lamp will rarely afford a clear and bright light. Another thing to be noticed by those who wish the lamp to perform its duty in the best possible manner is, that the wick be not of such size as, by its *length*, as well as its thickness, to fill the cup, and thereby leave no room for the oil. It must also be remembered that, although the wick when first adjusted may be of the proper size, the glutinous matter of the oil, filling its capillary tubes, causes the wick to swell, and thereby become too large for the tube, producing the same difficulty as has already been noticed in cases where the wick is too large to allow the free operation of capillary attraction.

* Endosmose, from ειδον, *within*, and ωομος, *impulsion*. Exosmose, from εξ, *outward*, and ωομος, *impulsion*.

Fluids of different specific gravity will arrange themselves in the order of their density, each preserving its own equilibrium.

425. Thus, if a quantity of mercury, water, oil and air, be put into the same vessel, they will arrange themselves in the order of their specific gravity. The mercury will sink to the bottom, the water will stand above the mercury, the oil above the water, and the air above the oil; and the surface of each fluid will partake of the spherical form of the earth, to which they all respectively gravitate.

What is a Spirit Level, or a Water Level? 426. A Water or Spirit Level is an instrument constructed on the principle of the equilibrium of fluids. It consists of a glass tube, partly filled with water, and closed at both ends. When the tube is not perfectly horizontal,— that is, if one end of the tube be lower than the other,— the water will run to the lower end. By this means the level of any line to which the instrument is applied may be ascertained.

Explain Fig. 61. 427. Fig. 61 represents a Water Level. A B is a glass tube partly filled with water. C is a bubble of air occupying the space not filled by the water. When both ends of the tube are on a level, the air-bubble will remain in the centre of the tube; but, if either end of the tube be depressed, the water will descend and the air-bubble will rise. The glass tube, when used, is generally set in a wooden or a brass box. It is an instrument much used by carpenters, masons, surveyors, &c.

Fig. 61.

[N. B. The tube is generally filled with spirit, instead of water, on account of the danger that the water will freeze and burst the glass. Hence the instrument is called indifferently the Spirit Level or the Water Level.]

Why do falling fluids do less damage than falling solids? 428. Effect of the Peculiar Gravitation of Fluids.— Solid bodies gravitate in masses, their parts being so connected as to form a whole, and their weight may be regarded as concentrated in a point, called the centre of gravity; while each

10*

particle of a fluid may be considered as a separate mass, gravitating independently.

It is for this reason that a body of water, in falling, does less injury than a solid body of the same weight. But, if the water be converted into ice, the particles losing their fluid form, and being united by cohesive attraction, gravitate unitedly in one mass.

In what direction do fluids press, on account of their weight?

429. PRESSURE OF FLUIDS.— Fluids not only press downwards like solids, but also upwards, sidewise,* and in every direction.

430. So long as the equality of pressure is undisturbed, every particle will remain at rest. If the fluid be disturbed by agitating it, the equality of pressure will be disturbed, and the fluid will not rest until the equilibrium is restored.

How are the downward, lateral and upward pressure of fluids shown?

431. The downward pressure of fluids is shown by making an aperture in the bottom of a vessel of water. Every particle of the fluid above the aperture will run downwards through the opening.

432. The lateral pressure is shown by making the aperture at the side of the vessel. The fluid will then escape through the aperture at the side.

433. The upward pressure is shown by taking a glass tube, open at both ends, inserting a cork in one end (or stopping it with the finger), and immersing the other in the water. The water will not rise in the tube. But the moment the cork is taken out (or the finger removed), the fluid will rise in the tube to a level with the surrounding water.

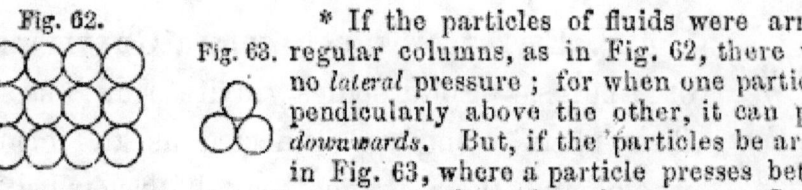

* If the particles of fluids were arranged in regular columns, as in Fig. 62, there would be no *lateral* pressure; for when one particle is perpendicularly above the other, it can press only *downwards*. But, if the particles be arranged as in Fig. 63, where a particle presses between two particles beneath, these last must suffer a *lateral* pressure. In whatever manner the particles are arranged, if they be globular, as is supposed, there must be spaces between them. [*See Fig.* 1, *page* 22.]

HYDROSTATICS.

What is the law of fluid pressure?

434. The pressure of a fluid is in proportion to the perpendicular distance from the surface; that is, the deeper the fluid, the greater will be the pressure. This pressure is exerted in every direction, so that all the parts at the same depth press each other with equal force.

435. A bladder, filled with air, being immersed in water, will be contracted in size, on account of the pressure of the water in all directions; and the deeper it is immersed, the more will it be contracted.*

436. An empty bottle, being corked, and, by means of a weight, let down to a certain depth in the sea, will either be broken by the pressure, or the cork will be driven into it, and the bottle be filled with water. This will take place even if the cork be secured with wire and sealed. But a bottle filled with water, or any other liquid, may be let down to any depth without damage, because, in this case, the internal pressure is equal to the external.†

---

* The weight of a cubic inch of water at the temperature of 62° of Fahrenheit's thermometer is 36066 millionths of a pound avoirdupois. The pressure of a column of water of the height of one foot will therefore be twelve times this quantity, or .4328 (making allowance for the repeating decimal), and the pressure upon a square foot by a column one foot high will be found by multiplying this last quantity by 144, the number of square inches in a square foot, and is therefore 62.3332.

Hence, at the depth of

		lbs.				lbs.
1 foot	the pressure on a square inch is	.4328,	on a square foot,			62.3232
2 feet	....	.8656,	" 2	"	feet,	124.6464
3 "	....	1.2984,	" 3	"	"	186.9696
4 "	....	1.7312,	" 4	"	"	249.2928
5 "	....	2.1640,	" 5	"	"	311.6160
6 "	....	2.5968,	" 6	"	"	373.9392
7 "	....	3.0296,	" 7	"	"	436.2624
8 "	....	3.4624,	" 8	"	"	498.5856
9 "	....	3.8952,	" 9	"	"	560.9088
10 "	....	4.3280,	" 10	"	"	623.2320
100 "	....	43.2800,	" 100	"	"	6232.3200

From this table, the pressure on any surface at any depth may easily be found.

It will thus be seen that there is a certain limit beyond which divers cannot plunge with impunity, nor fishes of any kind live. Wood that has been sunk to great depths in the sea will have its pores so filled with water, and its specific gravity so increased, that it will no longer float.

† "*Experiments at Sea.*— We are indebted to a friend, who has just arrived from Europe, says the *Baltimore Gazette*, for the following experiments made on board the Charlemagne:

"26th of September, 1836, the weather being calm, I corked an empty

437. *Questions for Solution.*

(1.) What pressure is sustained by the body of a fish having a surface of 9 square feet at the depth of 150 feet? *Ans.* 84136.32 *lb.*

(2.) What is the pressure on a square yard of the banks of a canal, at the depth of four feet? *Ans.* 2243.6352 *lb.*

(3.) What pressure is exerted on the body of a man, at the depth of 30 feet, supposing the surface of his body to be 2¼ *sq. yd.*? *Ans.* 42068.16 *lb.*

(4.) Suppose a whale to be at the depth of 200 feet, and that his body presents a surface of 150 yards. What is the pressure? *Ans.* 16827264 *lb.*

(5.) How deep may a glass vessel containing 18 inches of square surface be sunk without being broken, supposing it capable of resisting an equal pressure of 1500 lbs.? *Ans.* 192.54 *ft.* +

(6.) What is the pressure sustained on the sides of a cubical water-tight box at the depth of 150 feet below the surface, supposing the box to rest on the bed of the sea, and each side to be 8 feet square? *Ans.* 299151.36 *lb.*

(7.) How deep can a glass vessel be sunk without breaking, supposing that it be capable of resisting a pressure of 200 pounds on a square inch? *Ans.* 462.1 *ft.* +

What causes the lateral pressure of fluids?

438. The lateral pressure of a fluid proceeds entirely from the pressure downwards, or, in other words, from the weight of the liquid above; consequently, the lower an orifice is made in a vessel containing water or any other liquid, the greater will be the force and velocity with which the liquid will rush out.

wine-bottle, and tied a piece of linen over the cork; I then sank it into the sea six hundred feet; when drawn immediately up again, the cork was inside, the linen remained as it was placed, and the bottle was filled with water.

"I next made a noose of strong twine around the bottom of the cork, which I forced into the empty bottle, lashed the twine securely to the neck of the bottle, and sank the bottle six hundred feet. Upon drawing it up immediately, the cork was found inside, having forced its way by the twine, and in so doing had broken itself in two pieces; the bottle was filled with water.

"I then made a stopper of white pine, long enough to reach to the bottom of the bottle; after forcing this stopper into the bottle, I cut it off about half an inch above the top of the bottle, and drove two wedges, of the same wood, into the stopper. I sank it six hundred feet, and upon drawing it up immediately the stopper remained as I placed it, and there was about a gill of water in the bottle, which remained unbroken. The water must have forced its way through the pores of the wooden stopper, although wedged as aforesaid; and had the bottle remained sunk long enough, there is no doubt that it would have been filled with water." [*See also note on page* 109.]

It is the opinion of some philosophers that the pressure at very great depths of the sea is so great that the water is condensed into a solid state; and that at or near the centre of the earth, if the fluid could extend so deeply, this pressure would convert the whole into a solid mass of fire.

439. Fig. 64 represents a vessel of water, with orifices at the side at different distances from the surface. The different curves in the figure, described by the liquid in running out of the vessel, show the action of gravity, and the effects produced by the force of the pressure on the liquid at different depths. At A the pressure is the least, because there is less weight of fluid above. At B and C the fluid is driven outwards by the weight of that portion above, and the force will be strongest at C.

Explain Fig. 64.

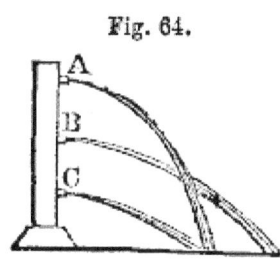

Fig. 64.

What effect has the length and the width of a body of fluid upon its lateral pressure?

440. As the lateral pressure arises solely from the downward pressure, it is not affected by the width nor the length of the vessel in which it is contained, but merely by its depth; for, as every particle acts independently of the rest, it is only the column of particles above the orifice that can weigh upon and press out the water.

To what is the lateral pressure equal?

441. The lateral pressure on one side of a cubical vessel will be equal only to half of the pressure downwards; for every particle at the bottom of a vessel is pressed upon by a column of the whole depth of the fluid, while the lateral pressure diminishes from the bottom upwards to the surface, where the particles have no pressure.

What causes the upward pressure of a fluid?

442. The upward pressure of fluids, although *apparently* in opposition to the principles of gravity, is but a necessary consequence of the operation of that principle; or, in other words, *the pressure upwards, as well as the pressure downwards,* is caused by gravity.

Explain Fig. 65.

443. When water is poured into a vessel with a spout (like a tea-pot, for instance), the water rises in the spout to a level with that in the body of the vessel. The particles of water at the bottom of the vessel are pressed upon by the particles above them, and to this pressure they will yield, if there is any mode of making way for the

particles above them. As they cannot descend through the bottom of the vessel, they will change their direction and rise in the spout. Fig. 65 represents a tea-pot, and the columns of balls represent the particles of water magnified.

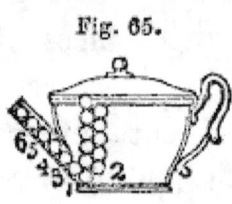

Fig. 65.

From an inspection of the figure, it appears that the particle numbered 1, at the bottom, will be pressed laterally by the particle numbered 2, and by this pressure forced into the spout, where, meeting with the particle 3, it presses it upwards, and this pressure will be continued from 3 to 4, from 4 to 5, and so on, till the water in the spout has risen to a level with that in the body of the vessel. If water be poured into the spout, the water will rise in the same manner in the body of the vessel,

Repeat the law of fluid pressure.

from which it appears that *the force of pressure depends entirely on the height, and not on the length or breadth, of the column of fluid.* [See No. 434.]

What is the Hydrostatic Paradox?

444. Any quantity of fluid, however small, may be made to balance any other quantity, however large. This is what is called the Hydrostatic Paradox.*

Explain Fig. 66.

445. The principle of what is called the hydrostatic paradox is illustrated by the hydrostatic bellows represented in Fig. 66. A B is a long tube, *one inch square.* C D E F are the bellows, consisting of two boards, *eight inches square,* connected by broad pieces of leather, or india-rubber cloth, in the manner of a pair of common bellows. One pound

---

* A paradox is something which is seemingly absurd, but true in fact. But in what is called the Hydrostatic Paradox there is in reality no paradox at all. It is true that a small quantity of fluid will balance any quantity, however large, but it is on the same principle as that with which the longer arm of the lever acts. In order to raise the larger quantity of fluid, the smaller quantity must be elevated to a height in proportion as the bulk of the larger quantity exceeds the smaller. Thus, to raise 500 lbs. of water by the descending force of one pound, the latter must descend 500 inches while the former is rising one inch; and hence, what is called the hydrostatic paradox is in strict conformity with the fundamental principle of Mechanics, that what is gained in power is lost in time, or in space.

of water poured into the tube will raise sixty-four pounds on the bellows. If a smaller tube be used, the same quantity of water will fill it higher, and, consequently, will raise a greater weight; but, if a larger tube be used, it will, of course, not fill it so high, and, consequently, will not raise so great a weight, because it is the *height, not the quantity, which causes the pressure.*

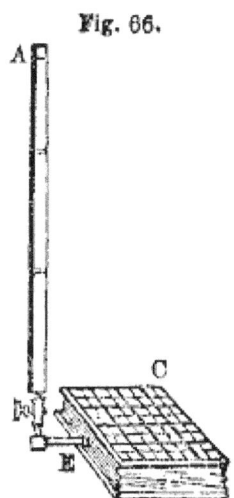

Fig. 66.

The hydrostatic bellows may be constructed in a variety of forms, the simplest of which consists, as in the figure, of two boards connected together by broad pieces of leather, or india-rubber cloth, in such a manner as to allow the upper board to rise and fall like the common bellows. A perpendicular tube is so adjusted to this apparatus that water poured into the tube, passing between the boards, will separate them by its upward pressure, even although the upper board is loaded with a considerable weight.

[N. B. A small quantity of water must be poured into the bellows to separate the surfaces before they are loaded with the weight.]

How is the force of pressure on the hydrostatic bellows estimated?

446. The force of pressure exerted on the bellows by the water poured into the tube is estimated by the comparative size of the tube and the bellows. Thus, if the tube be one inch square, and the top of the bellows twelve inches, thus containing 144 square inches, a pound of water poured into the tube will exert a pressure of 144 pounds on the bellows. Now, it will be clearly perceived that *this pressure is caused by the height of the column of water in the tube.* A pound, or a pint, of water will fill the tube 144 times as high as the same quantity would fill the bellows. To raise a weight of 144 pounds on the bellows to the height of one inch, it will be necessary to pour into the tube as much water as would fill the tube were it 144 inches long. It will thus be perceived that *the*

What fundamental law of *fundamental principle of the laws of motion is*

120 NATURAL PHILOSOPHY.

Mechanics applies also to hydrostatic pressure?

here also in full force, namely, that what is gained in power is lost either in time or in space; for, while the water in the bellows is rising to the height of one inch, that in the tube passes over 144 inches.

Explain Fig. 67.

447. Another form of apparatus, by means of which it can be proved that fluids press in proportion to their perpendicular height, and not their quantity, is seen in Fig. 67. This apparatus unites simplicity with convenience. Instead of two boards, connected with leather, an india-rubber bag is placed between two boards, connected by crossed bars with a board below, loaded with weights, and the upper boards are made to rise or fall as the water runs into or out of the bag. It is an apparatus easily repaired, and the bag may also be used for gas, or for experiments in Pneumatics.

A and B are two vessels of unequal size, but of the same length. These may successively be screwed to the apparatus, and filled with water. Weights may then be added to the suspended scale until the pressure is counterbalanced. It will then be perceived that, although A is ten times larger than B, the water will stand at the same height in both, *because they are of the same length.* If C be used instead of A or B, the apparatus may be used as the hydrostatic bellows.*

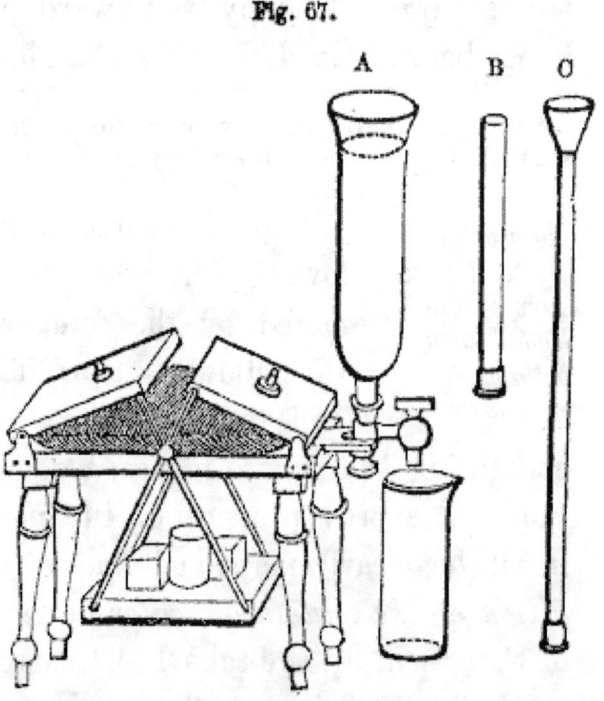

Fig. 67.

---

* If a cask be filled with water, and a long pipe be fitted to it, by pouring water into the pipe it will exert so great a pressure as to burst the cask.

In the same manner a mountain would be rent asunder by hydrostatic pressure, if a deep crevice, communicating with a small fountain below, be filled with water by the rain.

HYDROSTATICS. 121

In what manner may hydrostatic pressure be employed as a Mechanical Power?

448. HYDROSTATIC PRESSURE USED AS A MECHANICAL POWER. — If water be confined in any vessel, and a pressure to any amount be exerted on a square inch of that water, a pressure to an equal amount will be transmitted to every square inch of the surface of the vessel in which the water is confined.

449. This property of fluids seems to invest us with a power of increasing the intensity of a pressure exerted by a comparatively small force, without any other limit than that of the strength of the materials of which the engine itself is constructed. It also enables us with great facility to transmit the motion and force of one machine to another, in cases where local circumstances preclude the possibility of instituting any ordinary mechanical connexion between the two machines. Thus, merely by means of water-pipes, the force of a machine may be transmitted to any distance, and over inequalities of ground, or through any other obstructions.

On what principle is Bramah's hydrostatic press constructed? Explain Fig. 68.

450. It is on the principle of hydrostatic pressure that Bramah's hydrostatic press, represented in Fig. 68, is constructed. The main features of this apparatus are as follows: a is a narrow, and A a large metallic cylinder, having communication one with the other. Water stands in both the cylinders. The piston S carries a strong head P, which works in a frame opposite to a similar plate R. Between the two plates the substance W to be compressed is placed. In the narrow tube, a is a piston p, worked by a lever $a\,b\,d$, its short arm

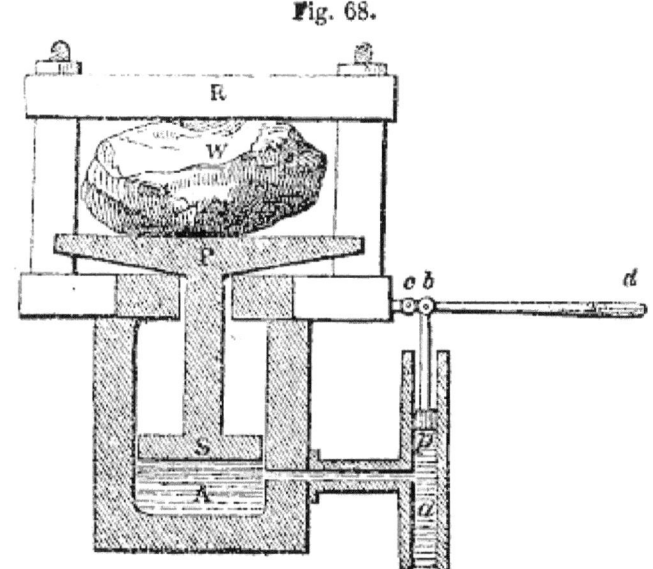

Fig. 68.

c b driving the piston, while the power is applied at *d*. The pressure exerted by the small piston *p* on the water at *a* is transmitted with equal force throughout the entire mass of the fluid, while the surface at A presses up the piston S with a force proportioned to its area. For instance, if the cylinder *a*, of the force-pump has an area of half an inch, while the great cylinder has an area of 200 inches, then the pressure of the water in the latter on the piston S will be equal to 400 times that on *p*.

Next, suppose the arms of the lever to be to each other as 1 to 50, and that at *d*, the extremity of the longer arm, a man works with a force of 50 pounds, the piston *p* will consequently descend on the water with a force of 2500 pounds. Deducting one-fourth for the loss of power caused by the different impediments to motion, and one man would still be able to exert a force of three-quarters of a million of pounds by means of this machine. This press is used in pressing paper, cloth, hay, gunpowder, &c.; also in uprooting trees, testing the strength of ropes, &c.

When will one fluid float on the surface of another fluid?

451. A fluid specifically lighter than another fluid will float upon its surface.*

[N. B. This is but another way of stating the law mentioned in Nos. 409 and 410.]

452. If an open bottle, filled with any fluid specifically lighter than water, be sunk in water, the lighter fluid will rise from the bottle, and its place will be supplied with the heavier water.

When will a body rise, sink or float, in a fluid?

453. Any substance whose specific gravity is greater than any fluid will sink to the bottom of that fluid, and a body of the same specific gravity with a fluid will neither rise nor fall in the fluid, but will remain in whatever portion of the fluid it is placed.

* The slaves in the West Indies, it is said, steal rum by inserting the long neck of a bottle, full of water, through the top aperture of the rum cask. The water falls out of the bottle into the cask, while the lighter rum ascends in its stead.

But a body whose specific gravity is less than that of a fluid will float.

This is the reason why some bodies will sink and others float, and still others neither sink nor float.*

How deep will a body sink in a fluid?   454. A body specifically lighter than a fluid will sink in the fluid until it has displaced a portion of the fluid equal in weight to itself.

455. If a piece of cork is placed in a vessel of water, about one-third part of the cork will sink below, and the remainder will stand above, the surface of the water; thereby displacing a portion of water equal in bulk to about a third part of the cork, and this quantity of water is equal in weight to the whole of the cork, because the specific gravity of water is about three times as great as that of cork.

456. It is on the same principle that boats, ships, &c., although composed of materials heavier than water, are made to float. From their peculiar shape, they are made to rest lightly on the water. The extent of the surface presented to the water counterbalances the weight of the materials, and the vessel sinks to such a depth as will cause it to displace a portion of water equal in weight to the whole weight of the vessel. From a knowledge of the specific gravity of water, and the materials of which a vessel is composed, rules have been formed by which to estimate the tonnage of vessels; that is to say, the weight which the vessel will sustain without sinking.

What is the standard for estimating the specific gravity of bodies?   457. The standard which has been adopted to estimate the specific gravity of bodies is rain or distilled water, at the temperature of 60°.†

---

* The bodies of birds that frequent the water, or that live in the water, are generally much lighter than the fluid in which they move. The feathers and down of water-fowl contribute much to their buoyancy, but fishes have the power of dilating and contracting their bodies by means of an internal air-vessel, which they can contract or expand at pleasure.

The reason that the bodies of persons who have been drowned first sink, and, after a number of days, will float, is, that when first drowned the air, being expelled from the lungs, makes the body specifically heavier than water, and it will of course sink; but, after decomposition has taken place, the gases generated within the body distend it, and render it lighter than water, and they will cause it to rise to the surface.

† As heat expands and cold condenses all metals, their specific gravity cannot be the same in summer that it is in winter. For this reason, they will not serve as a standard to estimate the specific gravity of other bodies. The reason that *distilled* water is used is, that spring, well, or river water is seldom perfectly pure, and the various substances mixed with it affect its

This is found to be a very convenient standard, because a cubic foot of water at that temperature weighs exactly one thousand ounces.

458. Taking a certain quantity of rain or distilled water, we find that a *quantity* of gold, *equal in bulk*, will weigh nearly twenty times as much as the water; of lead, nearly twelve times as much; while oil, spirit, cork, &c., will weigh less than water.*

---

weight. The cause of the ascent of steam or vapor may be found in its specific gravity. It may here be stated that rain, snow and hail, are formed by the condensation of the particles of vapor in the upper regions of the atmosphere. Fine, watery particles, coming within the sphere of each other's attraction, unite in the form of a drop, which, being heavier than the air, falls to the earth. Snow and hail differ from rain only in the different degrees of temperature at which the particles unite. When rain, snow, or hail falls, part of it reäscends in the form of vapor and forms clouds, part is absorbed by the roots of vegetables, and part descends into the earth and forms springs. The springs form brooks, rivulets, rivers, &c., and descend to the ocean, where, being again heated by the sun, the water, rising in the form of vapor, again forms clouds, and again descends in rain, snow, hail, &c. The specific gravity of the watery particles which constitute vapor is less than that of the air near the surface of the earth; they will, therefore, ascend until they reach a portion of the atmosphere of the same specific gravity with themselves. But the constant accession of fresh vapor from the earth, and the loss of heat, cause several particles to come within the sphere of each other's attraction, as has been stated above, and they unite in the form of a drop, the specific gravity of which being greater than that of the atmosphere, it will fall in the form of rain. Water, as it descends in rain, snow or hail, is perfectly pure; but, when it has fallen to the earth, it mixes with the various substances through which it passes, which gives it a species of flavor, without affecting its transparency.

* TABLE OF SPECIFIC GRAVITIES.
Temperature about 40° Fahrenheit.

Distilled Water,	1.	Palladium,	11.500
Mercury,	13.596	Iridium,	18.650
Sulphuric Acid,	1.841	Copper,	8.850
Nitric Acid,	1.220	Lead,	11.250
Prussic Acid,	.696	Bismuth,	9.822
Alcohol (pure),	.792	Tellurium,	6.240
Ether,	.715	Antimony,	6.720
Spirits of Turpentine,	.869	Chromium,	5.900
Essence of Cinnamon,	1.010	Tungsten,	17.500
Sea Water,	1.026	Nickel,	8.270
Milk,	1.030	Cobalt,	7.810
Wine,	.993	Tin,	7.293
Olive Oil,	.915	Cadmium,	8.687
Naphtha,	.847	Zinc,	7.190
Iodine,	4.946	Steel,	7.820
Platinum,	22.050	Iron,	7.788
Gold,	19.360	Cast-iron,	7.200
Silver,	10.500	Manganese,	8.012
Rhodium,	11.000	Sodium,	.972

HYDROSTATICS. 125

How is the specific gravity of a body ascertained when it is greater than that of water?

459. The specific gravity of bodies that will sink in water is ascertained by weighing them first in water, and then out of the water, and dividing the weight out of the water by the loss of weight in water.

Potassium,	.875	Elm,	.800
Diamond,	3.530	Yew,	.807
Arsenic,	5.670	Apple Tree,	.733
Graphite,	2.500	Yellow Fir,	.657
Phosphorus,	1.770	Cedar,	.561
Sulphur,	2.086	Sassafras,	.482
Lime,	3.150	Poplar,	.383
Galena,	7.580	Cork Tree,	.240
Marble,	2.850	Flint Glass,	3.330
White Lead,	6.730	Pearls,	2.750
Plaster of Paris,	2.330	Coral,	2.680
Nitrate of Potash,	1.930	China-ware,	2.380
Emerald,	2.700	Porcelain Clay,	2.210
Garnet,	3.350	Flint,	2.600
Feldspar,	2.500	Granite,	2.700
Serpentine,	2.470	Slate,	2.825
Alum,	1.700	Alabaster,	2.700
Topaz,	3.500	Brass,	8.300
Bituminous Coal,	1.250	Ice,	.865
Anthracite,	1.800	Common Air,	.001
Pulverized Charcoal,	1.500	Hydrogen Gas,	.000105
Woody Fibre,	1.500	Living Men,	.891
Lignum Vitæ,	1.350	Brandy,	.820
Boxwood,	1.320	Mahogany,	1.003
Beech,	.852	Chalk,	1.793
Ash,	.845	Carbonic Acid Gas,	.001527

By means of this table the weight of any mass of matter can be ascertained, if we know its cubical contents. A cubic foot of water weighs exactly 1000 ounces. If we multiply this by the number annexed to any substance in this table, the product will be the weight of a cubic foot of that substance. Thus anthracite coal has a specific gravity of 1.800. A thousand ounces, multiplied by this sum, produces 1800 ounces, which is the weight of a cubic foot of anthracite coal.

The bulk of any given weight of a substance may also readily be ascertained by dividing that weight in ounces by the number of ounces there are in a cubic foot. The result will be the number of cubic feet. The cube root of the number of cubic feet will give the length, depth and breadth, of the inside of a square box that will contain it.

It is to be understood that all substances whose specific gravity is greater than water will sink when immersed in it, and that all whose specific gravity is less than that of water will float in it. Let us, then, take a quantity of water which will weigh exactly one pound; a quantity of the substances specified in the table, of the same bulk, will weigh as follows:

Platinum,	23. lbs.	Silver,	11.091 lbs.
Fine Gold,	19.640 "	Copper,	9.000 "
Mercury,	14.019 "	Iron,	7.645 "
Lead,	11.525 "	Glass,	3.000 "

Describe the scales used for finding the specific gravity of a body.

460. Fig. 69 represents the scales for ascertaining the specific gravity of bodies. One scale is shorter than the other, and a hook is attached to the bottom of the scale, to which substances whose specific gravity is sought may be attached and sunk in water.

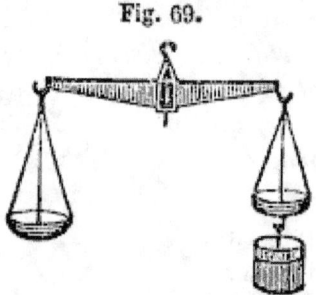

Fig. 69.

461. Suppose a cubic inch of gold weighs nineteen ounces when weighed out of the water, and but eighteen ounces * when weighed

Marble,	2.705 lbs.	Brandy,	.820 lbs.
Chalk,	1.793 "	Living Men,	.891 "
Coal,	1.250 "	Ash,	.800 "
Mahogany,	1.063 "	Beech,	.700 "
Milk,	1.034 "	Elm,	.600 "
Boxwood,	1.030 "	Fir,	.500 "
Rain Water,	1.000 "	Cork,	.240 "
Oil,	.920 "	Common Air,	.0011 "
Ice,	.908 "	Hydrogen Gas,	.000105 "

A cubic foot of water weighs one thousand avoirdupois ounces. By multiplying the number opposite to any substance in the above table by one thousand, we obtain the weight of a cubic foot of that substance in ounces. Thus, a cubic foot of platinum is 23,000 ounces in weight.

In the above table it appears that the specific gravity of *living men* is about one-ninth less than that of common water. So long, therefore, as the lungs can be kept free from water, a person, although unacquainted with the art of swimming, will not completely sink, provided the hands and arms be kept under water.

The specific gravity of sea-water is greater than that of the water of lakes and rivers, on account of the salt contained in it. On this account, the water of lakes and rivers has less buoyancy, and it is more difficult to swim in it.

* The gold will weigh less in the water than out of it, on account of the upward pressure of the particles of water, which in some measure supports it, and, by so doing, diminishes its weight. Now, as the upward pressure of these particles is exactly sufficient to balance the downward pressure of a quantity of water of exactly the same dimensions with the gold, it follows that the gold will lose exactly as much of its weight in water as a quantity of water of the same dimensions with the gold will weigh. And this rule applies to all bodies, heavier than water, that are immersed in it. *They will lose as much of their weight in water as a quantity of water of their own dimensions weighs.* All bodies, therefore, of the same size, lose the same quantity of their weight in water. Hence, *the specific gravity of a body is the weight of it compared with that of water.* As a body loses a quantity of its weight when immersed in water, it follows that when the body is lifted from the water that portion of its weight which it had lost will be restored. This is the reason that a bucket of water, drawn from a well, is heavier when it rises above the surface of the water in the well than it is while it remains below the surface. For the same reason our limbs feel heavy in leaving a bath

in water, the loss in water is one ounce. The weight out of water, nineteen ounces, being divided by one (the loss in water), gives nineteen. The specific gravity of gold, then, would be nineteen; or, in other words, gold is nineteen times heavier than water.

How is the specific gravity of a body lighter than water found?

462. The specific gravity of a body that will not sink in water is ascertained by dividing its weight by the sum of its weight added to the loss of weight which it occasions in a heavy body previously balanced in water.*

463. If a body lighter than water weighs six ounces, and, on being attached to a heavy body, balanced in water, is found to occasion it to lose twelve ounces of its weight, its specific gravity is determined by dividing its weight (six ounces) by the sum of its weight added to the loss of weight it occasions in the heavy body; namely, 6 added to 12, which, in other words, is 6 divided by 18, or $\frac{6}{18}$, which is $\frac{1}{3}$.

464. *Questions for Solution.*

(1.) A body lighter than water caused the loss of 10 lbs. to a heavier body immersed in water. In air the same body weighed 30 lbs. What was its specific gravity?

Solution. — 30 lbs., its weight, divided by (30+10=) 40 (the sum of its weight added to the loss of weight which it caused in another body previously balanced in the water). *Ans.* .75.

(2.) A body that weighed 15 lbs. in air weighed but 12 in water. What was its specific gravity? *Ans.* 5.

(3.) If a cubic foot of water weigh 1000 ounces, what is the weight of an equal bulk of gold? *Ans.* 1227 lb. 8 oz.

(4.) The weight of an equal bulk of lead? *Ans.* 720 lb. 5 oz.

(5.) The weight of an equal bulk of cork? *Ans.* 15 lb.

---

* The method of ascertaining the specific gravities of bodies was discovered accidentally by Archimedes. He had been employed by the King of Syracuse to investigate the metals of a golden crown, which he suspected had been adulterated by the workmen. The philosopher labored at the problem in vain, till, going one day into the bath, he perceived that the water rose in the bath in proportion to the bulk of his body. He instantly perceived that any other substance of equal size would raise the water just as much, though one of *equal* weight and *less* bulk could not produce the same effect. He then obtained two masses, one of gold and one of silver, each equal in weight to the crown, and having filled a vessel very accurately with water, he first plunged the silver mass into it, and observed the quantity of water that flowed over; he then did the same with the gold, and found that a less quantity had passed over than before. Hence he inferred that, though of equal weight, the bulk of the silver was greater than that of the gold, and that the quantity of water displaced was, in each experiment, equal to the bulk of the metal. He next made trial with the crown, and found that it displaced more water than the gold, and less than the silver, which led him to conclude that it was neither pure gold nor pure silver.

(6.) The weight of an equal bulk of iron? *Ans.* 477 *lb.* 13 *oz.*
(7.) What is the weight of a cubic foot of mahogany? *Ans.* 66 *lb.* 7 *oz.*
(8.) The weight of a cubic foot of marble? *Ans.* 169 *lb.* 1 *oz.*
(9.) What is the weight of an iceberg 6 miles long, ½ mile wide, and 400 feet thick? *Ans.* 949,259,520 *tons.*
(10.) What is the weight of a marble statue, supposing it to be exactly a yard and half of cubic measure? *Ans.* 6847.03 *lb.* +
(11.) If a cubical body of cork exactly 9 inches on each side be placed in water, how deep will it sink? *Ans.* 2.16 *in.*
(12.) Suppose that 4 boats were made each out of one of the following kinds of wood, namely, ash, beech, elm and fir, which would carry the greatest weight without sinking? *Ans. That of elm.*

What is an Hydrometer? and on what principle is it constructed?

465. An Hydrometer is an instrument to ascertain the specific gravity of liquids.

466. The hydrometer is constructed on the principle that the greater the weight of a liquid the greater will be its buoyancy.

How is an hydrometer constructed?

467. The hydrometer is made in a variety of forms, but it generally consists of a hollow ball of silver, glass, or other material, with a graduated scale rising from the upper part. A weight is attached below the ball. When the instrument thus constructed is immersed in a fluid, the specific gravity of the fluid is estimated by the portion of the scale that remains above the surface of the fluid. The greater the specific gravity of the fluid, the less will the scale sink.

Of what use is the hydrometer?

468. The hydrometer is a very useful instrument for ascertaining the purity of many articles in common use. It sinks to a certain determinate depth in various fluids, and if the fluids be adulterated the hydrometer will expose the cheat. Thus, for instance, the specific gravity of sperm oil is less than that of whale oil, and of course has less buoyancy. If, therefore, the hydrometer does not sink to the proper mark of sperm oil, it will at once be seen that the article is not pure.

Of what does Hydraulics treat?

469. HYDRAULICS. — Hydraulics treats of fluids in motion, and the instruments by which their motion is guided or controlled.

470. This branch of Hydrodynamics describes the effects of liquids issuing from pipes and tubes, orifices or apertures, the motion of rivers and canals, and the forces developed in the action of fluids with solids.

What quantity of a liquid will be discharged from an orifice or pipe of a given size?

471. The quantity of a liquid discharged in a given time through a pipe or orifice is equal to a column of the liquid having for its base the orifice or the area of the bore of the pipe, and a height equal to the space through which the liquid would pass in the given time.

472. Hence, when a fluid issues from an orifice in a vessel, it is discharged with the greatest rapidity when the vessel from which it flows is kept constantly full.* This is a necessary consequence of the law that pressure is proportioned to the height of the column above.

From what orifice will a fluid spout to the greatest distance?

473. When a fluid spouts from several orifices in the side of a vessel, it is thrown with the greatest random from the orifice nearest to the centre.

474. A vessel filled with any liquid will discharge a greater quantity of the liquid through an orifice to which a short pipe of peculiar shape is fitted, than through an orifice of the same size without a pipe.

This is caused by the cross-currents made by the rushing of the water from different directions towards the sharp-edged orifice. The pipe smooths the passage of the liquid. But, if the pipe project into the vessel, the quantity discharged will be diminished, instead of increased, by the pipe.

475. The quantity of a fluid discharged through a pipe or an orifice is increased by heating the liquid; because heat diminishes the cohesion of the particles, which exists, to a certain degree, in all liquids.

What part of a current of water flows most rapidly, and why?

476. Water, in its motion, is retarded by the friction of the bottom and sides of the channel through which it passes. For this reason, the velocity of the surface of a running stream is always greater than that of any other part.

---

* The velocity with which a liquid issues from an infinitely small orifice in the bottom or sides of a vessel that is kept full is equal to that which a heavy body would acquire by falling from the level of the surface to the level of the orifice. — [*Brande.*]

477. In consequence of the friction of the banks and beds of rivers, and the numerous obstacles they meet in their circuitous course, their progress is slow. If it were not for these impediments, the velocity which the waters would acquire would produce very disastrous consequences.* An inclination of three inches in a mile, in the bed of a river, will give the current a velocity of about three miles an hour.

478. To measure the velocity of a stream at its surface, hollow floating bodies are used; as, for example, a glass bottle filled with a sufficient quantity of water to make it sink just below the level of the current, and having a small flag projecting from the cork. A wheel may also be caused to revolve by the current striking against boards projecting from the circumference of the wheel, and the rapidity of the current may be estimated by the number of the revolutions in a given time.

How may the velocity of a current at any depth be ascertained?

479. The velocity of a current of water at any portion of its depth may be ascertained by immersing in it a bent tube, shaped like a tunnel at the end which is immersed.

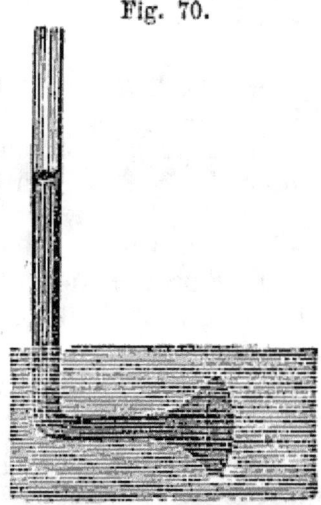

Fig. 70.

480. Fig. 70 is a tube shaped like a tunnel, with the larger end immersed in an opposite direction to the current. The rapidity of the current is estimated by the height to which the water is forced into the tube, above the surface of the current. By such an instrument the comparative velocity of different streams, or the same stream at different times, may be estimated.

How are waves caused?

481. Waves are caused, first, by the friction between air and water, and secondly, and on a much grander scale, by the attraction of the sun and moon exerted on the surface of the ocean, producing the phenomena of the tides.

482. The contriving hand of a benevolent Creator is seen more clearly in nothing, than in the laws and operations of the material world. Were it not for the almost ceaseless motion of the water the ocean itself would become a putrid mass. Decayed and decay-

* See what is stated with regard to friction in Nos. 373 and 374.

...g matter would be constantly emitting pestilential vapors, poisoning the atmosphere, and spreading contagion and death to every breathing inhabitant of the earth. The "ceaseless motion" mixes up the poisonous ingredients, and prevents their floating on the surface.*

483. The equilibrium of a fluid, according to recent discoveries, cannot be disturbed by waves to a greater depth than about three hundred and fifty times the altitude of the wave.

484. When oil is poured on the windward side of a pond, the whole surface will become smooth. The oil protects the water from the friction of the wind or air. It is said that boats have been preserved in a raging surf, in consequence of the sailors having emptied a barrel of oil on the water.

What are the principal hydraulic instruments or machines?

485. The instruments or machines for raising or drawing water are the common pump, the forcing-pump, the chain-pump, the siphon, the hydraulic ram, and the screw of Archimēdes.

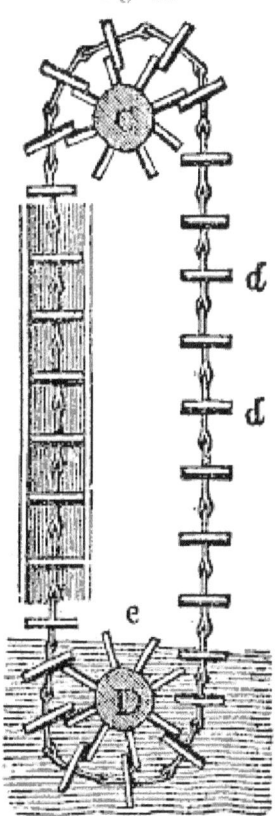

Fig. 71.

[The common pump and the forcing-pump will be noticed in connexion with Pneumatics, as their operation is dependent upon principles explained in that department of Philosophy. The fire-engine is nothing more than a double forcing-pump, and will be noticed in the same connexion.]

What is the Chain-pump?

486. The Chain-pump is a machine by which the water is lifted through a box or channel, by boards fitted to the channel and attached to a chain. It has been used principally on board of ships.

Explain Fig. 71.

487. Fig. 71 represents a Chain-pump. It consists of a square box through which a number of square boards or buckets, connected by a chain, is made to pass. The chain passes over the wheel C and under the wheel D, which is under water. The buckets are made to fit the box,

* The undulations of large bodies of water have also produced material changes on the face of the globe, purposely designed by Creative Wisdom working by secondary causes, the uses of which are described in the science of Geology.

132   NATURAL PHILOSOPHY.

so as to move with little friction. The upper wheel C is turned by a crank (not represented in the Fig.), which causes the chain with the buckets attached to pass through the box. Each bucket, as it enters the box, lifts up the water above it, and discharges it at the top.

What is the Screw of Archimēdes? 488. The screw of Archimēdes is a machine said to have been invented by the philosopher Archimēdes, for raising water and draining the lands of Egypt, about two hundred years before the Christian era.

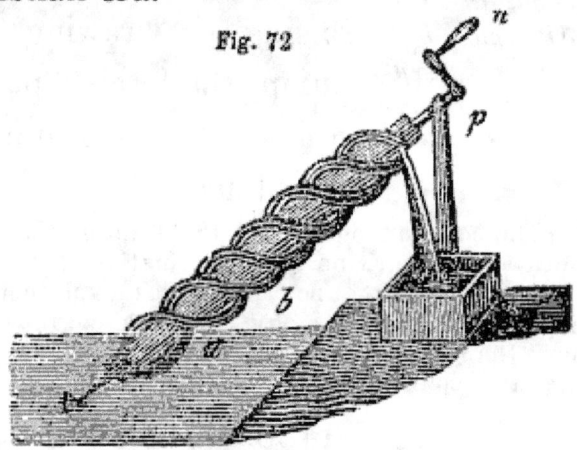

Fig. 72

Explain Fig. 72. Fig. 72 represents the screw of Archimēdes. A single tube, or two tubes, are wound in the form of a screw around a shaft or cylinder, supported by the prop and the pivot A, and turned by the handle *n*. As the end of the tube dips into the water, it is filled with the fluid, which is forced up the tube by every successive revolution, until it is discharged at the upper end.

What is the Siphon? 489. The Siphon is a tube bent in the form of the letter U, one side being a little longer than the other, to contain a longer column of the fluid.

Explain Fig. 73. 490. Fig. 73 represents a Siphon. A siphon is used by filling it with water or some other fluid, then stopping both ends, and in this state immersing the shorter leg or side into a vessel containing a liquid. The ends being then unstopped, the liquid will run through the siphon until the vessel is emptied. In performing this experiment, *the end of the siphon which is out of the water must always be below the surface of the water*

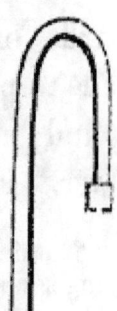

On what principle does the siphon act?

491. The principle on which the siphon acts is, that the longer column having the greater hydrostatic pressure, the fluid will run down in the direction of that column. The upward pressure in the smaller column will supply a continued stream so long as that column rests below the surface of the water.

[N. B. This principle will be better understood after the principle is explained on which the operation of the common pump depends; for the upward and downward pressure both depend on the pressure of the atmosphere.]

492. The siphon may be used in exemplifying the equilibrium of fluids; for, if the tube be inverted and two liquids of different density poured into the legs, they will stand at a height in an inverse proportion to their specific gravity. Thus, as the specific gravity of mercury is thirteen times greater than that of water, a column of mercury in one leg will balance a column of water in the other thirteen times higher than itself. But, if but one fluid be poured into both legs, that fluid will stand at equal height in both.

Explain the toy called Tantalus' Cup.

493. The toy called Tantalus' * Cup consists of a goblet containing a wooden figure, with a siphon concealed within. The water being poured into the cup until it is above the bend of the siphon, rises in the shorter leg, which opens into the cup, and runs out at the longer end, which pierces the bottom.

Fig. 74.

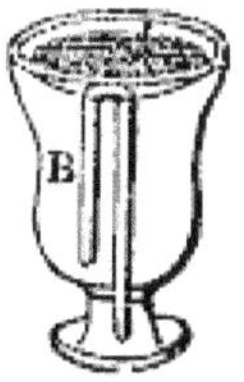

494. Fig. 74 represents the cup with the siphon, the figure of the man being omitted, in order that the position of the siphon may be seen.

What is the Hydraulic Ram?

495. THE HYDRAULIC RAM† is an ingenious machine, constructed for the purpose of raising water by means of its own impulse or momentum.

---

* Tantalus, in Heathen mythology, is represented as the victim of perpetual thirst, although placed up to the chin in a pool of water; for, as soon as he attempts to stoop to drink, the water flows away from his grasp; hence our English word *tantalize* takes its origin. In the toy described above, the siphon carries the water away before it reaches the mouth of the figure.

† The Hydraulic Ram, sometimes called by its French name, *Belier Hy-*

12

496. In the construction of an hydraulic ram, there must be, in the first place, a spring or reservoir elevated at least four or five feet above the horizontal level of the machine.*

Secondly, a pipe must conduct the water from the reservoir to the machine with a descent at least as great as one inch for every six feet of its length.

Thirdly, a channel must be provided by which the superfluous water may run off.

497. The ram itself consists of a pipe having two apertures, both guarded by valves of sufficient gravity to fall by their own weight, one of which opens downwards, the other opening upwards into an air-tight chamber. An air-vessel is generally attached to the chamber, for the purpose of causing a steady stream to flow from the chamber, through another pipe, to the desired point where the water is to be discharged.

Explain the construction of the Hydraulic Ram by Fig. 75.

498. Fig. 75 represents the hydraulic ram. A B represents the tube, or body of the ram, having two apertures, C and D, both guarded by valves; C opening downwards, D opening up-

---

draulique, in its present form, was invented by Montgolfier, of Montpelier. An instrument or machine of a similar construction had been previously constructed by Mr. Whitehurst, at Chester, but much less perfect in its mode of action, as it required to be opened and shut by the hand by means of a stop-cock. Montgolfier's machine, on the contrary, is set in motion by the action of the water itself.

* Such an elevation may easily be obtained in any brook or stream of running water by a dam at the upper part of the stream, to form a reservoir. It has been calculated that for every foot of fall in the pipe running from the reservoir to the ram sufficient power will be obtained to raise about a sixth part of the water to the height of ten feet. With a fall of only four feet and a half, sixty-three hundred gallons of water have been raised to the height of one hundred and thirty-four feet. But, the higher the reservoir, the greater the force with which the hydraulic ram will act. The operation of the principle by which the hydraulic ram acts is familiar to those who obtain water for domestic purposes by means of pipes from an elevated reservoir, as is the case in many of our large cities. A sudden stoppage of the flow, by turning the cock too quickly, causes a jarring of the pipes, which is distinctly perceived, and often loudly heard all over the building. This is due to the sudden change from a state of rapid motion to a state of rest. The *inertia* of the fluid, or its resistance to a change from a state of rapid motion to a state of rest, a property which it possesses in common with all other kinds of matter, explains the cause of the violent jarring of the pipes, the stopping of which arrests the motion of the fluid; and the violence, which is in exact proportion to the momentum of the fluid, is sometimes so great as to burst the pipes

wards, and both falling by their own weight. Let us now suppose the valve C to be open and D shut. The water, descending through the tube A B with a force proportionate to the height of the

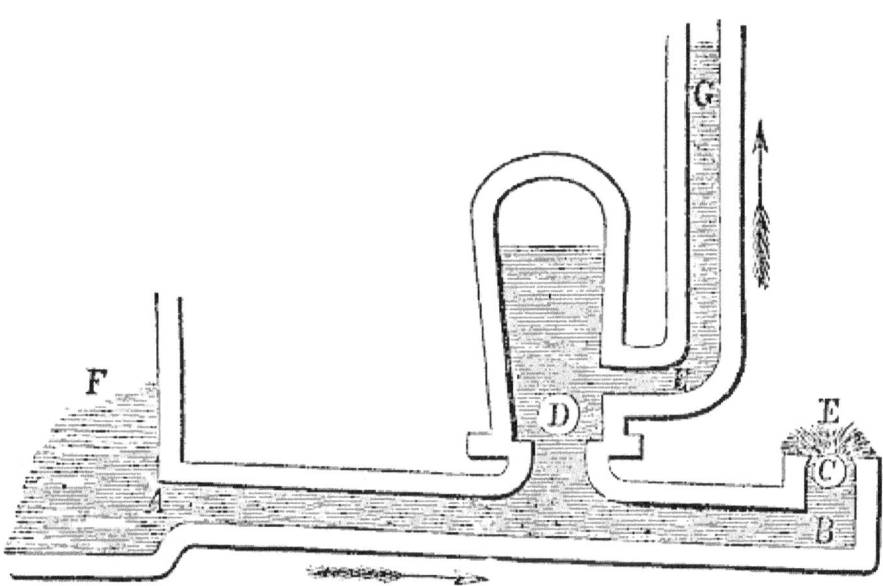

Fig. 75.

reservoir, forces up the valve C and closes the aperture, thus suddenly arresting the current, and causing, by its reäction, a pressure throughout the whole length of the pipe; this pressure forces up the valve D, and causes a portion of the water to enter the chamber above D. The current having thus spent its force, the valve C immediately falls by its own weight, by which means the current is again permitted to flow towards the aperture C. The pressure at D thereby being removed, that valve immediately falls, and closes the aperture. When this takes place, everything is in the same state in which it was at first. The water again begins to flow through the aperture at C, again closing that valve, and again opening D; and the same effects are repeated at intervals of time, which, for the same ram, undergo but little variation.

The water being thus forced into the chamber E, as it cannot return through the valve D, it must proceed upwards through the pipe G, and is thus carried to any desired point of discharge. An air-vessel is frequently attached to the chamber of the ram, which performs the same office as it does in the forcing-pump, namely, to cause a steady stream to flow from the pipe G. The action, both of the ram and the forcing-pump, without the air-vessel, would be spasmodic.*

How are Springs and Rivulets formed?

499. SPRINGS AND RIVULETS.—Springs and Rivulets are formed by the water from rain, snow, &c., which penetrates the earth, and descends until it meets a substance which it cannot penetrate. A reservoir is then formed by the union of small streams under ground, and the water continues to accumulate until it finds an outlet.

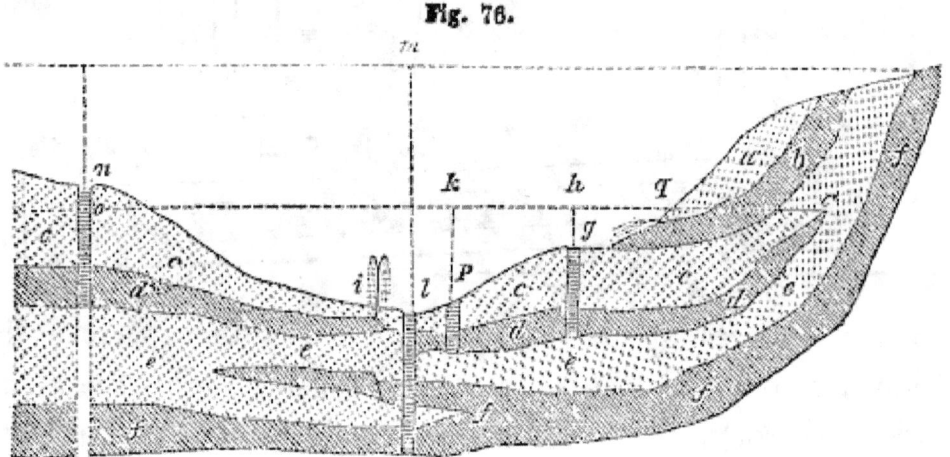

Fig. 76.

Fig. 76 represents a vertical section of the crust of the earth. a, c, and e are strata, either porous, or full of cracks, which permit the water to flow through, while b, d and f, are impervious to the water. Now, according to the laws of hydrostatics, the water at b will descend and form a natural spring at g: at i it will run with considerable force, forming a natural jet; and at l, p and g, artesian wells may be dug, in which the water will rise to the respective heights $g\,h$, $p\,k$, and $l\,m$, the water not

---

* The simplicity and economy of this mode of raising water have caused it to be quite extensively adopted in the Northern States. When well constructed, an hydraulic ram will last for years, involving no additional trouble and expense, more than occasionally leathering the valves when they have been too much worn by friction. The origin of the name will be readily perceived from the mode of its action.

" Et potum pastas age, Tityre et inter agendum,
   Occursare capro, *cornu ferit ille*, caveto."—*Virg.* Bucolic 9, v. 25

being allowed to come in contact with the porous soil through which the bore is made, but being brought in pipes to the surface; at *n* the water will ascend to about *o*, and there will be no fountain. This explains, also, the manner in which water is obtained by digging wells.

How high will the water of a spring rise? 500. A spring will rise nearly as high, but cannot rise higher than the reservoir from whence it issues.

Friction prevents the water from rising quite as high as the reservoir.

To what height may water be conveyed in tubes? 501. Water may be conveyed over hills and valleys in bent pipes and tubes, or through natural passages, to any height which is not greater than the level of the reservoir from whence it flows.

502. The ancient Romans, ignorant of this property of fluids, constructed vast aqueducts across valleys, at great expense, to convey water over them. The moderns effect the same object by means of wooden, metallic, or stone pipes.

How are fountains formed? 503. Fountains are formed by water carried through natural or artificial ducts from a reservoir. The water will spout from the ducts to nearly the height of the surface of the reservoir.

Explain the fountain by Fig. 76. 504. In Fig. 76 a fountain is represented at *i*, issuing from the reservoir, the height of which is represented by *a c*. The jet at *i* will rise nearly as high as *c*.

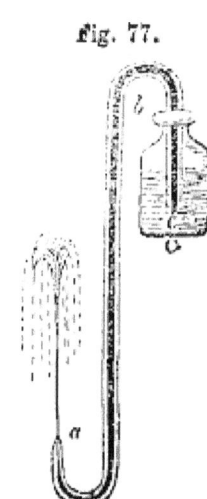

Fig. 77.

505. A simple method of making an artificial fountain may be understood by Fig. 77. A glass siphon *a b c* is immersed in a vessel of water, and the air being exhausted from the siphon, a jet will be produced at *a*, proportioned to the fineness of the bore and the length of the tube.

[N. B. The force of this kind of artificial jet is in a great measure dependent on a pneumatic principle.]

506. HERO'S FOUNTAIN.—The hydraulic instrument called Hero's Fountain is an apparatus for projecting water by means of the pressure of confined air.

Fig. 78 represents Hero's Fountain. 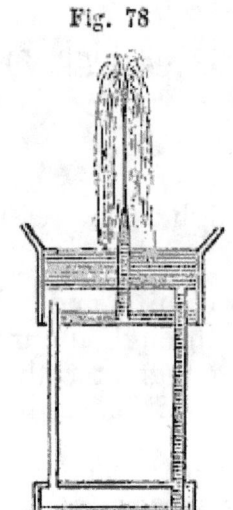 It consists of two vessels, both air-tight, and communicating by a pipe, which, being inserted into the top of the lower vessel, reaches nearly to the top of the upper vessel, which is in two parts, the upper part being filled with water, which descends in a pipe seen on the right in the figure to the lower vessel, and, as it fills the lower vessel, condenses the air, forcing it up through the left-hand pipe, and causing it to press on the surface of the water in the lower part of the upper vessel. The water in the upper vessel is thus forced through the central pipe in a jet, to a height nearly as great as the length of the pipe on the right. The supply of water is furnished in the upper part of the upper vessel, which may always be kept full by any external supply.

How does water become a mechanical agent? 507. MECHANICAL AGENCY OF FLUIDS.— Water becomes a mechanical agent of great power by means of its weight, its momentum and its fluidity.

It is used as the moving power of presses, to raise portions of itself, and to propel or turn wheels of different constructions, which, being connected with machinery of various kinds, form mills and other engines, capable of exerting great force.

What is Pneumatics? 508. PNEUMATICS.— Pneumatics treats of the mechanical properties and effects of air and similar fluids, called elastic fluids and gases, or aëriform fluids.

What is meant by an aëriform fluid? 509. Aëriform fluids are those which have the form of air. Many of them are invisible,* or

---

* Gases are all invisible, except when colored, which happens only in a few instances.

nearly so, and all of them perform very important operations in the material world. But, notwithstanding that they are in most instances imperceptible to our sight, they are really material, and possess all the essential properties of matter. They possess, also, in an eminent degree, all the properties which have been ascribed to liquids in general, besides others by which they are distinguished from liquids.

What is the difference between a permanent gas and a vapor?

510. Elastic fluids are divided into two classes, namely, permanent gases and vapors. The gases cannot be converted into the liquid state by any known process of art;*but the vapors are readily reduced to the liquid form either by pressure or diminution of temperature. There is, however, no essential difference between the *mechanical* properties of both classes of fluids.

What subjects are embraced in the science of Pneumatics?

511. As the air which we breathe, and which surrounds us, is the most familiar of all this class of bodies, it is generally selected as the subject of Pneumatics. But it must be premised that *the same laws, properties and effects, which belong to air, belong in common, also, to all aëriform fluids or gaseous bodies.*

What are the two principal properties of air, and other gaseous bodies?

512. There are two principal properties of air, namely, gravity and elasticity. These are called the principal properties of this class of bodies, because they are the means by which their presence and mechanical agency are especially exhibited.

What degree of cohesive attraction have gaseous bodies?

513. Although the aëriform fluids all have weight, they appear to possess no cohesive attraction.

514. The great degree of elasticity possessed by all aëriform fluids, renders them susceptible of compression and expansion to an almost unlimited extent. The repulsion of their particles causes them to expand, while within certain limits they are easily com-

---

* Carbonic acid gas forms an exception to this remark. Water also is the union of oxygen and hydrogen gas.

pressed. This materially affects the state of density and **rarity** under which they are at times exhibited.*

What laws pertain to aëriform bodies in general? 515. It may here be stated, that all the laws and properties of liquids (which have been described under the heads of Hydrostatics and Hydraulics) belong also to aëriform fluids.

The chemical properties of both liquids and fluids belong peculiarly to the science of Chemistry, and are not, therefore, considered in this volume.

What is the air which we breathe? 516. The air which we breathe is an elastic fluid, surrounding the earth, and extending to an indefinite distance above its surface, and constantly decreasing upwards in density.

Where is the air in its most condensed form, and why? 517. It has already been stated that the air near the surface of the earth bears the weight of that which is above it. Being compressed, therefore, by the weight of that above it, it must exist in a condensed form near the surface of the earth, while in the upper regions of the atmosphere, where there is no pressure, it is highly rarefied. This condensation, or pressure, is very similar to that of water at great depths in the sea.†

518. As the air diminishes in density upwards, it follows that it must be more rare upon a hill than on a plain. In very elevated situations it is so rare that it is scarcely fit for respiration or breathing, and the expansion which takes place in the more dense air contained within the body is often painful. It

---

* The terms "*rarefaction*" and "*condensation,*" and "*rarefied*" and "*condensed,*" must be clearly understood in this connexion. They are applied respectively to the expansion and compression of a body.

† The air is necessary to animal and vegetable life, and to combustion. It is a very heterogeneous mixture, being filled with vapors of all kinds. It consists, however, of two principal ingredients, called oxygen and nitrogen, or azote; of the former of which there are twenty-one parts, and of the latter seventy-nine, in a hundred. The air is not visible, because it is perfectly transparent. It may be felt when it moves in the form of wind, or by swinging the hand rapidly backwards and forwards.

occasions distension, and sometimes causes the bursting, of the smaller blood-vessels in the nose and ears. Besides, in such situations we are more exposed both to heat and cold; for, though the atmosphere is itself transparent, its lower regions abound with vapors and exhalations from the earth, which float in it, and act in some degree as a covering, which preserves us equally from the intensity of the sun's rays and from the severity of the cold.

519. Besides the two principal properties, gravity* and elasticity, the operations of which produce most of the phenomena of Pneumatics, it will be recollected that as air, although an invisible is yet a material substance, possessing all the common properties of matter, it possesses also the common property of *impenetrability*. This will be illustrated by experiments.

Where is the pressure of the air felt?
What pressure does a man of common stature experience from the weight of the air?

520. The pressure of the atmosphere caused by its weight is exerted on all substances, internally and externally, and it is a necessary consequence of its fluidity. The body of a man of common stature has a surface of about 2000 square inches, whence the pressure, at 15 pounds per square inch, will be 30,000 pounds. The reason why this immense weight is not felt is, that the air within the body and its pores counterbalances the weight of the external air. When the external pressure is artificially removed from any part, it is immediately felt by the reäction of the internal air.

What effect has heat upon air and other elastic fluids?

521. Heat insinuates itself between the particles of bodies and forces them asunder, in opposition to the attraction of cohesion and of gravity; it therefore exerts its power against both the attraction of gravitation and the attraction of cohesion. But, as the attraction of cohesion does not exist in aëriform fluids, the expansive power of heat upon them has nothing to contend with

---

* It has been computed that the weight of the whole atmosphere is equal to that of a globe of lead sixty miles in diameter, or to five thousand billions of tons.

but gravity. Any increase of temperature, therefore, expands an elastic fluid prodigiously, and a diminution of heat condenses it.

What is the weight of a column of air with a base of a square inch?

522. A column of air, having a base an inch square, and reaching to the top of the atmosphere, weighs about fifteen pounds. This pressure, like the pressure of liquids, is exerted equally in all directions.

What is meant by the elasticity of air and other aëriform fluids?

523. The elasticity of air and other aëriform fluids is that property by which they are increased or diminished in extension, according as they are compressed.

What effect has an increase or a diminution of pressure upon an aëriform body?

524. This property exists in a much greater degree in air and other similar fluids than in any other substance. In fact, it has no known limit; for, when the pressure is removed from any portion of air, it immediately expands to such a degree that the smallest quantity will diffuse itself over an indefinitely large space. And, on the contrary, when the pressure is increased, it will be compressed into indefinitely small dimensions.

What is Mariotte's Law?

525. The elasticity or pressure of air and all gases is in direct proportion to their density; or, what is the same thing, inversely proportional to the space which the fluid occupies. This law, which was discovered by Mariotte, is called "*Mariotte's Law.*" This law may perhaps be better expressed in the following language; namely, *the density of an elastic fluid is in direct proportion to the pressure which it sustains.*

How does air become a mechanical agent?

526. Air becomes a mechanical agent by means of its weight, its elasticity, its inertia, and its fluidity.

With what power does

527. *The fluidity of air invests it*, as it invests all other liquids, *with the power of transmitting*

PNEUMATICS. 143

fluidity invest a fluid? *pressure.* But it has already been shown, under the head of Hydrostatics, that fluidity is a necessary consequence of the independent gravitation of the particles of a fluid. It may, therefore, be included among the effects of weight.

528. The inertia of air is exhibited in the resistance which it opposes to motion, which has already been noticed under the head of Mechanics.* This is clearly seen in its effects upon falling bodies, as will be exemplified in the experiments with the air-pump.

What is a Vacuum?

529. A Vacuum is a space from which air and every other substance have been removed.

What is the most perfect vacuum that has been obtained?

530. The Torricellian vacuum was discovered by Torricelli, and was obtained in the following manner: A tube, closed at one end, and about thirty-two inches long, was filled with mercury; the open end was then covered with the finger, so as to prevent the escape of the mercury, and the tube inverted and plunged into a vessel of mercury. The finger was then removed, and the mercury permitted to run out of the tube. It was found, however, that the mercury still remained in the tube to the height of about thirty inches, leaving a vacuum at the top of about two inches. This vacuum, called from the discoverer the Torricellian vacuum, is the most perfect that has been discovered.†

---

* The *fly*, as it is called, in the mechanism of a clock by which the hours are struck, is an instance of the application of the *inertia of the air* in Mechanics.

† Torricelli was a pupil of the celebrated Galileo. The Grand Duke of Tuscany having had a deep well dug, the workmen found that the water would rise no higher than thirty-two feet. Galileo was applied to for an explanation of the reason without success. Torricelli conceived the idea of substituting mercury for water, arguing that if it was the pressure of the atmosphere that would raise the water in the pump to the height of thirty-two feet, that it would sustain a column of mercury only one-fourteenth as high, or thirty inches only, on account of its greater specific gravity. He therefore determined to test it by experiment. He accordingly filled a small glass tube, about four feet long, with mercury, and, stopping the open end with his finger, he inverted it into a basin of mercury. On removing his finger, the mercury immediately descended in the tube, and stood at the height of about thirty inches; thus demonstrating the fact that it was the pressure of the air on the surface of the mercury in the one case, and of the water in the other, that sustained the column of mercury in the tube, and of the water in the pump.

144   NATURAL PHILOSOPHY.

531. As this is one of the most important discoveries of the science of Pneumatics, it is thought to be deserving of a labored explanation. The whole phenomenon is the result of the equilibrium of fluids. The atmosphere, pressing by its weight (fifteen pounds on every square inch) on the surface of the mercury in the vessel, counterpoised the column of mercury in the tube when it was about thirty inches high, showing thereby that a column of the atmosphere is equal in weight to a column of mercury of the same base, having a height of thirty inches. Any increase or diminution in the density of the air produces a corresponding alteration in its weight, and, consequently, in its ability to sustain a longer or a shorter column of mercury. Had water been used instead of mercury, it would have required a height of about thirty-three feet to counterpoise the weight of the atmospheric column. Other fluids may be used, but the perpendicular height of the column of any fluid, to counterpoise the weight of the atmosphere, must be as much greater than that of mercury as the specific gravity of mercury exceeds that of the fluid employed.

532. This discovery of Torricelli led to the construction of the barometer,* for it was reasoned that if it was the weight of the atmosphere which sustained the column of mercury, that on ascending any eminence the column of mercury would descend in proportion to the elevation.

What is a Barometer?

533. The Barometer is an instrument to measure the weight of the atmosphere, and thereby to indicate the variations of the weather.†

Fig. 79.

Explain Fig. 79.

534. Fig. 83 represents a barometer. It consists of a long glass tube, about thirty-three inches in length, closed at the upper end, and filled with mercury. The tube is then inverted in a cup or leather bag of mercury, on which the pressure of the atmosphere is exerted. As the tube is closed at the top, it is evident that the mercury cannot descend in the tube without producing a vacuum. The pressure of the atmosphere (which is capable of supporting a column of mercury of about thirty inches in height) prevents the descent of the mercury; and

---

* Among those to whom the world is indebted for the invention of the barometer, and its applications in science, may be mentioned the names of Descartes, Pascal, Morienne and Boyle. The original idea is due to Torricelli's experiment.

† The word *barometer* is from the Greek, and signifies "*a measure of the weight*," that is, of the atmosphere.

the instrument, thus constructed, becomes an implement for ascertaining the weight of the atmosphere. As the air varies in weight or pressure, it must, of course, influence the mercury in the tube, which will rise or fall in exact proportion with the pressure. When the air is thin and light, the pressure is less, and the mercury will descend; and, when the air is dense and heavy, the mercury will rise.* At the side of the tube there is a scale, marked inches and tenths of an inch, to note the rise and fall of the mercury.

535. The barometer, as thus constructed, only required the addition of an index and a weather-glass, as seen in Fig. 80, to give a fair and true announcement of the state and weight of the atmosphere. The instruments are now manufactured in several different forms. The different forms of the barometer in general use are the common Mercurial Barometer, the Diagonal, and the Wheel Barometer, all of which are constructed with a column of mercury. The Aneroid or Portable Barometer is a new instrument, in which confined air is substituted for mercury. This is a convenient form of the instrument for portable purposes. But the principle is the same in all, and repeated observations during the ascent of the loftiest mountains in Europe and America have confirmed the truth of barometrical announcements; for, by its indications, the respective heights of the acclivities in high regions can now be ascertained by means of this instrument better than by any other course, — with this advantage, too, that no proportionate height need be known to ascertain the altitude.†

Fig. 80.

---

* The elasticity of the air causes an increase or diminution of its bulk, according as it is affected by heat and cold; and this increase and diminution of bulk materially affect its specific gravity. The height of a column of mercury that can be sustained by a column of the atmosphere must, therefore, be affected by the state of the atmosphere.

† From the explanation which has now been given of the barometer, it

On what principle is the barometer constructed?

536. The pressure of the atmosphere on the mercury, in the bag or cup of a barometer, being exerted on the principle of the equilibrium of fluids, must vary according to the situation in which the barometer is placed. For this reason, it will be the greatest in valleys and low situations, and least on the top of high mountains. Hence the barometer is often used to ascertain the height of mountains and other places above the level of the sea.

When is the atmosphere heaviest?

537. The air is the heaviest in dry weather, and consequently the mercury will then rise highest. In wet weather the dampness renders

will readily be seen that a column of any other fluid will answer as well as mercury, provided the tube be extended in an inverse proportion to the specific gravity of the fluid. But mercury is the most convenient, because it requires the shortest tube.

In navigation the barometer has become an important element of guidance, and a most interesting incident is recounted by Captain Basil Hall, indicative of its value in the open sea. While cruising off the coast of South America, in the Medusa frigate, one day, when within the tropics, the commander of a brig in company was dining with him. After dinner, the conversation turned on the natural phenomena of the region, when Captain Hall's attention was accidentally directed to the barometer in the state-room where they were seated, and, to his surprise, he observed it to evince violent and frequent alteration. His experience told him to expect bad weather, and he mentioned it to his friend. His companion, however, only laughed, for the day was splendid in the extreme, the sun was shining with its utmost brilliance, and not a cloud specked the deep-blue sky above. But Captain Hall was too uneasy to be satisfied with bare appearances. He hurried his friend to his ship, and gave immediate directions for shortening the top hamper of the frigate as speedily as possible. His lieutenants and the men looked at him in mute surprise, and one or two of the former ventured to suggest the inutility of the proceeding. The captain, however, persevered. The sails were furled, the top-masts were struck; in short, everything that could oppose the wind was made as snug as possible. His friend, on the contrary, stood in under every sail.

The wisdom of Captain Hall's proceedings was, however, speedily evident; just, indeed, as he was beginning to doubt the accuracy of his instrument. For hardly had the necessary preparations been made, and while his eye was ranging over the vessel to see if his instructions had been obeyed, a dark, hazy hue was seen to rise in the horizon, a leaden tint rapidly overspread the sullen waves, and one of the most tremendous hurricanes burst upon the vessels that ever seaman encountered on his ocean home. The sails of the brig were immediately torn to ribbons, her masts went by the board, and she was left a complete wreck on the tempestuous surf which raged around her, while the frigate was driven wildly along at a furious rate, and had to scud under bare poles across the wide Pacific, full three thousand miles, before it could be said that she was in safety from the blast

the air less salubrious, and it appears, therefore, more heavy then, although it is, in fact, much lighter.

At what time of the day is the highest and lowest state of the barometer?
538. The greatest depression of the barometer occurs daily at about four o'clock, both in the morning and in the afternoon; and its highest elevation at about ten o'clock, morning and night. In summer these extreme points are reached an hour or two earlier in the morning, and as much later in the afternoon.

539. Rules have been proposed by which the changes of the weather may be predicted by means of the barometer. Hence the graduated edge of the instrument is marked with the words " rain," " fair," " changeable," " frost," &c. These expressions are predicated on the assumption that the changes of the weather may correctly be predicted by the absolute height of the mercury. But on this little reliance can be placed. The best authorities agree that it is rather the *change in the height* on which the predications must be made.

540. As the barometer is much used at the present day, it has been thought expedient to subjoin a few general and special rules, from different authorities, by which some knowledge of the uses of the instrument may be acquired.

541. *General Rules by which Changes of the Weather may be prognosticated by means of the Barometer.**

(1.) Generally the rising of the mercury indicates the approach of fair weather.

(2.) In sultry weather the fall of the mercury indicates coming thunder. In winter the rise of the mercury indicates frost. In frost, its fall indicates thaw, and its rise indicates snow.

(3.) Whatever change of weather suddenly follows a change in the barometer, may be expected to last but a short time. Thus, if fair weather follow immediately the rise of the mercury, there will be very little of it, and, in the same way, if foul weather follow the fall of the mercury, it will last but a short time.

(4.) If fair weather continue for several days, during which the mercury continually falls, a long succession of foul weather will probably ensue; and again, if foul weather continue for several days, while the mercury continually rises, a long succession of fair weather will probably succeed.

(5.) A fluctuating and unsettled state in the mercurial column indicates changeable weather. — *Lardner, page 75, Pneumatics.*

542. *Special Rules by which we may know the Changes of the Weather by means of the Barometer.*†

(1.) The barometer is highest of all during a long frost, and it generally rises with a north-west wind.

---

* These rules, says Dr. Lardner, from whose work they are extracted, may to some extent be relied upon, but they are subject to some uncertainty.
† These rules are from a different authority.

(2.) The barometer is lowest of all during a thaw which follows a long frost, and it generally falls with a south or east wind.

(3.) While the mercury in the barometer stands above 30°, the air must be very dry or very cold, or perhaps both, and no rain may be expected.

(4.) When the mercury stands very low indeed, there will never be much rain, although a fine day will seldom occur at such times.

(5.) In summer, after a long continuance of fair weather, the barometer will fall gradually for two or three days before rain falls; but, if the fall of the mercury be very sudden, a thunder-storm may be expected.

(6.) When the sky is cloudless and seems to promise fair weather, if the barometer is low, the face of the sky will soon be suddenly overcast.

(7.) Dark, dense clouds will pass over without rain when the barometer is high; but if the barometer be low it will often rain without any appearance of clouds.

(8.) The higher the mercury, the greater probability of fair weather.

(9.) When the mercury is in a rising state, fine weather is at hand; but when the mercury is in a falling state, foul weather is near.

(10.) In frosty weather, if snow falls, the mercury generally rises to 30, where it remains so long as the snow continues to fall; if after this the weather clears up, very severe cold weather may be expected.

It will be observed that the barometer varies more in winter than in summer. It is at the highest in May and August; then in June, March, September and April. It is the lowest in November and February; then in October, July, December and January.

[These rules are from Dr. Brewer's work called "The Science of Familiar Things."]

543. OF THE DIFFERENT STATES OF THE BAROMETER. — *Of the Fall of the Barometer.* — In very hot weather the fall of the Barometer indicates thunder. Otherwise, the sudden fall of the barometer leads to the expectation of high wind.

In frosty weather the fall of the barometer denotes a thaw.

If wet weather follow soon after the fall of the barometer, but little of such weather may be expected.

In wet weather, if the barometer falls, expect much wet.

In fair weather, if the barometer falls and remains low, expect much wet in a few days, and probably wind.

The barometer sinks lowest of all for wind and rain together; next to that for wind, except it be an east or north-east wind.

544. *Of the Rise of the Barometer.* — In winter the rise of the barometer presages frost.

In frosty weather, the rise of the barometer presages snow.

If fair weather happens soon after the rise of the barometer, expect but little of it.

In wet weather, if the mercury rises high and remains so, expect continued fine weather in a day or two.

In wet weather, if the mercury rises suddenly very high, fine weather will not last long.

The barometer rises highest of all for north and west winds; for all other winds, it sinks.

545. *The Barometer in an Unsettled State.* — If the motion of the mercury be unsettled, expect unsettled weather.

If it stand at "*much rain,*" and rise to "*changeable,*" expect fair weather of short continuance.

If it stand at "*fair,*" and fall to "*changeable,*" expect foul weather.

Its motion upwards indicates the approach of fine weather; its motion downward indicates the approach of foul weather.

What is the Thermometer, and on what principle is it constructed?

546. THE THERMOMETER. — The Thermometer * is an instrument to indicate the temperature of the atmosphere. It is constructed on the principle that heat expands and cold contracts most substances.

547. The thermometer consists of a capillary tube, closed at the top and terminating downwards in a bulb. It is filled with mercury, which expands and fills the whole length of the tube or contracts altogether into the bulb, according to the degree of heat or cold to which it is exposed. Any other fluid may be used which is expanded by heat and contracted by cold, instead of mercury.

Fig. 81.

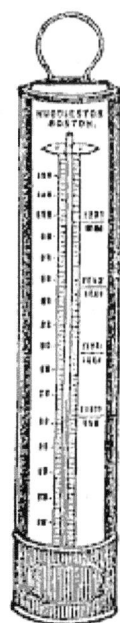

548. On the side of the thermometer is a scale to indicate the rise and fall of the mercury, and consequently the temperature of the weather.

What scale is adopted for the thermometer in this country?

549. There are several different scales applied to the thermometer, of which those of Fahrenheit, Reaumur, Delisle and Celsius, are the principal. The thermometer in common use in this country is graduated by Fahrenheit's scale, which, commencing with 0, or zero, extends upwards to 212 degrees, the boiling point of water, and downwards to 20 or 30 degrees. The scales of Reaumur and Celsius fix zero at the freezing point of water; and that of Delisle at the boiling point.

What is the Hygrometer?

550. THE HYGROMETER. — The Hygrometer is an instrument for showing the degree of moisture in the atmosphere.

---

* The word "*Thermometer*" is from the Greek, and means "*a measure of heat.*" "*Hygrometer*" means "*a measure of moisture.*"

How is it constructed?

551. The hygrometer may be constructed of any material which dryness or moisture expands or contracts; such as most kinds of wood, catgut, twisted cord, the beard of wild oats, &c. It is sometimes also composed of a scale balanced by weights on one side, and a sponge, or other substance which readily imbibes moisture, on the other.

552. By the action of the sun's heat upon the surface of the earth, whether land or water, immense quantities of vapor are raised into the atmosphere, supplying materials for all the water which is deposited again in the various forms of dew, fog, rain, snow, and hail. Experiments have been made to show the quantity of moisture thus raised from the ground by the heat of the sun. Dr. Watson found that an acre of ground, apparently dry and burnt up by the sun, dispersed into the air sixteen hundred gallons of water in the space of twelve hours. His experiment was thus made: He put a glass, mouth downwards, on a grass-plot, on which it had not rained for above a month. In less than two minutes the inside was covered with vapor; and in half an hour drops began to trickle down its inside. The mouth of the glass was 20 square inches. There are 1296 square inches in a square yard, and 4840 square yards in an acre. When the glass had stood a quarter of an hour, he wiped it with a piece of muslin, the weight of which had been previously ascertained. When the glass had been wiped dry, he again weighed the muslin, and found that its weight had increased six grains by the water collected from 20 square inches of earth; a quantity equal to 1600 gallons, from an acre, in 12 hours. Another experiment, after rain had fallen, gave a much larger quantity.

553. When the atmosphere is colder than the earth, the vapor which arises from the ground, or a body of water, is condensed and becomes visible. This is the way that fog is produced. When the earth is colder than the atmosphere, the moisture in the atmosphere condenses in the form of dew, on the ground, or other surfaces. Clouds are nothing more than vapor condensed by the cold of the upper regions of the atmosphere. Rain is produced by the sudden cooling of large quantities of watery vapor. Snow and hail are produced in a similar manner, and differ from rain only in the degree of cold which produces them.

What is the Diving-bell, and on what principle is it constructed?

554. THE DIVER'S BELL OR DIVING-BELL. — The Diving-bell is a large vessel shaped like an inverted goblet, in which a person may safely descend to great depths in the water. It is constructed on the principle of the impenetrability of air.

555. It has already been stated that air, being a material substance, possesses all the given essential properties of matter, and among them the property of impenetrability. The weight of the air giving it a pressure in every direction, or the property of fluidity, it penetrates and fills all things around us, unless by mechanical means it be carefully excluded. An open vessel, of whatever kind, is always full either of air or of some other substance, and unless the air is first permitted to escape no other substance can take the place of the air.

556. If a tumbler be inverted and immersed in water, the water will not rise in the tumbler, because the air in the tumbler fills it. If the tumbler be inclined so as to let the air ascend in obedience to the laws of the equilibrium of fluids, the water will rush in and displace the air, while the lighter air, ascending, rises to the surface of the water. If this experiment be made with a bottle, the air will rise in bubbles with a gurgling sound. The same experiment may be made with a tube closed at one end by the finger; the water will not enter the tube until by the removal of the finger the air be permitted to escape. It is on this principle that the diving-bell is constructed.

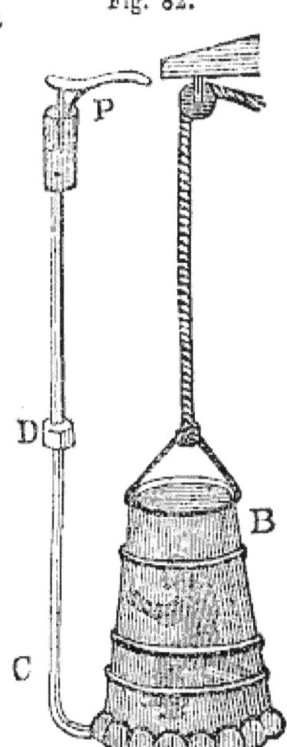

Fig. 82.

Explain the construction of the diving-bell by Fig. 82.

557. Fig. 82 represents a diving-bell. It consists of a large heavy vessel, formed like a bell (but may be made of any other shape), with the mouth open. It descends into the water with its mouth downwards. The air within it having no outlet, it is compelled by the order of specific gravities to ascend in the bell, and thus (as water and air cannot occupy the same space at the same time) prevents the water from rising in the bell. A person, therefore, may descend with safety in the bell to a great depth in the sea, and thus recover valuable articles that have been lost. A constant supply of fresh air is sent down, either by means of barrels, or by a forcing-pump. In the Fig. B represents the bell with the diver in it. C is a bent metallic tube attached to one side and reaching the air within; and P is the forcing-pump through which air is forced into the bell. The forcing-pump is attached to the tube by a joint at D. When the bell descends to a great depth, the pressure of the water

152   NATURAL PHILOSOPHY.

condenses the air within the bell, and causes the water to ascend in the bell. This is forced out by constant accessions of fresh air, supplied as above mentioned. Great care must be taken that a constant supply of fresh air is sent down, otherwise the lives of those within the bell will be endangered. The heated and impure air is allowed to escape through a stop-cock in the upper part of the bell.

How is water raised in a common pump?
How high may water be raised by a common pump?

558. THE COMMON WATER PUMP.—Water is raised in the common pump by means of the pressure of the atmosphere on the surface of the water. A vacuum being produced by raising the piston or pump-box,* the water below is forced up by the atmospheric pressure, on the principle of the equilibrium of fluids. On this principle the water can be raised only to the height of about thirty-three feet, because the pressure of the atmosphere will sustain a column of water of that height only.

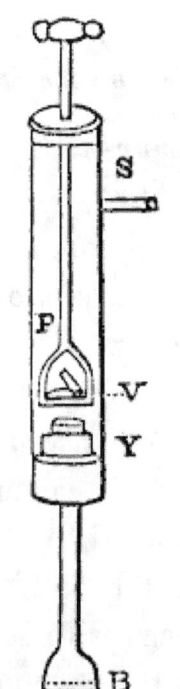

Fig. 83.

Explain Fig. 83.

559. Fig. 83 represents the common pump, improperly called the sucking-pump. The body consists of a large tube, or pipe, the lower end of which is immersed in the water which it is designed to raise. P is the piston, V a valve † in the piston, which, opening upwards, admits the water to rise through it, but prevents its return. Y is a similar valve in the body of the

* In order to produce such a vacuum, it is necessary that the piston or box should be accurately fitted to the bore of the pump; for, if the air above the piston has any means of rushing in to fill the vacuum, as it is produced by the raising of the piston, the water will not ascend. The piston is generally worked by a lever, which is the handle of the pump, not represented in the figure.

† A valve is a lid, or cover, so contrived as to open a communication in one way and close it in the other. Valves are made in different ways, according to the use for which they are intended. In the common pump they are generally made of thick leather partly covered with wood. In the air-pump they are made of oiled silk, or thin leather softened with oil. The clapper of a pair of bellows is a familiar specimen of a valve. **The valves of a pump are commonly called *boxes*.**

pump, below the piston. When the pump is not in action, the valves are closed by their own weight; but when the piston is raised it draws up the column of water which rested upon it, producing a vacuum between the piston and the lower valve Y. The water below immediately rushes through the lower valve and fills the vacuum. When the piston descends a second time, the water in the body of the pump passes through the valve V, and on the ascent of the piston is lifted up by the piston, and a vacuum is again formed below, which is immediately filled by the water rushing through the lower valve Y. In this manner the body of the pump is filled with water, until it reaches the spout S, where it runs out in an uninterrupted stream.

560. In the description here given of the common pump, as well as in the figure, it will be observed that the common form of the handle of the pump is not noticed. The handle of the pump is merely a lever of the first kind; the fulcrum is the pin which attaches it to the pump, and the iron rod connected with the upper valve of the pump is raised or depressed by means of the handle.

561. Although water can be raised by the atmospheric pressure only to the height of thirty-three feet above the surface, the common pump is so constructed that after the pressure of the atmosphere has forced the water through the valve in the body of the pump, and the descent of the piston has forced it through the valve in the piston, it is *lifted* up, when the piston is raised. For this reason, this pump is sometimes called the *lifting* pump. The distance of the upper valve from the surface of the water must never exceed thirty-two feet; and in practice it must be much less.

How does the Forcing-pump differ from the common pump?

562. THE FORCING-PUMP. The Forcing-pump differs from the common pump in having a forcing power added, to raise the water to any desired height.

Explain Fig. 84.

563. Fig. 84 represents the forcing-pump. The body and lower valve V are similar to those in the common pump. The piston P has no valve, but is solid; when, therefore, the vacuum is produced above the

154 NATURAL PHILOSOPHY.

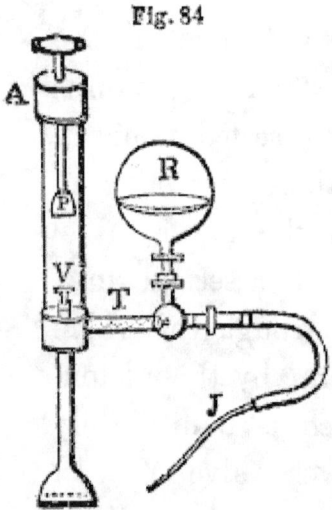

Fig. 84

lower valve, the water, on the descent of the piston, is forced through the tube into the reservoir or air-vessel R, where it compresses the air above it. The air, by its elasticity, forces the water out through the jet J in a continued stream, and with great force. It is on this principle that fire-engines are constructed.

Sometimes a pipe with a valve in it is substituted for the air-vessel; the water is then thrown out in a continued stream, but not with so much force.

How is the Fire-engine constructed?

564. THE FIRE-ENGINE consists of two forcing-pumps, worked successively by the elevation and depression of two long levers of the second kind, called "Brakes."

Fig. 85.

What is the Air-pump, and on what principle is it constructed?

565. THE AIR-PUMP.—The Air-pump is a machine constructed on the principle of the elasticity of the air, for the purpose of exhausting the air from a vessel prepared for the purpose. This vessel is called a receiver, and is made of glass, in order that the effects of the removal of the air may be seen.

566. Air-pumps are made in a great variety of forms; but all are constructed on the principle that, when any portion of confined

PNEUMATICS. 155

air is removed, the residue, immediately expanding, by its elasticity fills the space occupied by the portion that has been withdrawn.

Explain the construction of the air-pump by Fig. 86.

567. Fig. 86 represents a single-barrel air-pump, used both for condensing and exhausting. A D is the stand or platform of the instrument, which is screwed down to the table by means of a clamp, underneath, which is not represented in the figure. R is the glass vessel, or bulbed receiver, from which the air is to be exhausted. P is a solid piston, accurately fitted to the bore of the cylinder, and H the handle by which it is moved. The dotted line T represents the communication between the receiver R and the barrel B; it is a tube through which the air, entering at the opening I, on the plate of the pump, passes into the barrel through the exhausting valve E V. C V is the condensing valve, communicating with the barrel B by means of an aperture near E, and opening outwards through the condensing pipe p.

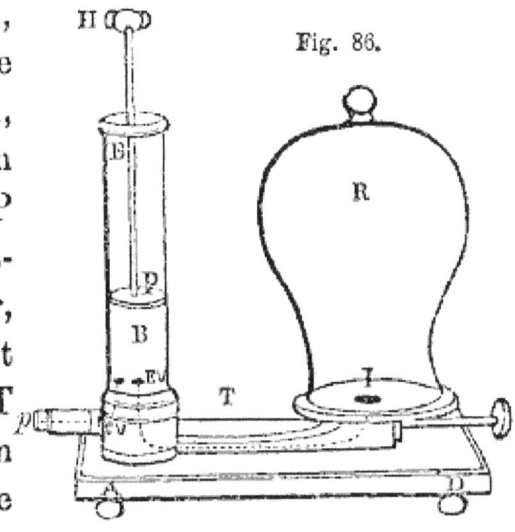

Fig. 86.

Explain the operation of the air-pump by Fig. 90.

568. *The operation of the pump is as follows;* The piston P being drawn upwards by the handle H, the air in the receiver R, expanding by its elasticity, passes by the aperture I through the tube T, and through the exhausting valve E V, into the barrel. On the descent of the piston, the air cannot return through that valve, because the valve opens *upwards* only: it must, therefore, pass through the aperture by the side of the valve, and through the condensing valve C V, into the pipe p, where it passes out into the open air. It cannot return through the condensing valve C V, because that valve opens *outwards* only. By continuing this operation, every ascent and descent of the piston P must render the air within the receiver R more and more

rare, until its elastic power is exhausted. The receiver is then said to be exhausted; and, although it still contains a small quantity of air, yet it is in so rare a state that the space within the receiver is considered a *vacuum*.

569. From this statement it will appear that a perfect vacuum can never be obtained by the air-pump as at present constructed. But so much of the air within a receiver may be exhausted that the residue will be reduced to such a degree of rarity as to subserve most of the practical purposes of a vacuum. The nearest approach made to a perfect vacuum is the famous experiment of Torricelli, which has been explained in No. 530. That would be a perfect vacuum, were there not vapor rising from the mercury.

How may the air be condensed by means of the pump which has been described?
570. From the explanation which has been given of the operation of this air-pump, it will readily be seen that, by removing the receiver R, and screwing any vessel to the pipe p, the air may be condensed in the vessel. Thus the pump is made to exhaust or to condense, without alteration.

What is a condensing syringe?
571. Air-pumps in general are not adapted for condensation; that office being performed by an instrument called "*a condensing syringe*," which *is an air-pump reversed, its valves being so arranged as to force air into a chamber, instead of drawing it out.* For this purpose, the valves open inwards in respect to the chamber, while in air-pumps they open outwards.

572. A guage, constructed on the principle of the barometer, is sometimes adjusted to the air-pump, for the purpose of exhibiting the degree of exhaustion.

How does the double air-pump differ from the single?
573. The double air-pump differs from the single air-pump, in having two barrels and two pistons; which, instead of being moved by the hand, are worked by means of a toothed wheel. playing in notches of the piston-rods.

Fig. 87 represents an air-pump of a different construction. In this pump the piston is stationary, while motion is given to the barrel by means of the lever H. The barrel is kept in a proper position by means of polished steel guides.

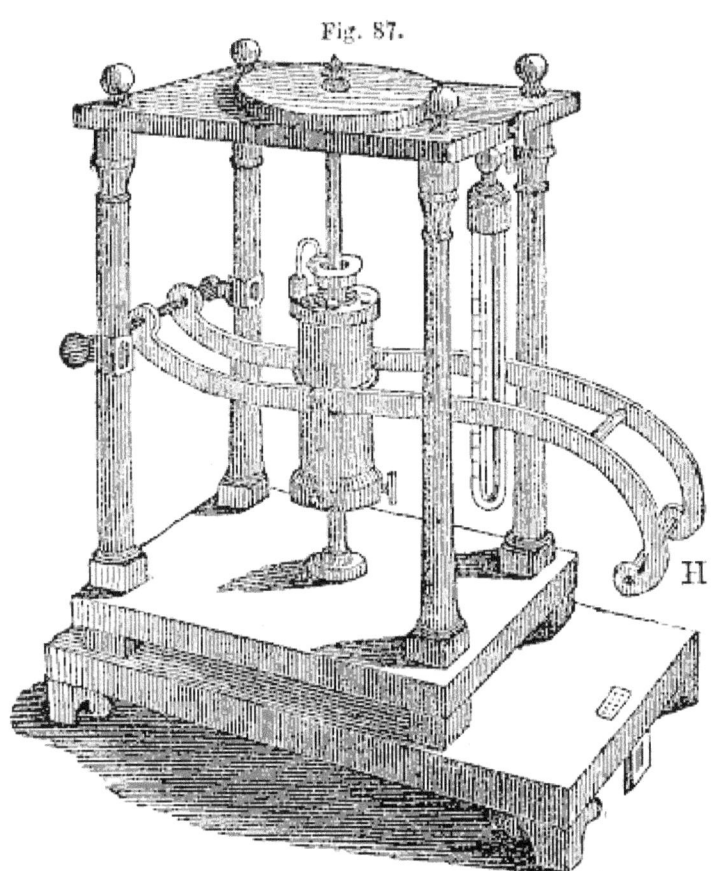

Fig. 87.

574. By means of the air-pump many interesting experiments may be performed, illustrating the gravity, elasticity, fluidity, and inertia of air.

575. EXPERIMENTS ILLUSTRATING THE GRAVITY OF AIR.—Having adjusted the receiver to the plate of the air-pump, exhaust the air and the receiver will be held firmly on the plate. The force which confines it is nothing more than the weight of the external air which, having no internal pressure to contend with, presses with a force of nearly fifteen pounds on every square inch of the external surface of the receiver.

576. The *exact* amount of pressure depends on the degree of exhaustion, being at its maximum of fifteen pounds when there is a perfect vacuum. On readmitting the air, the receiver may be readily removed.*

What are the Magdeburgh Cups, and what do they illustrate?

577. THE MAGDEBURGH CUPS, OR HEMISPHERES.— Fig. 88 represents the Magdeburgh Cups, or Hemispheres. They consist of two hollow brass cups, the edges of which are accurately fitted together. They each have a handle,

* The air is readmitted into the receiver by turning a screw which is inserted into the receiver, in which there is an aperture, through which the external air rushes with considerable force.

158 NATURAL PHILOSOPHY.

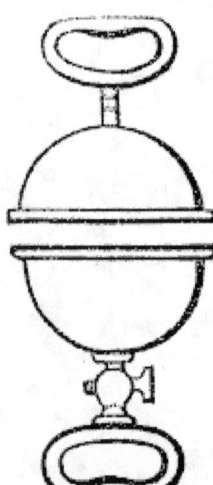

Fig. 88.

to one of which a stop-cock is fitted. The stop-cock, being attached to one of the cups, is to be screwed to the plate of the air-pump, and left open. Having joined the other cup to that on the pump, exhaust the air from within them, turn the stop-cock to prevent its readmission, and screw the handle that had been removed to the stop-cock. Two persons may then attempt to draw the cups asunder. It will be found that great power is required to separate them; but, on readmitting the air between them, by turning the cock, they will fall asunder by their own weight. When the air is exhausted from within them, the pressure of the surrounding air upon the outside keeps them united. This pressure being equal to a pressure of fifteen pounds on every square inch of the surface, it follows that the larger the cups, or hemispheres, the more difficult it will be to separate them.

578. The Magdeburgh Cups derive their name from the city where the experiment was first attempted. Otto Guericke constructed two hemispheres which, when the air was exhausted, were

Fig. 89.

held together by a force of about three-fourths of a ton. Fig. 89 shows the manner in which such an experiment may be tried.

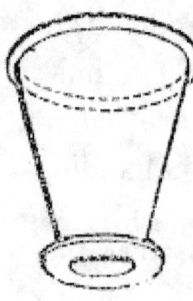

Fig. 90.

What principle does the Hand-glass illustrate?

579. THE HAND-GLASS.—Fig. 90 is nothing more than a tumbler, open at both ends, with the top and bottom ground smooth, so as to fit the brass plate of the air-pump. Placing it upon the plate, cover it closely with the palm of the hand, and work the pump. The air

within the glass being thus exhausted, the hand will be pressed down by the weight of the air above it: on readmitting the air, the hand may be easily removed.

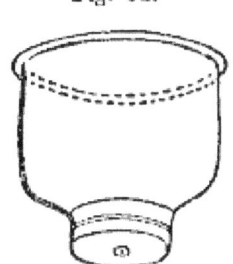

What principle is illustrated by the Bladder-glass?

580. THE BLADDER-GLASS.— Fig. 91 is a bell-shaped glass, covered with a piece of bladder, which is tied tightly around its neck. Thus prepared, it may be screwed to the plate of the air-pump, or connected with it by means of an elastic tube. On exhausting the air from the glass, the weight of the external air on the bladder will burst it inwards, with a loud explosion.

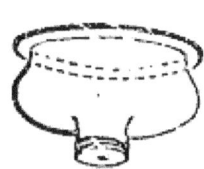

What does the India-rubber Glass show?

581. THE INDIA-RUBBER GLASS. — Fig. 92 is a glass similar to the one represented in the last figure, covered with india-rubber. The same experiments may be made with this as were mentioned in the last article, but with different results. Instead of bursting, the india-rubber will be pressed inwards the whole depth of the glass.

What is illustrated by means of the Fountain-glass and Jet?

582. THE FOUNTAIN-GLASS AND JET.— Fig. 93 represents the jet, which is a small brass tube. Fig. 94 is the fountain-glass. The experiment with these instruments is designed to show the pressure of the atmosphere on the surface of liquids. Screw the straight jet to the stop-cock, the stop-cock to the fountain-glass, with the straight jet inside of the fountain-glass, and the lower end of the stop-cock to the plate of the air-pump, and then open the stop-cock. Having exhausted the air from the fountain-glass, close the stop-cock, remove the glass from the pump, and, immersing it in a vessel of water, open the stop-cock. The pressure of the air on the surface of the water will cause it to rush up into the glass like a fountain.

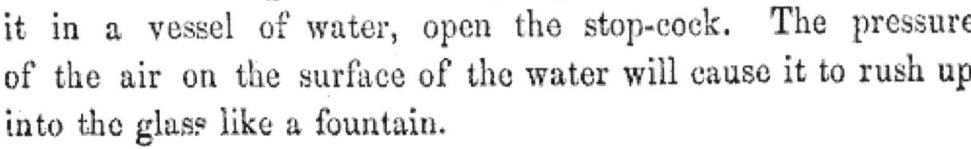

How are the Pneumatic Scales used?

583. Pneumatic Scales for Weighing Air.— Fig. 95 represents the flask, or glass vessel and scales for weighing air. Weigh the flask when full of air; then exhaust the air and weigh the flask again. The difference between its present and former weight is the weight of the air that was contained in the flask.

Fig. 95.

What principle does "the Sucker" illustrate?

584. The Sucker.— A circular piece of wet leather, with a string attached to the centre, being pressed upon a smooth surface, will adhere with considerable tenacity, when drawn upwards by the string. The string in this case must be attached to the leather, so that no air can pass under the leather.

What is the object of the Mercurial or Water Tube?

585. The Mercurial or Water Tube.— Exhaust the air from a glass tube three feet long fitted with a stop-cock at one end, and then immerse it in a vessel containing mercury or water. On turning the stop-cock, the mercury will rise to the height of nearly thirty inches; or, if immersed in water, the water will rise and fill the tube, and would fill it were it thirty feet long. This experiment shows the manner in which water is raised to the boxes or valves in common water-pumps.

How is the elasticity of the air illustrated?

586. Experiments showing the Elasticity of the Air.— Place an india-rubber bag, or a bladder, partly inflated, and tightly closed, under the receiver, and, on exhausting the air, the air within the bag or bladder, expanding, will fill the bag. On readmitting the air, the bag will collapse. The experiment may also be made with some kinds of shrivelled fruit, if the skin be sound. The internal air, expanding, will give the fruit a fresh and plump appearance, which will disappear on the readmission of the air.

587. The same principle may be illustrated by the india-

rubber and bladder glasses, if they have stop-cocks to confine the air.

588. A small bladder partly filled with air may be sunk in a vessel of water by means of a weight, and placed under the receiver. On exhausting the air from the receiver, the air in the bladder will expand, and, its specific gravity being thus diminished, the bladder with the weight will rise. On readmitting the air, the bladder will sink again.

How can the presence of air in wood be detected?

589. AIR CONTAINED IN WATER AND IN WOOD. — Place a vessel of water under the receiver, and, on exhausting the air from the receiver, the air in the water, previously invisible, will make its appearance in the form of bubbles, presenting the semblance of ebullition.

590. A piece of light porous wood being immersed in the water below the surface, the air will be seen issuing in bubbles from the pores of the wood.

Explain the principle of the Pneumatic Balloon.

591. THE PNEUMATIC BALLOON. — Fig. 96 represents a small glass balloon, with its car immersed in a jar of water, and placed under a receiver. On exhausting the air, the air within the balloon, expanding, gives it buoyancy, and it will rise in the jar. On readmitting the air, the balloon will sink.

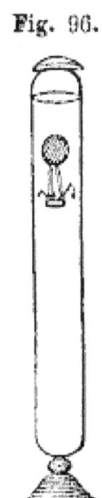

Fig. 96.

592. The experiment may be performed without the air-pump by covering the jar with some elastic substance, as india-rubber. By pressing on the elastic covering with the finger, the air will be condensed, the water will rise in the balloon, and it will sink. On removing the pressure, the air in the balloon, expanding, will expel part of the water, and the balloon will rise. This is the more convenient mode of performing the experiment, as it can be repeated at pleasure without resort to the pump.

593. The following is a full explanation: — The pressure on the top of the vessel first condenses the air between the cover

and the surface of the water; this condensation presses upon the water below, and, as this pressure affects every portion of the water throughout its whole extent, the water, by its upward pressure, compresses the air within the balloon, and makes room for the ascent of more water into the balloon, so as to alter the specific gravity of the balloon, and cause it to sink. As soon as the pressure ceases, the elasticity of the air in the balloon drives out the lately-entered water, and, restoring the former lightness to the balloon, causes it to rise. If, in the commencement of this experiment, the balloon be made to have a specific gravity too near that of water, it will not rise of itself, after once reaching the bottom, because the pressure of the water then above it will perpetuate the condensation of the air which caused it to descend. It may even then, however, be made to rise, if the perpendicular height of the water above it be diminished by inclining the vessel to one side.

594. This experiment proves many things; namely:

First. The *materiality of air*, by the pressure of the hand on the top being communicated to the water below through the air in the upper part of the vessel.

Secondly. The *compressibility of air*, by what happens in the globe before it descends.

Thirdly. The *elasticity*, or elastic force of air, when the water is expelled from the globe, on removing the pressure.

Fourthly. The *lightness of air*, in the buoyancy of the globe.

Fifthly. It shows that *the pressure of a liquid is exerted in all directions*, because the effects happen in whatever position the jar be held.

Sixthly. It shows *that pressure is as the depth*, because less pressure of the hand is required the further the globe has descended in the water.

Seventhly. It exemplifies many circumstances of *fluid support*. A person, therefore, who is familiar with this experiment, and can explain it, has learned the principal truths of Hydrostatics and Pneumatics.

595. The Pneumatic Balloon also exhibits the principle on which the well-known glass toy, called the Cartesian Devil, is constructed; and it may be thus explained: Several images of glass, hollow within, and each having a small opening at the heel by which water may pass in and out, may be made to manœuvre in a vessel of water. Place them in a vessel in the same manner with the balloon, but, by allowing different quantities of water to enter the

apertures in the images, cause them to differ a little from one another in specific gravity. Then, when a pressure is exerted on the cover, the heaviest will descend first, and the others follow in the order of their specific gravity; and they will stop or return to the surface in reverse order, when the pressure ceases. A person exhibiting these figures to spectators who do not understand them, while appearing carelessly to rest his hand on the cover of the vessel, seems to have the power of ordering their movements by his will. If the vessel containing the figures be inverted, and the cover be placed over a hole in the table, through which, unobserved, pressure can be made by a rod rising through the hole, and obeying the foot of the exhibiter, the most surprising evolutions may be produced among the figures, in perfect obedience to the word of command.

What is the use of the Condensing and Exhausting Syringe?

596. EXPERIMENTS WITH CONDENSED AIR.—THE CONDENSING AND EXHAUSTING SYRINGE.—The Condensing Syringe is the air-pump reversed. The Exhausting Syringe is the simple air-pump without its plate or stand. These implements are used respectively with such parts of the apparatus as cannot conveniently be attached to the air-pump, and as an addition to such pumps as do not perform the double office of exhaustion and condensation. In some sets of apparatus the condensing and exhausting syringes are united, and are made to perform each office respectively, by merely reversing the part which contains the valve.

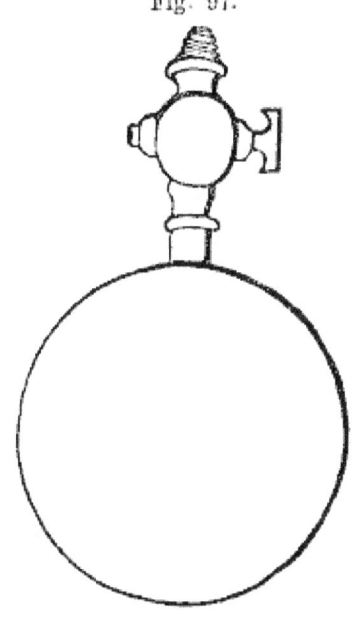

Fig. 97.

For what purpose is the Air-chamber used?

597. THE AIR-CHAMBER.—The air-chamber, Fig. 97, is a hollow brass globe prepared for the reception of a stop-cock, and is designed for the reception of condensed air. It is made in different forms in different sets, and is used by screwing it to a condensing pump or a condensing syringe.

What principle of Pneu-

598. STRAIGHT AND REVOLVING JETS FROM CONDENSED AIR.—Fill the air-chamber (Fig.

matics is illustrated by the straight and revolving jets?

97) partly with water, and then condense the air. Then confine the air by turning the cock; after which, unscrew it from the air-pump, and screw on the straight or the revolving jet. Then open the stop-cock, and the water will be thrown from the chamber in the one case in a straight continued stream, in the other in the form of a wheel. Figs. 98 and 99 represent a view of the straight and the revolving jets.

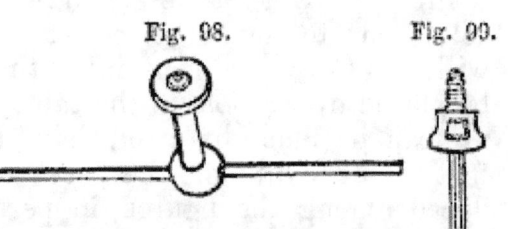

Fig. 98. Fig. 99.

In the revolving jet the water is thrown from two small apertures made at each end on opposite sides, to assist the revolution. The circular motion is caused by the reäction of the water on the opposite sides of the arms of the jets; for, as the water is forced into the tubes, it exerts an equal pressure on all sides of the tubes, and, as the pressure is relieved on one side by the jet-hole, the arm is caused to revolve in a contrary direction. This experiment, performed with the straight jet, illustrates the principle on which "Hero's ball" and "Hero's fountain" are constructed.

Explain the principle of the Air-gun.

599. THE PRINCIPLE OF THE AIR-GUN.—With the air-chamber, as in the last experiments, a small brass cylinder or gun-barrel, Fig. 100, may be substituted for the jets, and loaded with a small shot or paper ball. On turning the cock quickly, the condensed air, rushing out, will throw the shot to a considerable distance. In this way the air-gun operates, an apparatus resembling the lock of a gun being substituted for the stop-cock, by which a small portion only of the condensed air is admitted to escape at a time; so that the chamber, being once filled, will afford two or three dozen discharges. The force of the air-gun has never been equal to more than a fifteenth of the force of a common charge of powder, and the loudness of the report made in its discharge is always as great in proportion to its force as that of the common gun.

Fig. 100.

PNEUMATICS. 165

In weighing air what must always be taken into the account?

600. Condensed air may be weighed in the air-chamber, but, in estimating its weight, the temperature of the room must always be taken into consideration, as the density of air is materially affected by heat and cold.

What does the Guinea and Feather Drop illustrate?

601. EXPERIMENTS SHOWING THE INERTIA OF AIR.—THE GUINEA AND FEATHER DROP.—The inertia of air is shown by the guinea and feather drop, exhibiting the resistance which the air opposes to falling bodies. This apparatus is made in different forms, some having shelves on which the guinea and feather rest, and, when the air is exhausted, they are made to fall by the turning of a handle. A better form is that represented in Fig. 101, in which the guinea and feather (or a piece of brass substituted for the guinea) are enclosed, and the apparatus being screwed to the plate of the pump, the air is exhausted, a stop-cock turned to prevent the readmission of the air, and the apparatus being then unscrewed, the experiment may be repeatedly shown by one exhaustion of the air. It will then appear that every time the apparatus is inverted the guinea and the feather will fall simultaneously. The two forms of the guinea and feather drop are exhibited in Figs. 101 and 102, one of which, Fig. 101, is furnished with a stop-cock,* the other, Fig. 102, with shelves.

Fig. 101.

Fig. 102.

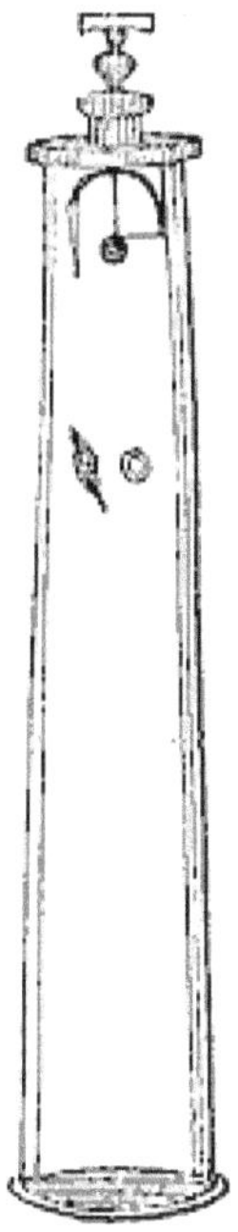

What principle is explained by means of the weight-lifter?

602. EXPERIMENTS SHOWING THE FLUIDITY OF AIR.—THE WEIGHT-LIFTER.—The upward pressure of the air, one of the properties of its fluidity, may be exhibited by an apparatus called the

---

* Most sets of philosophical apparatus are furnished with stop-cocks and elastic tubes, for the purpose of connecting the several parts with the pump, or with one another. In selecting the apparatus, it is important to have the screws of the stop-cocks and of all the apparatus of similar thread, in order that every article may subserve as many purposes as possible. This precaution is suggested by economy, as well as by convenience.

weight-lifter, made in different forms, but all on the same principle. The one represented in Fig. 103 consists of a glass tube, of large bore, set in a strong case or stand, supported by three legs. A piston is accurately fitted to the bore of the tube, and a hook is attached to the bottom of the piston, from which weights are to be suspended. One end of the elastic tube is to be screwed to the plate of the pump, and the other end attached to the top of this instrument.

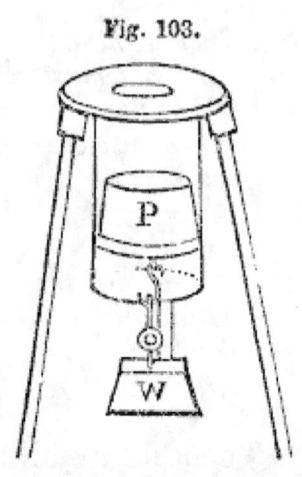

Fig. 103.

The air being then exhausted from the tube, the weights will be raised the whole length of the glass. The number of pounds' weight that can be raised by this instrument may be estimated by multiplying the number of square inches in the bottom of the piston by fifteen.

Explain the Pneumatic Shower-bath.

603. THE PNEUMATIC SHOWER-BATH. — On the principle of the upward pressure of the air the pneumatic shower-bath is constructed. It consists of a tin vessel perforated with holes in the bottom for the shower, and having an aperture at the top, which is opened or closed at pleasure by means of a spring-valve. [Instead of the spring-valve, a bent tube may be brought round from the top down the side of the vessel, with an aperture in the tube below the bottom of the vessel, which may be covered with the thumb.] On immersing the vessel thus constructed in a pail of water, with the valve open, and the tube (if it have one) on the outside of the pail, the water will fill the vessel. The aperture then being closed with the spring or with the thumb, and the vessel being lifted out of the water, the upward pressure of the air will confine the water in the vessel. On removing the thumb, or opening the valve, the water will descend in a shower, until the vessel is emptied.

What two properties of

604. MISCELLANEOUS EXPERIMENTS DEPENDING ON TWO OR MORE OF THE PROPERTIES OF AIR. —

air are illustrated by means of the Bolt-head and Jar?

THE BOLT-HEAD AND JAR.—Fig. 104, a glass globe with a long neck, called a bolt-head (or any long-necked bottle), partly filled with water, is inverted in a jar of water (colored with a few drops of red ink or any coloring matter, in order that the effects may be more distinctly visible), and placed under the receiver. On exhausting the air in the receiver, the air in the upper part of the bolt-head, expanding, expels the water, showing the elasticity of the air. On readmitting the air to the receiver, as it cannot return into the bolt-head, the pressure on the surface of the water in the jar forces the water into the bolt-head, showing the pressure of the air caused by its weight. The experiment may be repeated with the bolt-head without any water, and, on the readmission of the air, the water will nearly fill the bolt-head, affording an accurate test of the degree of exhaustion.

Fig. 104.

What two principles are concerned in the transfer of fluids from one vessel to another?

605. THE TRANSFER OF FLUIDS FROM ONE VESSEL TO ANOTHER.—The experiment may be made with two bottles tightly closed. Let one be partly filled with water, and the two connected by a bent tube, connecting the interior of the empty bottle with the water of the other, and extending nearly to the bottom of the water. On exhausting the air from the empty bottle, the water will pass to the other, and, on readmitting the air, the water will return to its original position, so long as the lower end of the bent tube is below the surface.

What experiments are performed with the siphon?

606. EXPERIMENTS WITH THE SIPHON.—Close the shorter end of the siphon with the finger or with a stop-cock, and pour mercury or water into the longer side. The air contained in the shorter side will prevent the liquid from rising in the shorter side. But, if the shorter end be opened, so as to afford free passage

outwards for the air, the fluid will rise to an equilibrium in both arms of the siphon.

607. Pour any liquid into the longer arm of the siphon until the shorter arm is filled. Then close the shorter end, to prevent the admission of the air; the siphon may then be turned in any direction and the fluid will not run out, on account of the pressure of the atmosphere against it. But, if the shorter end be unstopped, the fluid will run out freely.

What effect is produced on an animal placed under an exhausted receiver?

608. AIR ESSENTIAL TO ANIMAL LIFE. — If an animal be placed under the receiver, and the air exhausted, it will immediately droop, and, if the air be not speedily readmitted, it will die.

How is it shown that air is essential to combustion?

609. AIR ESSENTIAL TO COMBUSTION. — Place a lighted taper, cigar, or any other substance that will produce smoke, under the receiver, and exhaust the air; the light will be extinguished, and the smoke will fall, instead of rising. If the air be readmitted, the smoke will ascend.

What effect is produced on ether under an exhausted receiver?

610. THE PRESSURE OF THE AIR RETARDS EBULLITION.* — Ether, alcohol, and other distilled liquors, or warm water, placed under the receiver, will appear to boil when the air is exhausted.

What effect has the pressure of the air on the form of bodies?

611. The existence of many bodies in a liquid form depends on the weight or pressure of the atmosphere upon them. The same force, likewise, prevents the gases which exist in fluid and solid bodies from disengaging themselves. If, by rarefying the air, the pressure on these bodies be diminished, they either assume the form of vapors, or else the gas detaches itself altogether from the other body. The following experiment proves this: Place a quantity of lukewarm water, milk or alcohol, under a receiver, and exhaust the air, and the liquid

---

* EBULLITION. — The operation of boiling. The agitation of liquor by heat, which throws it up into bubbles.

will either pass off in vapor, or will have the appearance of boiling.

What experiment shows that the liquid form of some bodies is dependent on atmospheric pressure?

612. An experiment to prove that the pressure of the atmosphere preserves some bodies in the liquid form may thus be performed. Fill a long vial, or a tube closed at one end, with water, and invert it in a vessel of water. The atmospheric pressure will retain the water in the vial. Then, by means of a bent tube, introduce a few drops of sulphuric ether, which, by reason of their small specific gravity, will ascend to the top of the vial, expelling an equal bulk of water. Place the whole under the receiver, and exhaust the air, and the ether will be seen to assume the gaseous form, expanding in proportion to the rarefaction of the air under the receiver, so that it gradually expels the water from the vial, and fills up the entire space itself. On readmitting the air, the ether becomes condensed, and the water will reascend into the vial.

How may water be frozen under a receiver?

613. A simple and interesting experiment connected with the science of chemistry may thus be performed by means of the air-pump. A watch-glass, containing water, is placed over a small vessel containing sulphuric acid, and put under the bulbed receiver. When the air is exhausted, vapor will freely rise from the water, and be quickly absorbed by the acid. An intense degree of cold is thus produced, and the water will freeze.

614. In the above experiment, if ether be used instead of the acid, the ether will evaporate instead of the water, and, in the process of evaporation, depriving the water of its heat, the water will freeze. These two experiments, apparently similar in effects, namely, the freezing of the water, depend upon two different principles which pertain to the science of chemistry.

What is the Pneumatic Paradox?

615. THE PNEUMATIC PARADOX. — An interesting experiment, illustrative of the pneumatic

paradox, may be thus performed: Pass a small open tube (as a piece of quill) through the centre of a circular card two or three inches in diameter, and cement it, the lower end passing down, and the upper just even with the card. Then pass a pin through the centre of another similar card, and place it on the former, with the pin projecting into the tube to prevent the upper card from sliding off. It will then be impossible to displace the upper card by blowing through the quill, on account of the adhesion produced by the current passing between the discs. On this principle smoky chimneys have been remedied, and the office of ventilation more effectually performed.

What is Wind? 616. WIND. — Wind is air put in motion.

In what two ways may the motion of the air be explained? 617. There are two ways in which the motion of the air may arise. It may be considered as an absolute motion of the air, rarefied by heat and condensed by cold; or it may be only an apparent motion, caused by the superior velocity of the earth in its daily revolution.

618. When any portion of the atmosphere is heated it becomes rarefied, its specific gravity is diminished, and it consequently rises. The adjacent portions immediately rush into its place, to restore the equilibrium. This motion produces a current which rushes into the rarefied spot from all directions. This is what we call wind.

How is a north wind caused? 619. The portions north of the rarefied spot produce a north wind, those to the south produce a south wind, while those to the east and west in like manner, form currents moving in opposite directions. At the rarefied spot, agitated as it is by winds from all directions, turbulent and boisterous weather, whirlwinds, hurricanes, rain, thunder and lightning, prevail. This kind of weather occurs most frequently in the torrid zone, where the heat is greatest. The air, being more rarefied there than in any other part of the globe, is lighter, and, consequently, ascends; that about the polar regions is continually flowing from the poles to the equator, to restore the equilibrium; while the air rising from the equator flows in an upper current towards the poles, so that the polar regions may not be exhausted.

What wind prevails in the equatorial regions? 620. A regular east wind prevails about the equator, caused in part by the rarefaction of the air produced by the sun in his daily course from east to west. This wind, combining with that from the poles, causes a constant north-east wind for about thirty degrees north of the equator, and a south-east wind at the same distance south of the equator.

621. From what has now been said, it appears that there is a circulation of air in the atmosphere; the air in the lower strata flowing from the poles to the equator, and in the upper strata flowing back from the equator to the poles. It may here be remarked, that the periodical winds are more regular at sea than on the land; and the reason of this is, that the land reflects into the atmosphere a much greater quantity of the sun's rays than the water, therefore that part of the atmosphere which is over the land is more heated and rarefied than that which is over the sea. This occasions the wind to set in upon the land, as we find it regularly does on the coast of Guinea and other countries in the torrid zone. There are certain winds, called trade-winds, the theory of which may be easily explained on the principle of rarefaction, affected, as it is, by the relative position of the different parts of the earth with the sun at different seasons of the year, and at various parts of the day. A knowledge of the laws by which these winds are controlled is of importance to the mariner. When the place of the sun with respect to the different positions of the earth at the different seasons of the year is understood, it will be seen that they all depend upon the same principle. The reason that the wind generally subsides at the going down of the sun is, that the rarefaction of the air, in the particular spot which produces the wind, diminishes as the sun declines, and, consequently, the force of the wind abates. The great variety of winds in the temperate zone is thus explained. The air is an exceedingly elastic fluid, yielding to the slightest pressure: the agitations in it, therefore, caused by the regular winds, whose causes have been explained, must extend every way to a great distance, and the air, therefore, in all climates will suffer more or less perturbation, according to the situation of the country, the position of mountains, valleys, and a variety of other causes. Hence every climate must be liable to variable winds. The *quality* of winds is affected by the countries over which they pass; and

they are sometimes rendered pestilential by the heat of deserts or the putrid exhalations of marshes and lakes. Thus, from the deserts of Africa, Arabia and the neighboring countries, a hot wind blows, called *Samiel*, or *Simoon*, which sometimes produces instant death. A similar wind blows from the desert of Sahara, upon the western coast of Africa, called the *Harmattan*, producing a dryness and heat which is almost insupportable, scorching like the blasts of a furnace.

How is wind sometimes affected by the face of a country?

622. WHIRLWINDS AND WATERSPOUTS.—The direction of winds is sometimes influenced by the form of lofty and precipitous mountains, which, resisting their direct course, causes them to descend with a spiral and whirling motion, and with great force.

623. A similar effect is produced by two winds meeting at an angle, and then turning upon a centre. If a cloud happen to be between these two winds thus encountering each other, it will be condensed and rapidly turned round, and all light substances will be carried up into the air by the whirling motion thus produced.

What is supposed to be the cause of waterspouts?

624. The whirlwind, occurring at sea, occasions the singular phenomenon of the waterspout.

Fig. 105.

What does Fig. 105 represent? 625. Fig. 105 represents the several forms in which water-spouts are sometimes seen.

626. From a dense cloud a cone descends in the form of a trumpet, with the small end downwards. At the same time, the surface of the sea under it is agitated and whirled round, the waters are converted into vapor, and ascend with a spiral motion, till they unite with the cone proceeding from the cloud. Frequently, however, they disperse before the junction is effected. Both columns diminish towards their point of contact, where they are sometimes not more than three or four feet in diameter. In the centre of the water-spout there is generally a vacant space, in which none of the small particles of water ascend. In this, as well as around the outer edges of the water-spout, large drops of rain fall. Water-spouts sometimes disperse suddenly, and sometimes continue to move rapidly over the surface of the sea, continuing sometimes in sight for the space of a quarter of an hour. When the water-spout breaks, the water usually descends in the form of heavy rain. It is proper here to observe that by some authorities the phenomena of water-spouts are considered as due to electrical causes.

627. A notion has prevailed that water-spouts are dangerous to shipping. It is true that small vessels incur a risk of being overset if they carry much sail, because sudden gusts of wind, from all points of the compass, are very common in the vicinity of water-spouts; but large vessels, under but a small spread of canvas, encounter, as is said, but little danger.

628. Pneumatics forms a branch of physical science which has been entirely created by modern discoveries. Galileo first demonstrated that air possesses weight. His pupil, Torricelli, invented the barometer; and Pascal, by observing the difference of the altitudes of the mercurial column at the top and the foot of the *Puy de Dome*, proved that the suspension of the mercury is caused by the pressure of the atmosphere. Otto Guericke, a citizen of Magdeburg, invented the air-pump about the year 1654; and Boyle and Mariotte soon after detected, by its means, the principal mechanical properties of atmospheric air. Analogous properties have been proved to belong to all the other aëriform fluids. The problem of determining the velocity of their vibrations was solved by Newton and Euler, but more completely by Lagrange. The theoretical principles relative to the pressure and motion of elastic fluids, from which the practical formulæ are deduced, were established by Daniel Bernoulli in his *Hydrodynamica* (1738), but have been rendered more general by Navier.

What is Acoustics? 629. ACOUSTICS. — Acoustics is the science which treats of the nature and laws of sound. It includes the theory of musical concord or harmony.

What is sound? 630. Sound is the sensation produced in the organs of hearing by the vibrations or undulations transmitted through the air around.*

631. If a bell be rung under an exhausted receiver, no sound can be heard from it; but when the air is admitted to surround the bell, the vibrations immediately produce sound.

632. Again, if the experiments be made by enclosing the bell in a small receiver, full of air, and placing that under another receiver, from which the air can be withdrawn, though the bell, when struck, must then produce sound, as usual, yet it will not be heard if the outer receiver be well exhausted, and care be taken to prevent the vibrations from being communicated through any solid part of the apparatus, because there is no medium through which the vibrations of the bell in the smaller receiver can be communicated to the ear.†

Why is a sound louder in cold weather? 633. Sounds are louder when the air surrounding the sonorous body is dense than when it is in a rarefied state, and in general the intensity of sound increases with the density of the medium by which it is propagated.

634. For this reason the sound of a bell is louder in cold than in warm weather; and sound of any kind is transmitted to a greater distance in cold, clear weather, than in a warm, sultry day. On the top of mountains, where the air is rare, the human voice can be heard only at the distance of a few rods; and the firing of a gun produces a sound scarcely louder than the cracking of a whip.

What are Sonorous bodies? 635. Sonorous bodies are those which produce clear, distinct, regular, and durable sounds, such as a bell, a drum, wind instruments, musical strings and glasses. These vibrations can be communicated to a distance not only through the air, but also through liquids and solid bodies.

---

* "The sensation of sound is produced by the wave of air impinging on the membrane of the ear-drum, exactly as the momentum of a wave of the sea would strike the shore." — [*Lardner.*]

† In performing these experiments, the bell must be placed in such a manner that whatever supports it will rest on a soft cushion of wool, so as to prevent the vibrations from being communicated to the plate of the air pump, or any other of the solid parts of the apparatus.

ACOUSTICS.

To what do bodies owe their sonorous properties? 636. Bodies owe their sonorous property to their elasticity. But, although it is undoubtedly the case that all sonorous bodies are elastic, it is not to be inferred that all elastic bodies are sonorous.

637. The vibrations of a sonorous body give a tremulous or undulatory motion to the air or the medium by which it is surrounded, similar to the motion communicated to smooth water when a stone is thrown into it.

What are the best conductors of sound? 638. Sound is communicated more rapidly and with greater power through solid bodies than through the air, or fluids. It is conducted by water about four times quicker than by air, and by solids about twice as rapidly as by water.

639. If a person lay his head on a long piece of timber, he can hear the scratch of a pen at the other end, while it could not be heard through the air.

640. If the ear be placed against a long, dry brick wall, and a person strike it once with a hammer, the sound will be heard *twice*, because the wall will convey it with greater rapidity than the air, though each will bring it to the ear.

641. It is on the principle of the greater power of solid bodies to communicate sound that the instrument called the Stethoscope * is constructed.

What is the Stethoscope, and on what principle is it constructed? 642. The Stethoscope is a perforated cylinder, of light, fine-grained wood, with a funnel-shaped extremity, which is applied externally to the cavities of the body, to distinguish the sounds within.

What is the use of the stethoscope? 643. By means of the stethoscope the physician is enabled to form an opinion of the healthy action of the lungs, and other organs to which the ear cannot be directly applied.

---

* The word Stethoscope is derived from two Greek words, στεθος, the breast, and σκοπεω, to examine, and is given to this instrument because it is applied to the breast of a person for the purpose of ascertaining the condition of the lungs and other internal organs. Dr. Webster suggests that the term *Phonophorus*, or Sound-conductor, would be a preferable name for the instrument.

With what rapidity does sound move?

644. Sound passing through the air moves at the rate of 1120 feet in a second of time; and this rule applies to all kinds of sound, whether loud or soft.*

What kind of sounds move fastest?

645. The softest whisper, therefore, flies as fast as the loudest thunder; and the force and direction of the wind, although they affect the continuance of a sound, have but slight effect on its velocity.

646. Were it not for this uniform velocity of all kinds of sound, the music of a choir, or of an orchestra, at a short distance, would be but a strange confusion of discordant sounds; for the different instruments or voices, having different degrees of loudness, could not simultaneously reach the ear.

647. The air is a better conductor of sound when it is humid than when it is dry. A bell can be more distinctly heard just before a rain; and sound is heard better in the night than in the day, because the air is generally more damp in the night.

648. The distance to which sound may be heard depends upon various circumstances, on which no definite calculations can be predicated. Volcanoes, among the Andes, in South America, have been heard at the distance of three hundred miles; naval engagements have been heard two hundred; and even the watchword "*All's well,*" pronounced by the unassisted human voice, has been heard from Old to New Gibraltar, a distance of twelve miles. It is said that the cannon fired at the battle of Waterloo were heard at Dover.

649. A clear and frosty atmosphere is favorable to the transmission of sound, especially where the surface over which it passes is smooth and level. Conversation in the polar regions has been carried on between persons more than a mile apart. The cannon in naval engagements in the English Channel have been heard in the centre of England.

650. A blow struck *under the water* of the Lake of Geneva was heard across the whole breadth of the lake, a distance of nine miles. The earth itself is a good conductor of sound. The trampling of horses can be heard at a great distance by putting the ear to the ground, and the approach of railroad-cars can be ascertained when very far off by applying the ear to the rail.

---

* The velocity of sound has sometimes been estimated as much as eleven hundred and forty-two feet in a second. The state of the air must, however, be taken into consideration. The higher the temperature, the greater the velocity; and it has been ascertained that within certain limits the velocity is increased about one foot for every degree that the thermometer rises. Experiments made with a cannon at midnight by Arago, Gay Lassac, and others, when the thermometer stood at 61°, gave 1118.39 feet per second as the velocity of sound. The rate stated in No. 644 will not therefore be far from the truth. The experiments which gave a result of eleven hundred and forty-two feet in a second were probably made when the weather was extremely warm.

To what practical use is the velocity of sound applied? 651. This uniform velocity of sound enables us to ascertain, with some degree of accuracy, the distance of an object from which it proceeds.

If, for instance, the flash of a gun at sea is seen a half of a minute before the report is heard, the vessel must be at the distance of about six miles.

652. In the same manner the distance of a thunder-cloud may be estimated by counting the seconds that intervene between the flash of the lightning and the roaring of the thunder, and multiplying them by 1120.

What is the Acoustic Paradox? 653. THE ACOUSTIC PARADOX.—Sound, as has already been stated, is propagated by the undulations of the air. Now, as these undulations or waves are precisely analogous to the case of two series of waves formed upon the surface of a liquid, there is a point where the elevation of a wave, produced by one cause, will coincide with the depression of another wave produced by another cause, and the consequence will be neither elevation nor depression of the liquid.

Explain the acoustic paradox. 654. When, therefore, two sounds are produced in different places, there is a point between them where the undulations will counteract each other, and the *two sounds may produce silence.*

655. A simple illustration of this fact may be made with a tuning-fork. If this instrument be put into vibration and held up to the ear and rapidly turned, the sound, instead of being continuous, will appear to be pulsative or interrupted; but, if slowly caused to revolve at a distance from the ear, a position of the forks will be found at which the sound will be inaudible.

656. A similar experiment may be made with the tuning-fork held over a cylindrical glass vessel. Another glass vessel of similar kind being placed with its mouth at right angles to the first, no sound will be heard; but, if either cylinder be removed, the sound will be distinctly audible in the other. The *silence* produced in this way is due to the interference of the undulations.

This seeming paradox, when thus explained, like the paradox mentioned under the heads of Hydrostatics and Pneumatics, and another to be mentioned under the head of Optics, will be found to be perfectly consistent with the laws of sound.

What is an echo? 657. An echo is produced by the vibrations of the air meeting a hard and regular surface, such as a wall, a rock, a mountain, and being reflected back to the ear, thus producing the same sound a second, and sometimes a third and fourth time.

Why are there no echoes at sea, or on a plain?

658. For this reason, it is evident that no echo can be heard at sea, or on an extensive plain, where there are no objects to reflect the sound.

By what law is sound reflected?

659. Sound, as well as light and heat, is reflected in obedience to the same law that has already been stated in Mechanics, namely, the angles of incidence and of reflection are always equal.

660. It is only necessary, therefore, to know how sound strikes against a reflecting surface, to know how it will be reflected. It is related of Dionysius, the tyrant of Sicily, that he had a dungeon (called the ear of Dionysius) in which the roof was so constructed as to collect the words, and even the whispers, of the prisoners confined therein, and direct them along a hidden conductor to the place where he sat to listen; and thus he became acquainted with the most secret expressions of his unhappy victims.

On what principle are speaking-trumpets constructed?

661. Speaking-trumpets are constructed on the principle of the reflection of sound.

662. The voice, instead of being diffused in the open air, is confined within the trumpet; and the vibrations which spread and fall against the sides of the instrument are reflected according to the angle of incidence, and fall in the direction of the vibrations, which proceed straight forward. The whole of the vibrations are thus collected into a focus; and, if the ear be situated in or near that spot, the sound will be prodigiously increased.

How is a hearing trumpet constructed?

663. Hearing-trumpets, or the trumpets used by deaf persons, are also constructed on the same principle; but, as the voice enters the large end of the trumpet, instead of the small one, it is not so much confined, nor so much increased.*

664. The musical instrument called the trumpet acts also on the same principle with the speaking-trumpet, so far as its form tends to increase the sound.

665. The smooth and polished surface of the interior parts of certain kinds of shells, particularly if they be spiral or undulating,

---

* In this connexion the author cannot refrain from giving publicity to the value of a pair of acoustic instruments worn by one of the members of his family. They consist of two small hearing-trumpets of a peculiar construction, connected by a slender spring with an adjusting slide, which, passing over the head, keeps both trumpets in their place. They are concealed from observation by the head-dress, and enable the wearer to join in conversation of ordinary tone, from which without them she is wholly debarred. The instruments were made by B. S. Codman & Co. 57 Tremont st., Boston.

fit them to collect and reflect the various sounds which are taking place in the vicinity. Hence the Cyprias, the Nautilus, and some other shells, when held near the ear, give a continued sound, which resembles the roar of the distant ocean.

On what principle are whispering-galleries constructed? 666. Sound, like light, after it has been reflected from several surfaces may be collected into one point, as a focus, where it will be more audible than in any other part; and on this principle whispering-galleries may be constructed.

667. The famous whispering-gallery in the dome of St. Paul's church, in London, is constructed on this principle. Persons at very remote parts of the building can carry on a conversation in a soft whisper, which will be distinctly audible to one another, while others in the building cannot hear it; and the ticking of a watch may be heard from side to side.

668. There is a church in the town of Newburyport, in Massachusetts, which, as was accidentally discovered, has the same property as a whispering-gallery. Persons in opposite corners of the building, by facing the wall, may carry on a conversation in the softest whisper, unnoticed by others in any other part of the building. It is the building which contains in its cemetery the remains of the distinguished preacher, Whitefield.

What is an Acoustic Tube? 669. ACOUSTIC TUBES.—Sounds may be conveyed to a much greater distance through continuous tubes than through the open air. The tubes used to convey sounds are called Acoustic Tubes. They are much used in public houses, stores, counting-rooms, &c., to convey communications from one room to another.

670. The quality of sound is affected by the furniture of a room, particularly the softer kinds, such as curtains, carpets, &c.; because, having little elasticity, they present surfaces unfavorable to vibrations.

671. For this reason, music always sounds better in rooms with bare walls, without carpets, and without curtains. For the same reason, a crowded audience increases the difficulty of speaking.

672. As a general rule, it may be stated that *plane and smooth surfaces reflect sound without dispersing it; convex surfaces disperse it, and concave surfaces collect it.*

How is the sound of the human voice produced? 673. The sound of the human voice is produced by the vibration of two delicate membranes situated at the top of the windpipe, between which the air from the lungs passes.

674. The tones are varied from grave to acute, by opening or contracting the passage; and they are regulated by the muscles belonging to the throat, by the tongue, and by the cheeks. The management of the voice depends much upon cultivation; and although many persons can both speak and sing with ease, and with great power, without much attention to its culture, yet it is found that they who cultivate their voices by use acquire a degree of flexibility and ease in its management, which, in a great measure, supplies the deficiency of nature.*

What is Ventriloquism?

675. Ventriloquism † is the art of speaking in such a manner as to cause the voice to appear to proceed from a distance.

676. The art of ventriloquism was not unknown to the ancients, and it is supposed by some authors that the famous responses of the oracles at Delphi, at Ephesus, &c., were delivered by persons who possessed this faculty. There is no doubt that many apparently wonderful pieces of deception, which, in the days of superstition and ignorance, were considered as little short of miracles, were performed by means of ventriloquism. Thus houses have been made

---

* Dr. Rush's very valuable work on "The Philosophy of the Human Voice" contains much valuable matter in relation to the human voice. Dr. Barber's "Grammar of Elocution," and the "Rhetorical Reader," by the author of this volume, contain useful instructions in a practical form. To the work of Dr. Rush both of the latter works are largely indebted.

† The word Ventriloquism literally means, "*speaking from the belly*," and it is so defined in Chambers' Dictionary of Arts and Sciences. The ventriloquist, by a singular management of the voice, seems to have it in his power "*to throw his voice*" in any direction, so that the sound shall appear to proceed from that spot. The words are pronounced by the organs usually employed for that purpose, but in such a manner as to give little or no motion to the lips, the organs chiefly concerned being those of the throat and tongue. The variety of sounds which the human voice is capable of thus producing is altogether beyond common belief, and, indeed, is truly surprising. Adepts in this art will mimic the voices of all ages and conditions of human life, from the smallest infant to the tremulous voice of tottering age, and from the intoxicated foreign beggar to the high-bred, artificial tones of the fashionable lady. Some will also imitate the warbling of the nightingale, the loud tones of the whip-poor-will, and the scream of the peacock, with equal truth and facility. Nor are these arts confined to professed imitators; for in many villages boys may be found who are in the habit of imitating the brawling and spitting of cats in such a manner as to deceive almost every hearer.

The human voice is also capable of imitating almost every inanimate sound. Thus, the turning and occasional creaking of a grindstone, with the rush of the water, the sawing of wood, the trundling and creaking of a wheelbarrow, the drawing of corks, and the gurgling of the flowing liquor, the sound of air rushing through a crevice on a wintry night, and a great variety of other noises of the same kind, are imitated by the voice so exactly as to deceive any hearer who does not know whence they proceed.

to appear haunted, voices have been heard from tombs, and the dead have been made to appear to speak, to the great dismay of the neighborhood, by means of this wonderful art.

Ventriloquism is, without doubt, in great measure the gift of nature; but many persons can, with a little practice, utter sounds and pronounce words without opening the lips or moving the muscles of the face; and this appears to be the great secret of the art.

How is the sound of a musical string caused?

677. MUSICAL SOUNDS, OR HARMONY.—The sound produced by a musical string is caused by its vibrations; and the height or depth of the tone depends upon the rapidity of these vibrations. Long strings vibrate with less rapidity than short ones; and for this reason the low tones in a musical instrument proceed from the long strings, and the high tones from the short ones.

Explain Fig. 106.

678. Fig. 106. A B represents a musical string. If it be drawn up to G, its elasticity will not only carry it back again, but will give it a momen-

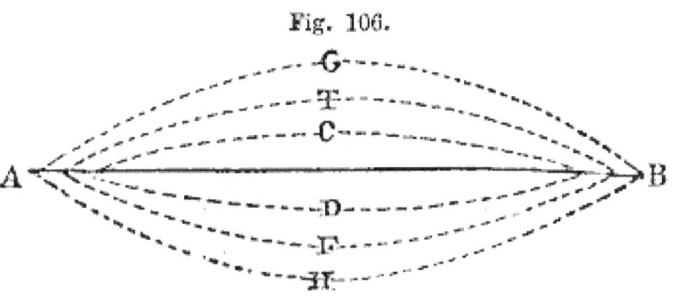
Fig. 106.

tum which will carry it to H, from whence it will successively return to T, F, C, D, &c., until the resistance of the air entirely destroys its motion.

On what does the quality of the tone of a string depend?

679. The quality of the sound produced by strings depends upon their length, thickness, weight, and degree of tension. The quality of the sound produced by wind instruments depends upon their size, their length, and their internal diameter.

680. When music is made by the use of strings, the air is struck by the body, and the sound is caused by the vibrations; when it is made by pipes, the body is struck by the air; but, as action and reaction are equal, the effect is the same in both cases.

681. Long and large strings, when loose, produce the lowest

tones; but different tones may be produced from the same string, according to the degree of tension. Large wind instruments, also, produce the lowest tones; but different tones may be produced from the same instrument, according to the distance of the aperture for the escape of the wind from the aperture where it enters.

How does the temperature of the weather affect the tone of a musical instrument?

682. The quality of the sound of all musical instruments is affected by the changes in the temperature and specific gravity of the atmosphere.

683. As heat expands and cold contracts the materials of which the instrument is made, it follows that the strings will have a greater degree of tension, and that pipes and other wind instruments will be contracted, or shortened, in cold weather. For this reason, most musical instruments are higher in tone (or sharper) in cold weather, and lower in tone (or more flat) in warm weather.

On what is the science of harmony founded?

684. The science of harmony is founded on the relation which the vibrations of sonorous bodies have to each other.

685. Thus, when the vibrations of one string are double those of another, the chord of an octave is produced. If the vibrations of two strings be as two to three, the chord of a fifth is produced. When the vibrations of two strings frequently coincide, they produce a musical chord; and when the coincidence of the vibrations is unfrequent, discord is produced.

686. A simple instrument, called a *monochord*, contrived for the purpose of showing the length and degree of tension of a string to produce the various musical tones, and to show their relations, may thus be made. A single string of catgut or wire, attached at one end to a fixed point, is carried over a pulley, and a weight is suspended to the other end of the string. The string rests on two bridges, between the fixed point and the pulley, one of which is fixed, the other movable. A scale is placed beneath the string by which the length of the vibrating part between the two bridges may be measured. By means of this simple instrument, the respective lengths required to produce the seven successive notes of the gamut will be as follows: it being premised that the longer the string the slower will be its vibrations.

687. Let the length of the string required to produce the note called C be 1; the length of the string to produce the successive notes will be

$$\begin{array}{cccccccc} C & D & E & F & G & A & B & C \\ 1 & \frac{8}{9} & \frac{4}{5} & \frac{3}{4} & \frac{2}{3} & \frac{3}{5} & \frac{8}{15} & \frac{1}{2}. \end{array}$$

ACOUSTICS.

688. Hence, the octave will require only half of the length of the fundamental note, and the vibrations that produce it will be as two to one. The vibrations of the string in producing the successive notes of the scale will be as follows:

$$\begin{array}{cccccccc} C & D & E & F & G & A & B & C \\ 1 & \tfrac{9}{8} & \tfrac{5}{4} & \tfrac{4}{3} & \tfrac{3}{2} & \tfrac{5}{3} & \tfrac{15}{8} & 2. \end{array}$$

That is, to produce the note D nine vibrations will be made in the same time that eight are made by C, five of E to four of C, four of F to three of C, three of G to two of C, five of A to three of C, fifteen of B to eight of C, and two of the octave C to one of the fundamental C.

689. The same relations exist in each successive octave throughout the whole musical scale.

690. As harmony depends upon the coincidence of vibrations, it follows that the octave produces the most perfect harmony; next in order is the fifth, as every third vibration of the fifth corresponds with every second vibration of the fundamental. Next to the fifth in the order of harmony follows the fourth, and after the fourth the third.

691. The following scale, containing three octaves, exhibits the proportions which exist between the fundamental and all the other notes within that compass.

692. In the lowest line of this scale the numbers show the intervals. The figures above express the number of vibrations of the fundamental or tonic, and the upper line of figures represents the proportionate vibrations of each successive interval.

693. It is found in practice that when two sounds are caused by vibrations which are in some simple numerical proportion to each other, such as 1 to 2, or 2 to 3, or 3 to 4, &c., they are pleasing to the ear; and the whole science of harmony is founded on such relations.

694. The principal harmonies are the octave, fifth, fourth, major third, and

minor third; and the relations between them and the fundamental or tonic are as follows

Octave,	2 to 1.
Fifth,	3 " 2.
Fourth,	4 " 3.
Major Third,	5 " 4.
Minor Third,	6 " 5.

695. *The following Rules may serve as the basis of interesting calculations.*

(1.) Strings of the same diameter and equal tension vibrate in times in an inverse proportion to their lengths.

(2.) The vibrations of strings of equal length and tension are in an inverse proportion to their diameters.

(3.) The vibrations of strings of the same length and diameter are as the square roots of the weights causing their tension.

(4.) The vibrations of cylindric tubes closed at one end are in an inverse proportion to their length.

(5.) The sound of tubes open at both ends is the same with that of tubes of half the length closed at one end.

[The limits of this work will not admit the further consideration of the subject of Harmony. It constitutes of itself a science, involving principles which require deep study and investigation; and they who would pursue it advantageously will scarcely expect, in the pages of an elementary work of this kind, that their wants will be supplied.]

696. *Questions for Solution.*

(1.) A rocket was seen to explode, and in two seconds the sound of the explosion was heard; how far off was the rocket? *Ans.* 2240 *ft.*

(2.) The flash from a cloud was seen, and in five seconds the thunder was heard; what was the distance of the cloud? *Ans.* 5600 *ft.*

(3.) A musical string four feet long gave a certain tone; what must be the length of a string of similar size and tension to produce the note of a fifth? *Ans.* 2 *ft.* 8 *in.*

(4.) A certain string vibrates 100 times in a second; how many times must a string of the same kind vibrate to produce the octave? the fifth? the minor third? the major third? *Ans.* 200; 150; 120; 125.

(5.) Supposing that two sounds, with all their attending circumstances similar, reach an ear situated at the point of interference of the undulations, — what will be the consequence? [*See Nos.* 653 *and* 654.]

(6.) The length of a string being 36, what will be length of its octave? fifth? fourth? major and minor thirds? *Ans.* 18; 24; 27; 28.8; 30.

(7.) A stone, being dropped into a pit, is heard to strike the bottom in 7 seconds; how deep was the pit? *Ans.* By Algebra, 660 *ft.*

[N. B. In estimating the velocity of sound, it is generally reckoned in practice as only at 1090 feet per second, supposing the thermometer at the freezing point; and one foot per second is added for every degree of temperature above the freezing point, or 32°. The average rate of 1120 feet has been assumed in the text.]

(8.) Suppose the length of a music-string to be five feet; what will be the length of the 15th, all other circumstances being equal? *Ans. 4 in.*

(9.) The length of the fifth being four, what will be the length of the fundamental, or tonic? *Ans. 6.*

(10.) What must be the length of a pipe of an open diapason to produce the same tone with four foot C of the stopped diapason? *Ans. 4 ft.*

[N. B. The open diapason consists of pipes open at both ends; the stopped diapason has its pipes closed at one end. [See No. 695 (5).]

(11.) In what proportion are the vibrations of two strings of equal length and diameter, one stretched with a weight of twenty-five pounds, the other with a weight of fifty pounds? [See No. 695 (8).] *Ans. 1 to 1.41 +*

(12.) In what proportion are the vibrations of two strings of equal length and tension, but having diameters in the proportion of 3 to 5?
Ans. 5 to 3.

What is Pyronomics? 697. PYRONOMICS, OR THE LAWS OF HEAT. — Pyronomics is the science which treats of the laws, the properties and operations of heat.

What is heat, and what is its weight? 698. The nature of heat is unknown, but it has been proved that the addition of heat to any substance produces no perceptible alteration in the weight of that substance. Hence it is inferred that heat is imponderable.

What is cold? 699. Heat is undoubtedly a positive substance, or quality. Cold is merely negative, being only the absence of heat.

What effect has heat on all bodies? 700. Heat pervades all bodies, insinuating itself, more or less, between their particles, and forcing them asunder. Heat and the attraction of cohesion constantly act in opposition to each other; hence, the more a body is heated, the more its particles will be separated.

701. Heat causes most substances to dilate or expand, while cold (which is merely the absence of heat) causes them to contract.* Since there is a continual change in the temperature of all bodies on the surface of the earth, it necessarily follows that there will be a constant corresponding change in their magnitude as they are affected by heat and cold. They expand their bulk in a warm day, and contract it in a cold one. In warm weather the flesh swells,

---

* The exceptions to this remark are *water* and *clay*. Water *expands* when it *freezes;* clay *contracts* when *heated.*

the blood-vessels are well filled, the hands and the feet, as well as other parts of the body, expand or acquire a degree of plumpness, and the skin is distended; while, on the contrary, in cold weather the flesh appears to contract, the vessels shrink, and the skin appears shrivelled. Hence a glove or a shoe which is too tight in the summer will often be found to be easy in cold weather.

702. The effect of heat in separating the particles of different kinds of substances is seen in the melting of solids, such as metals, wax, butter, &c. The heat insinuates itself between the particles, and forces them asunder. These particles then are removed from that degree of proximity to each other within which cohesive attraction exists, and the body is reduced to a fluid form. When the heat is removed, the bodies return to their former solid state.

What kind of bodies arrest the progress of heat?

703. Heat passes through some bodies with more difficulty than through others, but there is no kind of matter which can completely arrest its progress.

What is steam?

704. Of all the effects of heat, that produced upon water is, perhaps, the most familiar. The particles are totally separated, and converted into steam or vapor, and their extension is wonderfully increased. The steam which arises from boiling water is nothing more than portions of the water heated. The heat insinuates itself between the particles of the water, and forces them asunder. When deprived of the heat, the particles will unite in the form of drops of water.

This fact can be seen by holding a cold plate over boiling water. The steam rising from the water will be condensed into drops on the bottom of the plate. The air which we breathe generally contains a considerable portion of moisture. On a cold day this moisture condenses on the glass in the windows, and becomes visible. We see it also collected into drops on the outside of a tumbler or other vessel containing cold water in warm weather. Heat also produces most remarkable effects upon air, causing it to expand to a wonderful extent, while the absence of heat causes it to shrink or contract into very small dimensions.

How is rain produced?

705. The attraction of cohesion causes the small watery particles which compose mist or vapor to unite together in the form of drops of water. It is thus that rain is produced. The clouds consist of

mist or vapor expanded by heat. They rise to the cold regions of the skies, where the particles of vapor lose their heat, and then, uniting in drops, fall to the earth. But so long as they retain their heat the attraction of cohesion can have no influence upon them, and they will continue to exist in the form of steam, vapor or mist.

706. The thermometer, an instrument designed to measure degrees of heat, has already been described, in connexion with the barometer, under the head of Pneumatics. Heat, under the name of *caloric*, is properly a subject of consideration in the science of Chemistry. It exists in two states, called, respectively, free heat and latent heat. Free heat, or free caloric, is that which is perceptible to the senses, as the heat of a fire, the heat of the sun, &c. Latent heat is that which exists in most kinds of substances, but is not perceptible to the senses until it is brought out by mechanical or chemical action. Thus, when a piece of cold iron is hammered upon an anvil, it becomes intensely heated; and when a small portion of sulphuric acid, or vitriol, is poured into a vial of cold water, the vial and the liquid immediately become hot. A further illustration of the existence of latent or concealed heat is given at the fir side every day. A portion of cold fuel is placed upon the grate or hearth, and a spark is applied to kindle the fire which warms us. It is evident that the heat given out by the fuel, when ignited, does not all proceed from the spark, nor can we perceive it in the fuel; it must, therefore, have existed somewhere in a latent state. It is, however, the effects of free heat, or free caloric, which are embraced in the science of Pyronomics. The subject of latent heat belongs more properly to the science of Chemistry.

707. The terms heat and cold, as they are generally used, are merely relative terms; for a substance which in one person would excite the sensation of heat might, at the same time, seem cold to another. Thus, also, to the same individual the same thing may be made to appear, relatively, both warm and cold. If, for instance, a person were to hold one hand near to a warm fire, and the other on a cold stone, or marble slab, and then plunge both into a basin of lukewarm water, the liquid would appear cold to the warm hand and warm to the cold one.

What are the principal sources of heat?

708. SOURCES OF HEAT. — The four principal sources of the development of heat are the Sun, Electricity, Chemical Action and Mechanical Action. The heat produced by fire and flame is due to chemical action.

What is the source of the greatest degree of heat?

709. But, of all the sources from which heat has been developed by human agency, that produced by electrical action, and especially the galvanic battery, is by far the most eminent in its degree and in its effects. It can reduce the most refractory substances to a fluid state, or convert them to their original elements.

710. The heat generally ascribed to the sun is attended by peculiar phenomena, but imperfectly understood. It may, perhaps, be questioned whether there be any absolute heat in the rays of that luminary, for we find that the heat is not in all cases proportionate to his proximity. Thus, on the tops of high mountains, and at great elevation, it is not found that the heat is increased, but, on the contrary, diminished. But there are other phenomena which lead to the conclusion that his rays are accompanied by the development of heat, if they are not the cause and the source of it.

711. All mechanical operations are attended by heat. Friction, sudden compression, violent extension, are all attended by heat. The savage makes his fire by the friction of two pieces of dry wood. Air, suddenly and violently compressed, ignites dry substances;* and India-rubber especially, when suddenly extended, shows evident signs of heat; and an iron bar may be made red hot by beating it quickly on an anvil. Even water, when strongly compressed, gives out heat.

What are the principal effects of heat?

712. The principal effects of heat are three, namely:

(1.) Heat expands most substances.

(2.) It converts them from a solid to a fluid state.

(3.) It destroys their texture by combustion.

713. There are many substances on which ordinary degrees of heat, and, indeed, heat of great intensity, seems to produce no sensible effects; and they have, therefore, received the name of incombustible bodies. Bodies usually called incombustible are generally mineral substances, such as stones, the earths, &c. All vegetable substances, and most animal substances, are highly combustible. The metals also all yield to the electrical or galvanic battery. But there is sufficient evidence that all bodies were once in a fluid or gaseous state, and that the solid forms that they have assumed are due to the loss of heat. Could the same degree of

---

* Syringes have been constructed on this principle. A solid piston being forcibly driven downward on dry tinder, ignites it.

intensity of heat be restored, it is presumed that they would resume their liquid or gaseous form.

What is the first law of heat?

714. Heat tends to diffuse itself equally through all substances.

If a heated body be placed near a cold one, the temperature of the former will be lowered, while that of the latter will be raised. All substances contain a certain quantity of heat; but, on account of its tendency to diffuse itself equally, and the difference in the power of different substances to conduct it, bodies of the same absolute temperature appear to possess different degrees of heat.

Thus, if the hand be successively applied to a woollen garment, a mahogany table, and a marble slab, all of which have been for some time in the same room, the woollen garment will appear the warmest, and the marble slab the coldest, of the three articles; but, if a thermometer be applied to each, no difference in the temperature will be observed.

What is the reason that some substances feel warm and others cold in the same room?

715. From this it appears that *some substances conduct heat readily, and others with great difficulty.* The reason that the marble slab seems the coldest is, that marble, being a good conductor of heat, receives the heat from the hand so readily that the loss is instantly felt by the hand; while the woollen garment, being a bad conductor of heat, receives the heat from the hand so slowly that the loss is imperceptible.

What is the cause of the difference in the warmth of different garments?

716. The different power of receiving and conducting heat, possessed by different substances, is the cause of the difference in the warmth of various substances used for clothing.

Why are woollen garments warm, and linen, cool?

717. Thus, woollen garments are warm garments, because they part slowly with the heat which they acquire from the body, and, consequently, they do not readily convey the warmth of the body to the air; while, on the contrary, a linen garment is a cool one, because it parts with its heat readily, and as readily receives fresh heat from the body. It is, therefore, constantly receiving heat from the body and throwing it out into

the air, while the woollen garment retains the heat which it receives, and thus encases the body with a warm covering.

718. For a similar reason, ice in summer is wrapped in woollen cloths. It is then protected from the heat of the air, and will not melt.

How is heat propagated? 719. Heat is propagated in two ways, namely, by conduction and by radiation. Heat is propagated by conduction when it passes from one substance to another in contact with it. Heat is propagated by radiation when it passes through the air, or any other elastic fluid.

What are the best conductors of heat? 720. Different bodies conduct heat with different degrees of facility. The metals are the best conductors; and with regard to their conducting power, stand in the following order, namely: Gold, platinum, silver, copper, iron, zinc, tin, lead.

721. Any liquid, therefore, may be more readily heated in a silver vessel than in any other of the same thickness, except one of gold, or of platinum, on account of its great conducting power.

Why are the handles of tea and coffee pots made of wood? 722. Metals, on account of their conducting power, cannot be handled when raised to a temperature above 120 degrees of Fahrenheit. For this reason, the handles of metal tea-pots and coffee-pots are commonly made of wood; since, if they were made of metal, they would become too hot to be grasped by the hand, soon after the vessel is filled with heated fluid.

723. Wood conducts heat very imperfectly. For this reason, wooden spoons and forks are preferred for ice. Indeed, so imperfect a conductor of heat is wood, that a stick of wood may be grasped by the hand while one end of the stick is a burning coal. But an iron bar, being a good conductor of heat, cannot be handled near the heated end.

724. Animal and vegetable substances, of a loose texture, such as fur, wool, cotton, &c., conduct heat very imperfectly; hence their efficacy in preserving the warmth of the body. Water becomes scalding hot at 150 degrees; but air, heated far beyond the temperature of boiling water, may be applied to the skin without much pain. Sir Joseph Banks, with several other gentlemen, remained some time in a room when the heat was 52 degrees above the boiling point; but, though they could bear the contact of the heated *air* they could not touch any metallic substance, as their watch-chains

money, &c. Eggs, placed on a tin frame, were roasted hard in twenty minutes; and a beef-steak was overdone in thirty-three minutes.

725. Chantrey, the celebrated sculptor, had an oven which he used for drying his plaster cuts and moulds. The thermometer generally stood at 300 degrees in it, yet the workmen entered, and remained in it some minutes without difficulty; but a gentleman once entering it with a pair of silver-mounted spectacles on, had his face burnt where the metal came in contact with the skin.

726. The air, being a bad conductor, never radiates heat, nor is it ever made hot by the direct rays of the sun. The air which comes in contact with the surface of the earth ascends, and warms the air through which it passes in its ascent. Other air, heated in the same way, also ascends, carrying heat, and this process is repeated till all the air is made hot.

727. In like manner, in cold weather, the air resting on the earth is made cold by contact. This cold air makes the air above it cold, and cold currents (or wind) agitate the mass together till a uniform temperature is produced.

How is heat reflected? 728. Heat is reflected by bright substances, and the angle of reflection will be equal to the angle of incidence.

729. Advantage has been taken of this property of heat in the construction of a simple apparatus for baking. It is a bright tin case, having a cover inclined towards the fire in such a manner as to reflect the heat downwards. In this manner use is made both of the direct heat of the fire, and the reflected heat, which would otherwise pass into the room. The whole apparatus, thus connected with the culinary department, is called, in New England, "*The Connecticut baker.*"

730. This power of reflecting heat, possessed by bright substances, is the reason why andirons and other articles that are kept bright, although standing very near the fire, never become hot; while other darker substances, further from the fire, become hot. But, if they are not bright, heat will penetrate them.

731. The reflecting power of bright and light-colored substances accounts also for the superior coolness of white and light-colored fabrics for clothing.

Why are dark garments warmer than light ones? 732. Black and dark-colored surfaces absorb heat. This is the reason why black and dark-colored fabrics are warmer when made into garments than those of light color.

733. Snow or ice will melt under a piece of black cloth, while it would remain perfectly solid under a white one. The farmers in some of the mountainous parts of Europe are accustomed to spread

black earth, or soot, over the snow, in the spring, to hasten its melting. and enable them to commence ploughing.

What effect has heat upon the density of substances? 734. The density of all substances is augmented by cold, and diminished by heat.

There is a remarkable exception to this remark, and that is in the case of water; which, instead of contracting, expands at the freezing point, or when it is frozen. This is the reason why pitchers, and other vessels, containing water and other similar fluids, are so often broken when the liquid freezes in them. For the same reason, ice floats instead of sinking in water; for, as its density is diminished, its specific gravity is consequently diminished. Were it not for this remarkable property of water, large ponds and lakes, exposed to intense cold, would become solid masses of ice; for, if the ice, when formed on the surface, were more dense (that is, more heavy) than the water below, it would sink to the bottom, and the water above, freezing in its turn, would also sink, until the whole body of the water would be frozen. The consequence would be the total destruction of all creatures in the water. But the specific gravity of ice causes it to continue on the surface, protecting the water below from congelation.

What is cold? 735. *Cold* is merely the absence of heat; or rather, more properly speaking, inferior degrees of heat are termed *cold*.

736. The effect of heat and cold, in the expansion and contraction of glass, is an object of common observation; for it is this expansion and contraction which cause so many accidents with glass articles. Thus, when hot water is suddenly poured into a cold glass of any form, the glass, if it have any thickness, will crack; and, on the contrary, if cold water be poured into a heated glass vessel, the same effect will be produced. The reason of which is this; Heat makes its way but slowly through glass; the inner surface, therefore, when the hot water is poured into it, becomes heated, and, of course, distended before the outer surface, and the irregular expansion causes the vessel to break. There is less danger of fracture, therefore, when the glass is thin, because the heat readily penetrates it, and there is no irregular expansion.

737. The glass chimneys, used for oil and gas burners, are often broken by being suddenly placed, when cold, over a hot flame. The danger of fracture may be prevented (it is said) by making a minute notch on the bottom of the tube with a diamond. This precaution has been used in an establishment where six lamps were lighted every day, and not a single glass has been broken in nine years.

What bodies retain heat the longest? 738. Different bodies require different quantities of heat to raise them to the same temperature; and those which are heated with most difficulty retain their heat the longest.

Thus, oil becomes heated more speedily than water, and it likewise cools more quickly.

739. The most obvious and direct effect of heat on a body is to increase its extension in all directions.

740. Coopers, wheelwrights and other artificers, avail themselves of this property in fixing iron hoops on casks, and the tires or irons on wheels. The hoop or tire, having been heated, expands, and, being adapted in that state to the cask or the wheel, as the metal contracts in cooling it clasps the parts very firmly together.

741. From what has been stated above, it will be seen that an allowance should be made for the alteration of the dimensions in metallic beams or supporters, caused by the dilatation and contraction effected by the weather. In the iron arches of Southwark Bridge, over the Thames, the variation of the temperature of the air causes a difference of height, at different times, amounting to nearly an inch.

A happy application of the expansive power of heat to the mechanic arts was made some years ago, at Paris. The weight of the roof of a building, in the Conservatory of Arts and Trades, had pressed outwards the side walls of the structure, and endangered its security. The following method was adopted to restore the perpendicular direction of the structure. Several apertures were made in the walls, opposite to each other, through which iron bars were introduced, which, stretching across the building, extended beyond the outside of the walls. These bars terminated in screws, at each end, to which large broad nuts were attached. Each alternate bar was then heated by means of powerful lamps, and their lengths being thus increased, the nuts on the outside of the building were screwed up close to it, and the bars were suffered to cool. The powerful contraction of the bars drew the walls of the building closer together, and the same process being repeated on all the bars, the walls were gradually and steadily restored to their upright position.

What is the Pyrometer? 742. The Pyrometer is an instrument to show the expansion of bodies by the application of heat.

It consists of a metallic bar or wire, with an index connected with one extremity. On the application of heat, the bar expands, and turns the index to show the degree of expansion.

743. Wedgewood's pyrometer, the instrument commonly used for high temperatures, measures heat by the contraction of clay.

What effect has heat on bodies respectively, in the solid, liquid and aëriform state?

744. The expansion caused by heat in solid and liquid bodies differs in different substances; but aëriform fluids all expand alike, and undergo uniform degrees of expansion at various temperatures.

745. The expansion of solid bodies depends, in some degree, on the cohesion of their particles; but, as gases and vapors are destitute of cohesion, heat operates on them without any opposing power.

What effect has heat on the form of liquid bodies?

746. When heat is applied to water or other liquids, it converts them into steam or vapor. The deprivation of heat reconverts them into the liquid form. It is on this principle that distillation takes place.

What is a Still?

747. The vessel employed for distillation is called a Still.*

Fig. 107.

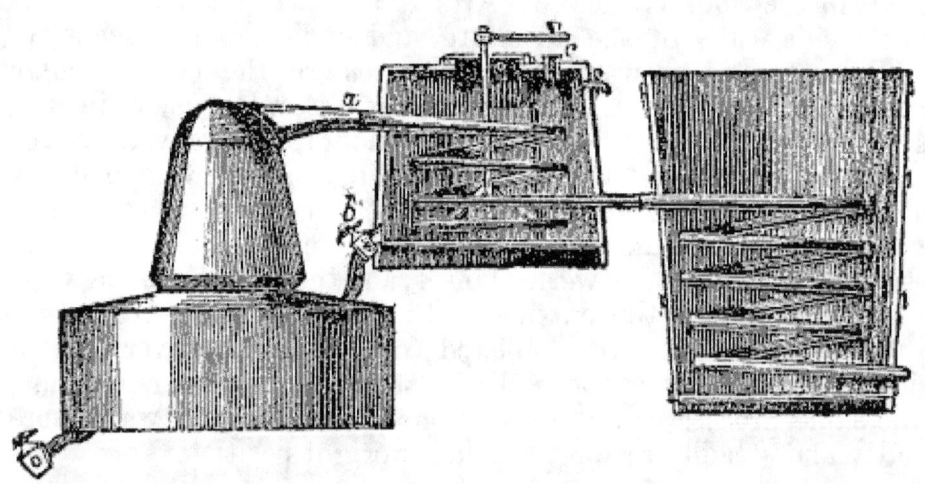

Explain Fig. 107.

748. Fig. 107 represents a *Still*. A liquid being poured into the large vessel *a*, heat is applied below, which converts the liquid gradually into steam or vapor, which, having no other outlet, passes through the spiral tube, called the worm, in vessel *b*, and from *b* through another worm, in *c*. The worm, being surrounded with cold water, condenses the vapor in the tube or worm, and reconverts it to a fluid state, and it flows out

* The subject of distillation properly belongs to the science of Chemistry, but it is here introduced for the benefit of those who cannot readily refer to a treatise on that subject.

at *e* in a tepid stream. The worm is of different lengths, and its only use is to present a large extent of surface to the cold water, so that the vapor may readily be condensed.

749. The process of distillation is sometimes used to purify a liquid, as the vapors which rise are unmixed with the impurities of the fluid. Important changes are thus made, and the still becomes highly useful in the arts.

At what temperature is water converted into steam? 750. When water is raised to the temperature of 212° of Fahrenheit's thermometer, it is converted into steam. It is then highly elastic and compressible.

What effect has heat upon steam? 751. The elastic force of steam is increased by heat; and decrease of heat diminishes it. The amount of pressure which steam will exert depends, therefore, on its temperature.

What is the temperature of confined steam? 752. The temperature of steam is always the same with that of the liquid from which it is formed, *while it remains in contact with that liquid;* and when heated to a great degree, its elastic force will cause the vessel in which it is contained to burst, unless it is made sufficiently strong to resist a prodigious pressure.

753. It has already been stated that water is converted into steam at the temperature of 212°. When closely confined it may be raised to a higher temperature, and it will then emit steam of greatly increased elastic force.

How is steam condensed? 754. When any portion of steam comes in contact with water, it instantly parts with its heat to the water, and becomes condensed into water. The whole mass then becomes water, increased in temperature by the amount of heat which the steam has lost.

On what property do the mechanical agencies of steam depend? 755. This is the great and peculiar property of steam, on which its mechanical agencies depend, namely, *its power of exerting a high degree of elastic force, and losing it instantaneously.*

How may the mechanical force of steam be instantly destroyed?

756. There are two ways in which steam is made instantly to lose its mechanical force; namely, *first*, by suddenly opening a passage for its escape into the open air, where it immediately becomes visible,* by a sudden loss of part of its heat, which it gives to the air; and *secondly*, by conveying it to a vessel called a condenser, where it comes directly into contact with a stream of water, to which it instantly gives up its heat and is condensed into water.

What space does steam occupy?

757. Steam occupies a space about seventeen hundred times larger than when it is converted into water. But the space that a given quantity of water converted into steam will occupy depends upon the temperature of the steam. The more it is heated the greater space it will fill, and the greater will be its expansive force.

What is the Steam-engine?

758. THE STEAM-ENGINE. — The Steam-engine is a machine moved by the expansive force of steam.

In what manner is steam made to act?

759. The mode in which steam is made to act is by causing its expansive force to raise a solid piston accurately fitted to the bore of a cylinder, like that in the forcing-pump. The piston rises by the impulse of expanding steam, admitted into the cylinder below. When the piston is thus raised, if the steam below it be suddenly condensed by the admission of cold water, or withdrawn from under

---

* Steam in a highly elastic state — that is, when at a high temperature — is perfectly dry and invisible. The reason that we are able to see it *after it has performed its work* and issues from the steam-engine is, that as soon as it comes in contact with the air it immediately parts with a portion of its heat (and, because air is not a good conductor, only a portion), and is condensed into small vesicles, which present a visible form, resembling smoke. Its expansive force, however, is not wholly destroyed; for the vesicles themselves expand as they rise, and soon become invisible, mingling with other vapors in the air. Could we look into the cylinder, *filled* with highly elastic steam, we should be able to see nothing. But, that the steam is there, and in its invisible form exerting a prodigious force, we know by the movements of the piston

it, a vacuum will be formed, and the pressure of the atmosphere on the piston above will drive it down. The admission of more steam below will raise it again, and thus a continued motion of the piston, up and down, will be produced. This motion of the piston is communicated to wheels, levers, and other machinery, in such a manner as to produce the effect intended.

How was the steam-engine of Newcomen and Savery constructed? 760. This is the mode in which the engine of Newcomen and Savery, commonly called the atmospheric engine, was constructed. It was called the atmospheric engine because half of the work was done by the pressure of the atmosphere, namely, the downward motion of the piston.

What improvements did Watt make in the steam-engine? 761. The celebrated Mr. James Watt introduced two important improvements into the steam-engine. Observing that the cooling of the cylinder by the water thrown into it to condense the steam lessened the expansibility of the steam, he contrived a method to withdraw the steam from the principal cylinder, after it had performed its office, into a condensing-chamber, where it is reconverted into water, and conveyed back to the boiler. The other improvement, called the *double action*, consists in substituting the expansive power of steam for the atmospheric pressure. This was performed by admitting the steam into the cylinder *above* the raised piston, at the same moment that it is removed from *below* it; and thus the power of steam is exerted in the *descending* as well as in the ascending stroke of the piston; and a much greater impetus is given to the machinery than by the former method. From the *double action* of the steam *above*, as well as *below* the piston, and from the condensation of the steam after it has performed its office, this engine is called Watt's *double-acting condensing* steam-engine. [*See also, No.* 766.]

Explain Fig. 108. 762. Fig. 108 represents that portion of the steam-engine in which steam is made to act, and propel such machinery as may be connected with it. It also exhibits **two**

17*

improvements of Mr. Watt. The principal parts are the boiler, the cylinder and its piston, the condenser, the air-pump, the steam-pipe, the eduction-pipe, and the cistern. In this figure, A represents the boiler, C the cylinder, with H the

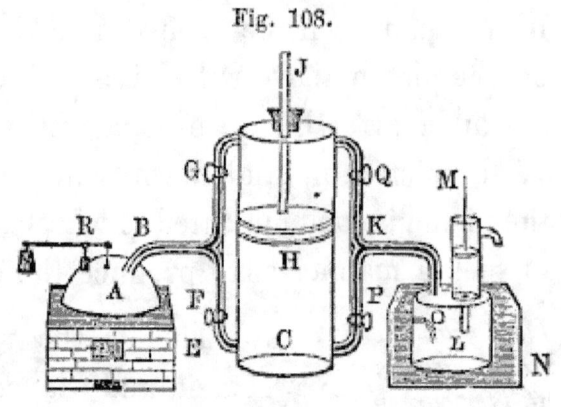

Fig. 108.

piston, B the steam-pipe, with two *branches** communicating with the cylinder, the one above and the other below the piston. This pipe has two valves, F and G, which are opened and closed alternately by machinery connected with the piston. The steam is carried through this pipe by the valves, when open, to the cylinder, both above and below the piston. K is the eduction-pipe, having two branches, like the steam-pipe, furnished with valves, &c., which are opened and shut by the same machinery. By the eduction-pipe the steam is led off from the cylinder, as the piston ascends and descends.

L is the condenser, and O a stop-cock for the admission of cold water. M is the pump. N is the cistern of cold water in which the condenser is immersed. R is the safety-valve. When the valves are all open, the steam issues freely from the boiler, and circulates through all the parts of the machine, expelling the air. This process is called blowing out, and is heard when a steamboat is about starting.

Now, the valves F and Q being closed, and G and P remaining open, the steam presses upon the piston and forces it down. As it descends, it draws with it the end of the working-beam, which is attached to the piston-rod J (but which is not represented in the figure). To this working-beam (which is a lever of the first kind) bars or rods are attached, which, rising and falling with the beam and the piston, open the stop-cock O, ad-

---

* The steam and the eduction pipes are sometimes made in forms differing from those in the figure, and they differ much in different engines.

mitting a stream of cold water, which meets the steam from the cylinder and condenses it, leaving no force below the piston to oppose its descent. At this moment the rods attached to the working-beam close the stop-cocks G and P, and open F and Q. The steam then flows in below the piston, and rushes from above it into the condenser, by which means the piston is forced up again with the same power as that with which it descended. Thus the steam-cocks G and P and F and Q are alternately opened and closed; the steam passing from the boiler drives the piston alternately upwards and downwards, and thus produces a regular and continued motion. This motion of the piston, being communicated to the working-beam, is extended to other machinery, and thus an engine of great power is obtained.

The pump M, the rod of which is connected with the working-beam, carries the water from the condenser back into the boiler, by a communication represented in Fig. 109.

The safety-valve R, connected with a lever of the second kind, is made to open when the pressure of the steam within the boiler is too great. The steam then rushing through the aperture under the valve, removes the danger of the bursting of the boiler.

How is the power of a steam-engine estimated? 763. The power of a steam-engine is generally expressed by the power of a horse, which can raise 33,000 lbs. to the height of one foot in a minute. An engine of 100 horse power is one that will raise 3,300,000 lbs. to the height of one foot in one minute.

What are the two kinds of steam-engines, and how do they differ? 764. The steam-engine is constructed in various forms, and no two manufacturers following exactly the same pattern; but the two principal kinds are the high and the low pressure engines, or, as they are sometimes called, the non-condensing and the condensing engines. The non-condensing or high-pressure engines differ from the low-pressure or condensing engines in having no condenser. The steam, after having moved the piston, is let off into the open air. As this kind of engine occupies less space, and is much less complicated, it is generally used on railroads. In the low-pressure or condensing engines, the steam, after having moved the piston, is condensed, or converted into water, and then conducted back into the boiler.

Who were the principal improvers of the steam-engine?

765. The steam-engine, as it is constructed at the present day, is the result of the inventions and discoveries of a number of distinguished individuals, at different periods. Among those who have contributed to its present state of perfection, and its application to practical purposes, may be mentioned the names of Somerset, the Marquis of Worcester, Savery, Newcomen, Fulton, and especially Mr. James Watt.

766. To the inventive genius of Watt the engine is indebted for the *condenser, the appendages for parallel motion,* the application of the *governor*, and for the *double action*. In the words of Mr. Jeffrey, it may be added, that, " by his admirable contrivances, and those of Mr. Fulton, it has become a thing alike stupendous for its force and its flexibility; for the prodigious power it can exert, and the ease and precision and ductility with which it can be varied, distributed, and applied. The trunk of an elephant, that can pick up a pin, or rend an oak, is as nothing to it. It can engrave a seal, and crush masses of obdurate metal before it; draw out, without breaking, a thread as fine as gossamer, and lift up a ship of war like a bauble in the air. It can embroider muslin, and forge anchors; cut steel into ribands, and impel loaded vessels against the fury of the winds and waves."

Explain Fig. 109.

767. Fig. 109 represents Watt's double-acting condensing steam-engine, in which A represents the boiler, containing a large quantity of water, which is constantly replaced as fast as portions are converted into steam. B is the steam-pipe, conveying the steam to the cylinder, having a steam-cock *b* to admit or exclude the steam at pleasure.

C is the cylinder, surrounded by the jacket *c c*, a space kept constantly supplied with hot steam, in order to keep the cylinder from being cooled by the external air. D is the eduction-pipe, communicating between the cylinder and the condenser. E is he condenser, with a valve *e*, called the injection-cock, admitting a jet of cold water, which meets the steam the instant that the steam enters the condenser. F is the air-pump, which is a common suction-pump, but is here called the air-pump because it removes from the condenser not only the water, but also the air, and the steam that escapes condensation. G G is a cold-water cistern, which surrounds the condenser, and supplies it with cold water, being filled by the cold-water pump, which is represented

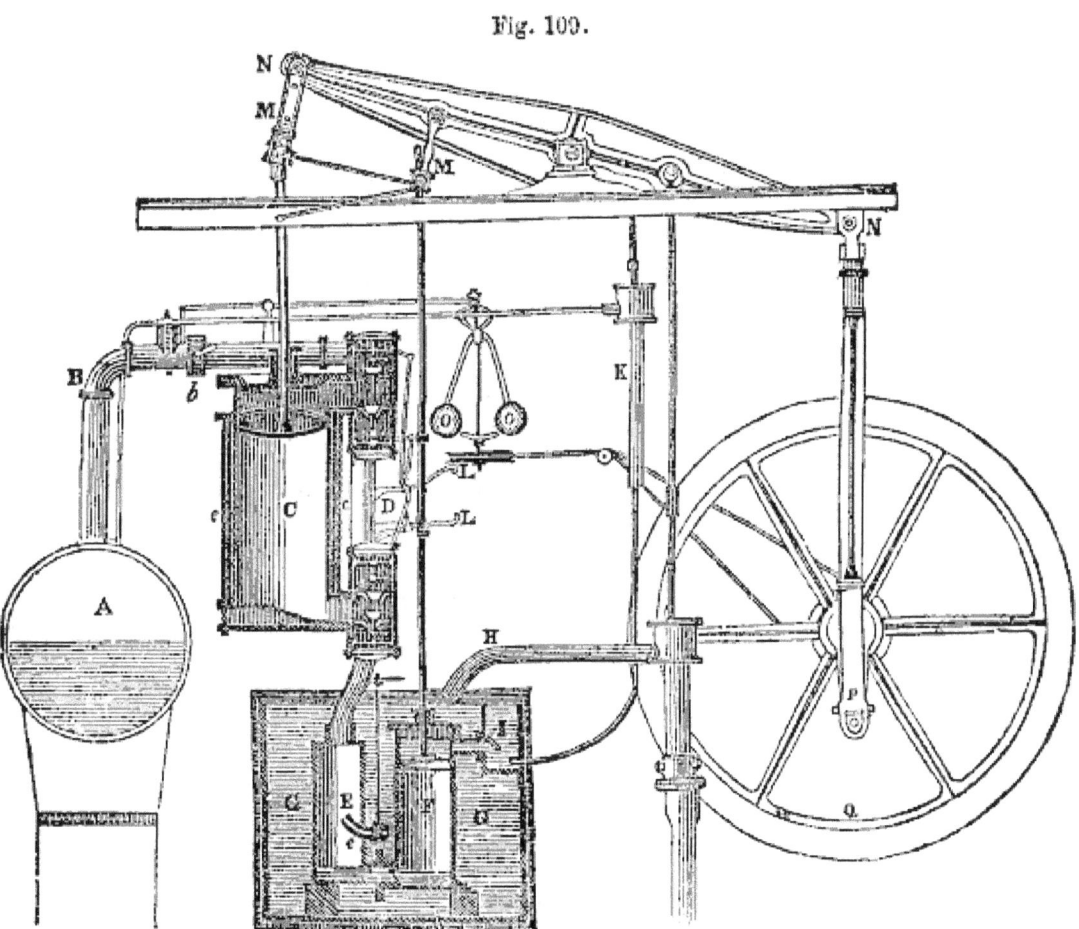

Fig. 109.

by H. I is the hot well, containing water from the condenser. K is the hot-water pump, which conveys back the water of condensation from the hot well to the boiler.

L L are levers, which open and shut the valves in the channel between the steam-pipe, cylinder, eduction-pipe, and condenser; which levers are raised or depressed by projections attached to the piston-rod of the pump. M M is an apparatus for changing the circular motion of the working-beam into par-

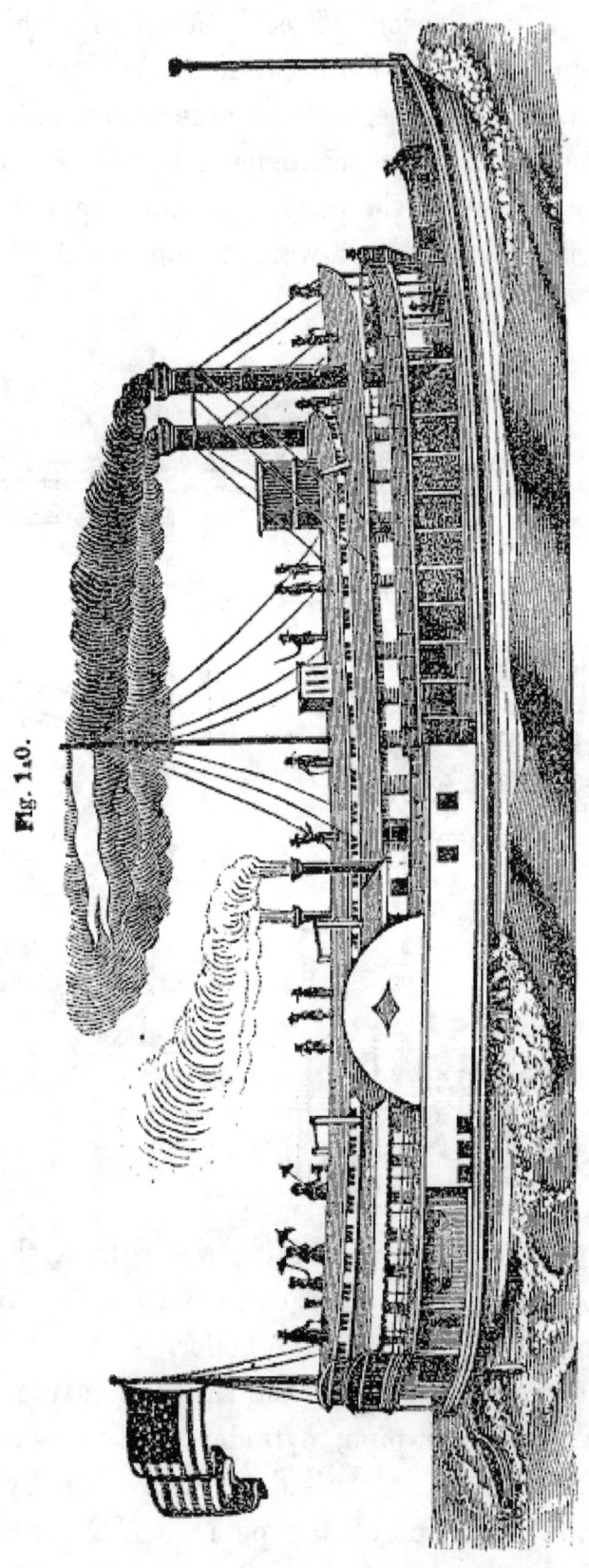

Fig. 140.

allel motion, so that the piston-rods are made to move in a straight line. N N is the working-beam, which, being moved by the rising and falling of the piston attached to one end, communicates motion to the fly-wheel by means of the crank P, and from the fly-wheel the motion is communicated by bands, wheels or levers, to the other parts of the machinery. O O is the governor.

The governor, being connected with the fly-wheel, is made to participate the common motion of the engine, and the balls will remain at a constant distance from the perpendicular shaft so long as the motion of the engine is uniform; but, whenever the engine moves faster than usual, the balls will recede further from the shaft, and by partly closing a valve connected with the boiler, will diminish the supply of steam to the cylinder, and thus reduce the speed to the rate required.

The steam-engine thus constructed is applied to boats to turn wheels having paddles attached to their circumference, which answer the purpose of oars. [*See Fig.* 110.] It is used also in work-shops, factories, &c.; and different directions and velocities may be given to the motion produced by the action of the steam on the piston, by connecting the piston to the beam with wheels, axles and levers, according to the principles stated under the head of Mechanics.

Steamboats are used principally on rivers, in harbors, bays, and on the coast. They are made of all sizes, and carry engines of different power, proportioned to the size of the boat.

The steamship [*See Fig.* 111], in addition to its steam-engines

Fig. 111.

and paddles, is rigged with masts and sails to increase the speed, or to make progress if the engines get out of order.

The Propeller differs from a steam-boat or steam-ship, by having an immense screw projecting from under the stern of the ship, instead of paddle-wheels. The screw is caused to revolve by means of steam-engines, and forces the vessel forward by its action on the water.

What is the locomotive steam-engine? 768. The locomotive engine is a high-pressure steam-engine, mounted on wheels, and used to draw loads on a railroad, or other level road. It is usually accompanied by a large wagon, called a *tender*, in which the wood and water used by the engine are carried.

Explain Fig. 112. 769. Fig. 112 represents a side view of the internal construction of a locomotive steam-engine; in which F represents the fire-box, or place where the fire is kept; D the door through which the fuel is introduced. The spaces marked B are the interior of the boiler, in which the water stands at the height indicated by the dotted line. The boiler is closed on all sides, all its openings being guarded by valves. The tubes marked $p\ p$ conduct the smoke and flame of the fuel through the boiler to the chimney C C, serving, at the same time, to communicate the heat to the remotest part of the boiler. By this arrangement, none of the heat is lost, as these tubes are all surrounded by the water. S S S is the steam-pipe, open at the top V S, having a steam-tight cock, or regulator, V, which is opened and shut by the lever L, extending outside of the boiler, and managed by the engineer.

The operation of the machine is as follows: The steam being generated in great abundance in the boiler, and being unable to escape out of it, acquires a considerable degree of elastic force. If at that moment the valve V be opened, by the handle L, the steam, entering the pipe S, passes in the direction of the arrow, through the tube, and enters the valve-box at X. There a sliding-valve, which moves at the same time with the machine, opens for the steam a communication successively with each end of the cylinder below. Thus, in the figure, the entrance on the right hand of the sliding-valve is represented as being open, and

STEAM-ENGINE.

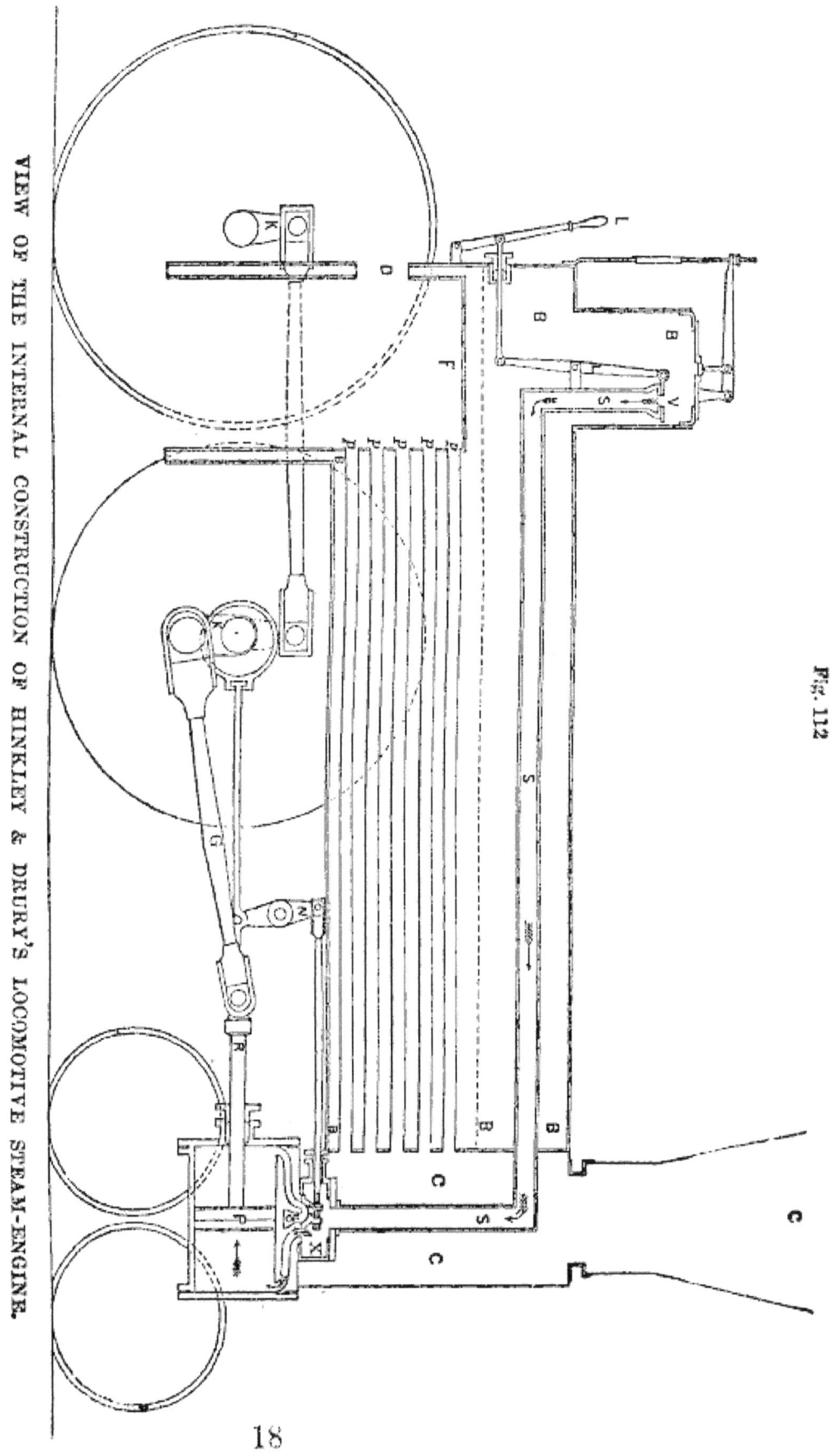

Fig. 112

VIEW OF THE INTERNAL CONSTRUCTION OF HINKLEY & DRURY'S LOCOMOTIVE STEAM-ENGINE.

206 NATURAL PHILOSOPHY.

Fig. 113.

STEAM-ENGINE. 207

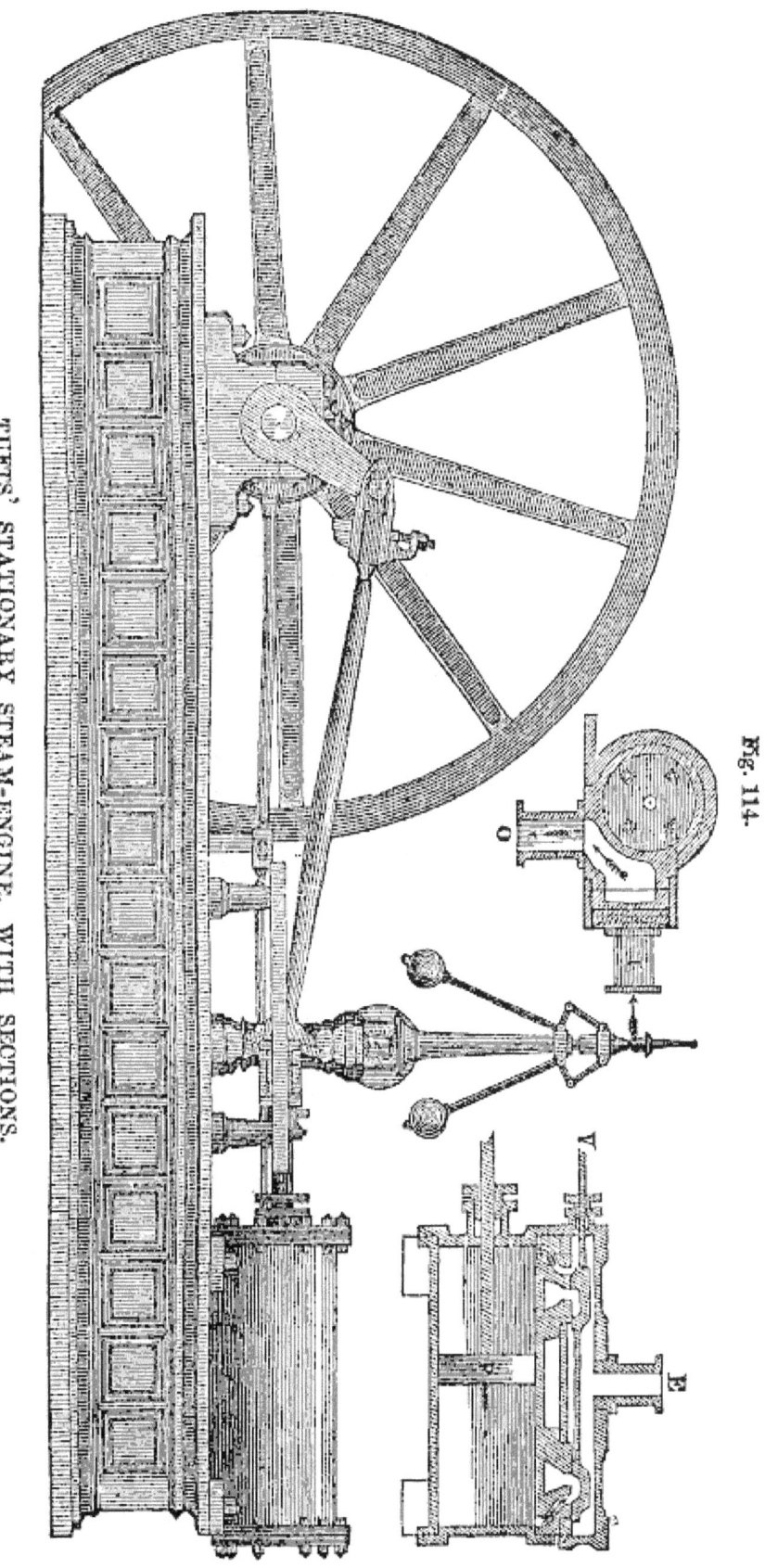

Fig. 114.

"TUFTS' STATIONARY STEAM-ENGINE, WITH SECTIONS.

Fig. 115.

the steam follows in the direction of the arrows into the cylinder, where its expansive force will move the piston P in the direction of the arrow. The steam or air on the other side of the piston passes out in the opposite direction, and is conveyed by a tube passing through C C into the open air.

The motion of the piston in the direction of the arrow causes the lever N to close the sliding-valve on the right, and open a communication for the steam on the opposite side of the piston P, where it drives the piston back towards the arrow, at the same time affording a passage for the steam on the right of the piston to pass into the open air.

Motion being thus given to the piston, it is communicated, by means of the rod R and the beam G, to the cranks K K, which, being connected with the axle of the wheel, causes it to turn, and thus move the machine.

Thus constructed, and placed on a railroad, the locomotive steam-engine is advantageously used as a substitute for horse power, for drawing heavy loads.

The apparatus of safety-valves, and other appliances for the management of the power produced by the machine, are the same in principle. though differing in form, with those used in other steam-engines; for a particular description of which, the student is referred to practical treatises upon the subject.

What is the best form of the steam-engine?

770. THE STATIONARY STEAM-ENGINE.— This engine is generally a high-pressure or non-condensing engine, used to propel machinery in work-shops and factories. As it is designed for a labor-saving machine, it is desirable to combine simplicity and economy with safety and durability in its construction; and that form of this engine is to be preferred which in the greatest degree unites these qualities.

Describe the Stationary Steam-engine.

771. The figure on page 207 represents Tufts' stationary steam-engine,* with sections of the interior. Like the double-acting condens-

* This engine was constructed by Mr. Otis Tufts, of East Boston, Mas-

ing engine of Mr. Watt, described in Fig. 109, it is furnished with a governor, by which the supply of steam is regulated; and, like the locomotive, Fig. 112, the cylinder, with its piston, has a horizontal position. The steam is admitted into the valve-box through an aperture at E, *in the section*, and from thence passes into the cylinder through a sliding-valve, alternately to each side of the piston P, as is represented by the direction of the arrows, the sliding-valve being moved by the rod V, communicating with an "eccentric" apparatus attached to the axis of the fly-wheel. The direction of the current of steam to the valve-box is represented by the arrow at I, and its passage outward from the cylinder, after it has moved the piston, is seen at O. In this engine there is no working-beam, as in Watt's engine, Fig. 109, but the motion is communicated from the piston-rod to a crank connected with the fly-wheel, which, turning the wheel, will move all machinery connected either with the axle or the circumference of that wheel.

Fig. 115 represents the Locomotive Steam-engine in one of its most perfect forms, as used on railways at the present day.

What is Optics?

772. OPTICS.—Optics is the science which treats of light, of colors, and of vision.

How are all substances considered in Optics?

773. The science of Optics divides all substances into the following classes: namely, luminous, transparent, and translucent; reflecting, refracting, and opaque.

What are luminous bodies?

774. Luminous bodies are those which shine by their own light; such as the sun, the stars, a burning lamp, or a fire.

---

sachusetts. It is the engine used to propel the machinery at a late Fair of the Massachusetts Mechanic Association, where it was very highly and justly commended for its beauty and simplicity of construction, and the perfectly "*noiseless tenor of its way.*" The figure which represents it is an *electrotype copy* of a steel plate, designed by Brown & Harbrys, under the direction of Mr. Tufts. The electrotype copy was taken by Mr. A. Wilcox, Washington-street, Boston. The electrotype process will be noticed in a subsequent page of this volume.

OPTICS.

What are transparent substances?
775. Transparent substances are those which allow light to pass through them freely, so that objects can be distinctly seen through them; as glass, water, air, &c.*

What are translucent bodies?
776. Translucent bodies are those which permit a portion of light to pass through them, but render the object behind them indistinct; as horn, oiled paper, colored glass, &c.

What are reflecting substances?
777. Reflecting substances are those which do not permit light to pass through them; but throw it off in a direction more or less oblique, according as it falls on the reflecting surface; as polished steel, looking-glasses, polished metal, &c.

What are refracting substances?
778. Refracting substances are those which turn the light from its course in its passage through them; and opaque substances are those which permit no light to pass through them, as metals, wood, &c.

What is light? What are the two theories respecting the nature of light?
779. It is not known what light is. Sir Isaac Newton supposed it to consist of exceedingly small particles, moving from luminous bodies; others think that it consists of the undulations of an elastic medium, which fills all space.† These undulations (as is supposed) produce the

---

* No substance that exists on our earth is *perfectly* transparent, and light must, therefore, necessarily be impaired in its passage through all transparent media, and the diminution it suffers will vary as the medium is more or less transparent, and as the passage it makes is of greater or less length. The exact ratio in which light is diminished has not yet been determined; it is, however, an established fact, that even those bodies which approach most nearly to perfect transparency become opaque when their thickness is considerably increased.

† These two theories of light are called respectively the *corpuscular* and the *undulatory* theory. By the former the reflection of light is supposed to take place in the same manner as the reflection of solid elastic bodies, as has been explained under the head of Mechanics [*see No.* 165, *page* 49]. By the latter the propagation of light takes place from every luminous point, by means of the undulatory movements of the ether. On this hypoth-

sensation of light to the eye, in the same manner as the vibrations of the air produce the sensation of sound to the ear. The opinions of philosophers at the present day are inclining to the undulatory theory.

What is a ray of light?
780. A ray of light is a single line of light proceeding from a luminous body.

When are rays said to diverge?
781. Rays of light are said to diverge when they separate more widely as they proceed from a luminous body.

Fig. 116.

Explain Fig. 116.
Fig. 116 represents the rays of light diverging as they proceed from the luminous body, from F to D.

782. It will be seen by this figure that, as light is projected in every direction, its intensity must decrease with the distance, and this decrease is determined by a fixed law. The light received upon any surface decreases as the square of the distance increases. Thus, if a portion of light fall on a surface at the distance of two feet from any luminary, a surface twice that distance will receive only one-fourth as much light; at three times that distance, one-ninth; at four times the distance, one-sixteenth, &c. Hence a person can see to read at a short distance from a single lamp much better than at twice the same distance with two lamps, &c.

When are rays of light said to converge?
783. Rays of light are said to converge when they approach each other. The point

esis, the *waves* of light follow the general laws of the reflection of all elastic *fluids*, and, accordingly, every wave from every point, when it impinges on any resisting object so as to be reflected, forms a new wave in its course back, having its centre as much on the other side of the obstacle as the centre of the original wave was on this side. In the case of light the centre of the original wave is, obviously, the luminous point. There is a remarkable similarity, therefore, between the reflection of light, and *echo*, or the reflection of sound. It has been shown, under the head of Acoustics, that when two waves meet under certain circumstances, the elevation of one wave exactly filling up the depression of another wave, produces what is called the *acoustic paradox*, namely, *two sounds producing silence*. It will readily be seen that the same undulatory movements in Optics will produce the same analogous effect; or, in other words, that *two rays of light may produce darkness;* and this may, with equal propriety, be termed the *optical paradox*. But a clear understanding of the principles involved in what is called respectively the hydrostatic, pneumatic, acoustic and optical paradox, shows that there is no paradox at all, but that each is the necessary result of certain fixed and determinate laws.

OPTICS.   213

at which converging rays meet is called the focus.

Fig. 117.

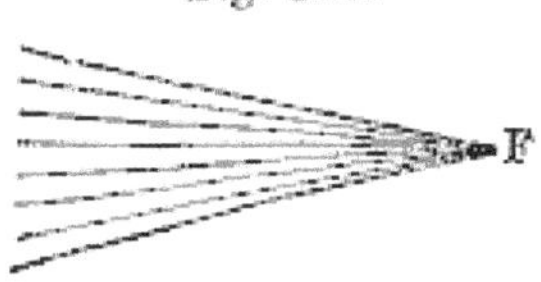

Explain Fig. 117.

Fig. 117 represents converging rays of light, of which the point F is the focus.

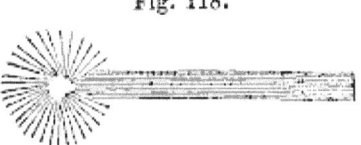

Fig. 118.

What is a beam of light?

784. A beam of light consists of many rays running in parallel lines.

Explain Fig. 118.

Fig. 118 represents a beam of light.

What is a pencil of rays?

785. A pencil of rays is a collection of diverging or converging rays. [*See Figs.* 116 and 117.]

In what direction, and with what rapidity, does light move?

786. Light proceeding from a luminous body is projected forward in straight lines in every possible direction. It moves with a rapidity but little short of two hundred thousand miles in a second of time.

From what part of a luminous body does light proceed?

787. Every point of a luminous body is a centre, from which light radiates in every direction. Rays of light proceeding from different bodies cross each other without interfering. The rays of light which issue from terrestrial bodies continually diverge, until they meet with a refracting substance; but the rays of the sun diverge so little, on account of the immense distance of that luminary, that they are considered parallel.

What is a shadow?

788. A shadow is the darkness produced by the intervention of an opaque body, which prevents the rays of light from reaching an object behind the opaque body.

Why are shadows of different

789. Shadows are of different degrees of darkness, because the light from other lumi-

degrees of darkness? nous bodies reaches the spot where the shadow is formed. Thus, if a shadow be formed when two candles are burning in a room, that shadow will be both deeper and darker if one of the candles be extinguished. The darkness of a shadow is proportioned to the intensity of the light, when the shadow is produced by the interruption of the rays from a single luminous body.

What produces the darkest shadow? 790. As the degree of light and darkness can be estimated only by comparison, the strongest light will appear to produce the deepest shadow. Hence, a total eclipse of the sun occasions a more sensible darkness than midnight, because it is immediately contrasted with the strong light of day. Hence, also, by causing the shadow of a single object to be thrown on a surface,—as, for instance, the wall,—from two or more lights, we can tell which is the brightest light, because it will cause the darkest shadow.

What is the shape of the shadow of an opaque body? 791. When a luminous body is larger than an opaque body, the shadow of the opaque body will gradually diminish in size till it terminates in a point. The form of the shadow of a spherical body will be that of a cone.

Explain Fig. 119. Fig. 119. A represents the sun, and B the moon. The sun being much larger than the moon, causes it to cast a converging shadow, which terminates at E.

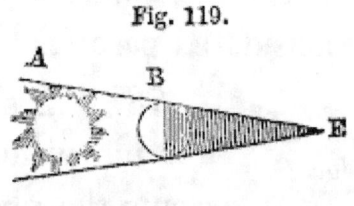

Fig. 119.

792. When the luminous body is smaller than the opaque body, the shadow of the opaque body will gradually increase in size with the distance, without limit.

In Fig. 120 the shadow of the object A increases in size at the different distances B, C, D, E; or, in other words, it constantly diverges.

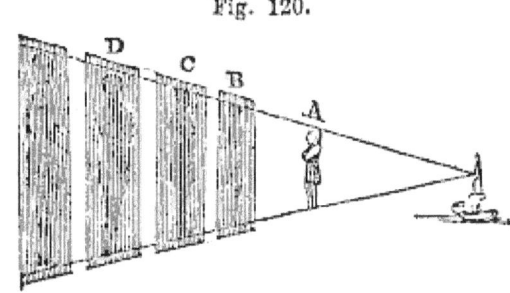

Fig. 120.

793. When several luminous bodies shine upon the same object, each one will produce a shadow.

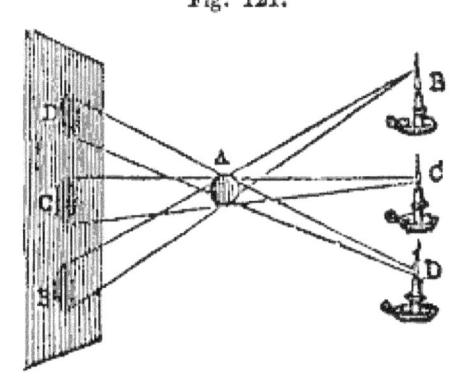

Fig. 121.

What is it the object of Fig. 121 to show?

Fig. 121 represents a ball A, illuminated by the three candles B, C, and D. The light B produces the shadow *b*, the light C the shadow *c*, and the light D the shadow *d*; but, as the light from each of the candles shines upon all the shadows except its own, the shadows will be faint.

What becomes of the light which falls on an opaque object?

794. When rays of light fall upon an opaque body, part of them are absorbed, and part are reflected.

When is light said to be reflected?

Light is said to be reflected when it is thrown off from the body on which it falls; and it is reflected in the largest quantities from the most highly polished surfaces. Thus, although most substances reflect it in a degree, polished metals, looking-glasses, or mirrors, &c., reflect it in so perfect a manner as to convey to our eyes, when situated in a proper position to receive them, perfect images of whatever objects shine on them, either by their own or by borrowed light.

What is Catoptrics?

795. That part of the science of Optics which relates to reflected light is called *Catoptrics*.

What is the fundamental law of Catoptrics?

796. The laws of reflected light are the same as those of reflected motion. Thus, when light falls perpendicularly on an opaque body, it is reflected back in the same line towards the point whence it proceeded. If it fall obliquely, it will be reflected obliquely in the opposite direction; and in all cases the angle of incidence will be equal to the angle of reflection. This is the fundamental law of Catoptrics, or reflected light.

797. The angles of incidence and reflection have already been described under the head of Mechanics [*see explanation of Fig.* 10, *No.* 162]; but, as all the phenomena of reflected light depend upon the law stated above, and a clear idea of these angles is necessary in order to understand the law, it is deemed expedient to repeat in this connection the explanation already given.

An incident ray is a ray proceeding *to* or falling *on* any surface; and a reflected ray is the ray which proceeds *from* any reflecting surface.

Explain Fig. 122.

Fig. 122 is designed to show the angles of incidence and of reflection. In this figure, M A M is a mirror, or reflecting surface. P is a line perpendicular to the surface. I A represents an incident ray, falling on the mirror in such a manner as to form, with the perpendicular P, the angle I A P. This is called the angle of incidence. The line R A is to be drawn on the other side of P A in such a manner as to have the same inclination with P A as I A has: that is, the angle R A P is equal to I A P. The line R A will then show the course of the reflected ray; and the angle R A P will be the angle of reflection.

Fig. 122.

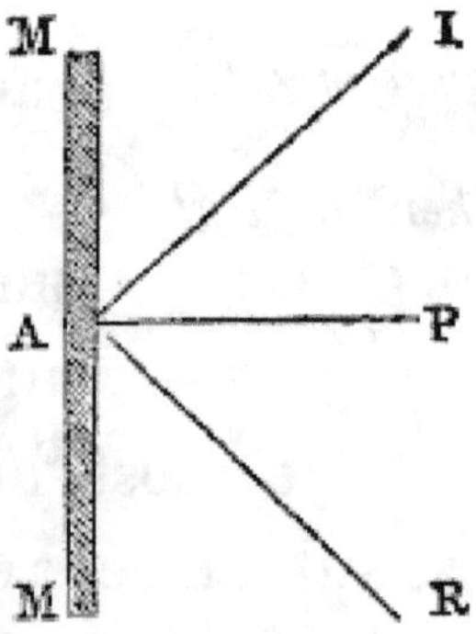

From whatever surface a ray of light is reflected,—whether it be a plain surface, a convex surface, or a concave surface,—this law invariably prevails; so that, if we notice the inclination of any incident ray, and the situation of the perpendicular to the surface on which it falls, we can always determine in what manner or to what point it will be reflected. This law explains the reason why, when we are standing on one side of a mirror, we can see the reflection of objects on the opposite side of the room, but not those on the same side on which we are standing. It also explains the reason why a person can see his whole figure in a mirror not more than half of his height. It also accounts for all the apparent peculiarities of the reflection of the different kinds of mirrors.

How are luminous and opaque bodies respectively seen? 798. Opaque bodies are seen only by reflected light. Luminous bodies are seen by the rays of light which they send directly to our eyes.

What effect has reflection on the intensity of light? 799. All bodies absorb a portion of the light which they receive; therefore the intensity of light is diminished every time that it is reflected.

What does every portion of a reflecting surface reflect? 800. Every portion of a reflecting surface reflects an entire image of the luminous body shining upon it.

Why do we not see many images of the same thing reflected by a reflecting surface? When the sun or the moon shines upon a sheet of water, every portion of the surface reflects an entire image of the luminary; but, as the image can be seen only by reflected rays, and as the angle of reflection is always equal to the angle of incidence, the image from any point can be seen only in the reflected ray prolonged.

Why do objects appear fainter by moonlight? 801. Objects seen by moonlight appear fainter than when seen by daylight, because the light by which they are seen has been twice reflected; for, the moon is not a luminous body, but its light is caused by the sun shining upon it. This light, reflected from the moon and falling upon any object, is again reflected by that object. It suffers, therefore, two reflections; and since a portion is absorbed by each surface that reflects it, the light must be proportionally fainter. In traversing the atmosphere, also, the rays, both of the sun and moon, suffer diminution; for, although pure air is a transparent medium, which transmits the rays of light freely, it is generally surcharged with vapors and exhalations, by which some portion of light is absorbed.

When is an object invisible?   802. All objects are seen by means of the rays of light emanating or reflected from them; and therefore, when no light falls upon an opaque body, it is invisible.

This is the reason why none but luminous bodies can be seen in the dark. For the same reason, objects in the shade or in a darkened room appear indistinct, while those which are exposed to a strong light can be clearly seen. We see the things around us, when the sun does not shine directly upon them, solely by means of reflected light. Everything on which it shines directly reflects a portion of its rays in all possible directions, and it is by means of this reflected light that we are enabled to see the objects around us in the day-time which are not in the direct rays of the sun. It may here also be remarked that it is entirely owing to the reflection of the atmosphere that the heavens appear bright in the day-time. If the atmosphere had no reflective power, only that part would be luminous in which the sun is placed; and, on turning our back to the sun, the whole heavens would appear as dark as in the night; we should have no twilight, but a sudden transition from the brightest sunshine to darkness immediately upon the setting of the sun.

How do rays of light enter a small aperture?   803. When rays of light, proceeding from any object, enter a small aperture, they cross one another, and form an inverted image of the object. This is a necessary consequence of the law that light always moves in straight lines.

Explain Fig. 123.   804. Fig. 123 represents the rays from an object, $a\ c$, entering an aperture. The ray from a passes

OPTICS. 219

lown through the aperture to *d*, and the ray from *c* passes up to *b*, and thus these rays, crossing at the aperture, form an inverted image on the wall. The room in which this experiment is made should be darkened, and no light permitted to enter, excepting through the aperture. It then becomes a camera obscura.

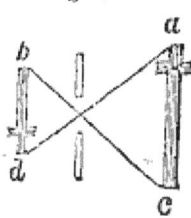
Fig. 123.

805. These words signify *a darkened chamber*. In the future description which will be given of *the eye*, it will be seen that the camera obscura is constructed on the same principle as the eye. If a convex lens be placed in the aperture, an inverted picture, not only of a single object, but of the entire landscape, will be found on the wall. A portable camera obscura is made by admitting the light into a box of any size, through a convex lens, which throws the image upon an inclined mirror, from whence it is reflected upwards to a plate of ground glass. In this manner a beautiful but diminished image of the landscape, or of any group of objects, is presented on the plate in an erect position.

What is the angle of vision?   806. The angle of vision is the angle formed at the eye by two lines drawn from opposite parts of an object.

What is the object of Figures 124 and 125?   807. The angle C, in Fig. 124, represents the angle of vision. The line A C, proceeding from one extremity of the object, meets the line B C from the opposite extremity, and forms an angle C at the eye;— this is the angle of vision.

808. Fig. 125 represents the different angles made by the same object at different distances. From an inspection of the figure, it is evident that the nearer the object is to the eye, the wider must be the opening of the lines to admit the extremities of the object, and, consequently, the larger the angle under which it is seen; and, on the contrary, that objects at a distance will form small angles of vision. Thus, in this figure, the three crosses F G, D E, and A B, are

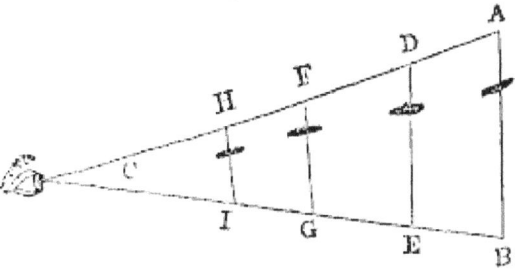
Fig. 124.

Fig. 125.

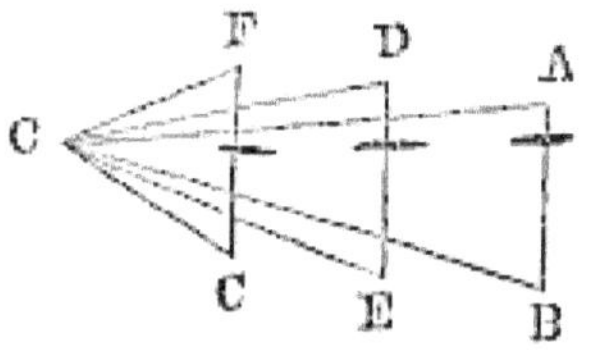

all of the same size; but A B, being the most distant, subtends the smallest angle A C B, while D E and F G, being nearer to the eye, situated at C, form respectively the larger angles D C E and F C G.

On what does the apparent size of an object depend?
809. The apparent size of an object depends upon the size of the angle of vision. But we are accustomed to correct, by experience, the fallacy of appearances; and, therefore, since we know that real objects do not vary in size, but that the angles under which we see them do vary with the distance, we are not deceived by the variations in the appearance of objects.

Thus, a house at a distance appears absolutely smaller than the window through which we look at it; otherwise we could not see it through the window; but our knowledge of the real size of the house prevents our alluding to its apparent magnitude. In Fig. 124 it will be seen that the several crosses, A B, D E, F G, and H I, although very different in size, on account of their different distances, subtend the same angle A C B; they, therefore, all appear to the eye to be of the same size, while, in Fig. 125, the three objects A B, D E, and F G, although of the same absolute size, are seen at a different angle of vision, and they, therefore, will seem of different sizes, appearing larger as they approach the eye.

It is to a correct observance of the angle of vision that the art of perspective drawing is indebted for its accuracy.

When is an object invisible on account of is distance?
810. When an object, at any distance, does not subtend an angle of more than two seconds of a degree, it is invisible.

At the distance of four miles a man of common stature will thus become invisible, because his height at that distance will not subtend an angle of two seconds of a degree. The size of the apparent diameter of the heavenly bodies is generally stated by the angle which they subtend.

When is motion imperceptible?
811. When the velocity of a moving body does not exceed twenty degrees in an hour, its motion is imperceptible to the eye.

It is for this reason that the motion of the heavenly bodies is invisible, notwithstanding their immense velocity.

812. The real velocity of a body in motion round a point depends on the space comprehended in a degree. The more dis-

tant the moving body from the centre, or, in other words, the larger the circle which it has to describe, the larger will be the degree.

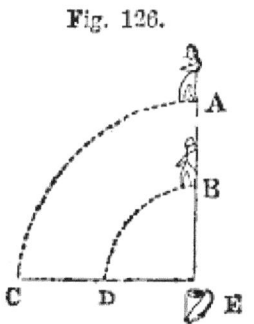

Fig. 126.

813. In Fig. 126, if the man at A, and the man at B, both start together, it is manifest that A must move more rapidly than B, to arrive at C at the same time that B reaches D, because the arc A C is the arc of a larger circle than the arc B D. But to the eye at E the velocity of both appears to be the same, because both are seen under the same angle of vision.

What are mirrors, and how are they made?

814. A mirror is a smooth and polished surface, that forms images by the reflection of the rays of light. Mirrors (or looking-glasses) are made of glass, with the back covered with an amalgam, or mixture of mercury and tin foil. It is the smooth and bright surface of the mercury that reflects the rays, the glass acting only as a transparent case, or covering, through which the rays find an easy passage. Some of the rays are absorbed in their passage through the glass, because the purest glass is not free from imperfections. For this reason, the best mirrors are made of an alloy of copper and tin, called speculum metal.

What are the different kinds of mirrors?

815. There are three kinds of mirrors, namely, the plain, the concave, and the convex mirror.

Plain mirrors are those which have a flat surface, such as a common looking-glass; and they neither magnify nor diminish the image of objects reflected from them.

By what law are objects reflected from a looking glass?

816. The reflection from plain mirrors is always obedient to the law that the angles of incidence and reflection are equal. For this reason, no person can see another in a looking-glass, if the other cannot see him in return.

How do looking-glasses make all objects appear?

817. Looking-glasses or plain mirrors cause everything to appear reversed. Standing before a looking-glass, if a person holds up his left hand it will appear in the glass to be the right.

818. A looking-glass, to reflect the whole person, needs be but half of the length of the person.

819. When two plain mirrors stand opposite to each other, the reflections of the one are cast upon the other, and to a person between them they present a long-continued vista.

820. When two reflecting surfaces are inclined at an angle, the reflected objects appear to have a common centre to an eye viewing them obliquely. It is on this principle that the kaleidoscope is constructed.

What is a Kaleidoscope?

821. The Kaleidoscope consists of two reflecting surfaces, or pieces of looking-glass, inclined to each other at an angle of sixty degrees, and placed between the eye and the objects intended to form the picture.

The two plates are enclosed in a tin or paper tube, and the objects, consisting of pieces of colored glass, beads, or other highly-colored fragments, are loosely confined between two circular pieces of common glass, the outer one of which is slightly ground, to make the light uniform. On looking down the tube through a small aperture, and where the ends of the glass plates nearly meet, a beautiful figure will be seen, having six angles, the reflectors being inclined the sixth part of a circle. If inclined the twelfth part or twentieth part of a circle, twelve or twenty angles will be seen. By turning the tube so as to alter the position of the colored fragments within, these beautiful forms will be changed; and in this manner an almost infinite variety of patterns may be produced.

The word Kaleidoscope is derived from the Greek language, and means "the sight of a beautiful form." The instrument was invented by Dr. Brewster, of Edinburgh, a few years ago.

822. A convex mirror is a portion of the external surface of a sphere. Convex mirrors have therefore a convex surface.

823. A concave mirror is a portion of the inner surface

of a hollow sphere. Concave mirrors have therefore a concave surface.

Explain Fig. 127. 824. In Fig. 127, M N represents both a convex and a concave mirror. They are both a portion of a sphere of which O is the centre. The outer part of M N is a convex, and the inner part is a concave mirror. Let A B, C D, E F, represent rays falling on the convex mirror M N. As the three rays are parallel, they would all be perpendicular to a plane or flat mirror; but *no ray can fall perpendicularly on a concave or convex mirror which is not directed towards the centre of the sphere of which the mirror is a portion.*

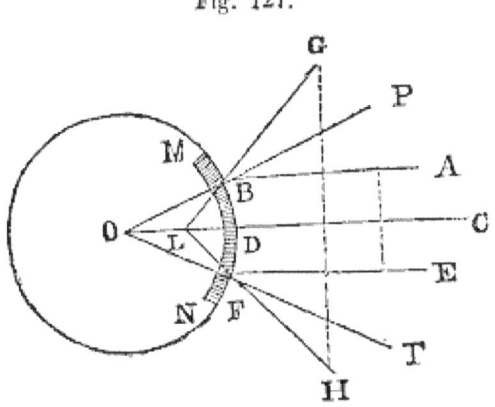

Fig. 127.

For this reason, the ray C D is perpendicular to the mirror, while the other rays, A B and E F, fall obliquely upon it. The middle ray therefore, falling perpendicularly on the mirror, will be reflected back in the same line, while the two other rays, falling obliquely, will be reflected obliquely; namely, the ray A B will be reflected to G, and the ray E F to H, and the angles of incidence A B P and E F T will be equal to the angles of reflection P B G and T F H; and, since *we see objects in the direction of the reflected rays,* we shall see the image at L, which is the point at which the reflected rays, if continued through the mirror, would unite and form the image. This point is equally distant from the surface and the centre of the sphere, and is called the imaginary focus of the mirror. It is called the *imaginary* focus, because the rays do not really unite at that point, but only appear to do so; for the rays do not pass through the mirror, since they are reflected by it.

825. The image of an object reflected from a convex mirror is smaller than the object.

What is the object of Fig. 128?

826. This is owing to the divergence of the reflected rays. *A convex mirror converts, by reflection, parallel rays into divergent rays; rays that fall upon the mirror divergent are rendered still more divergent* by reflection, and convergent rays are reflected either parallel, or less convergent. If, then, an object, A B, be placed before any part of a convex mirror, the two rays A and B, proceeding from the extremities, falling convergent on the mirror, will be reflected less convergent, and will not come to a focus until they arrive at C; then an eye placed in the direction of the reflected rays will see the image formed in (or rather behind) the mirror at *a b*; and, as the image is seen under a smaller angle than the object, it will appear smaller than the object.

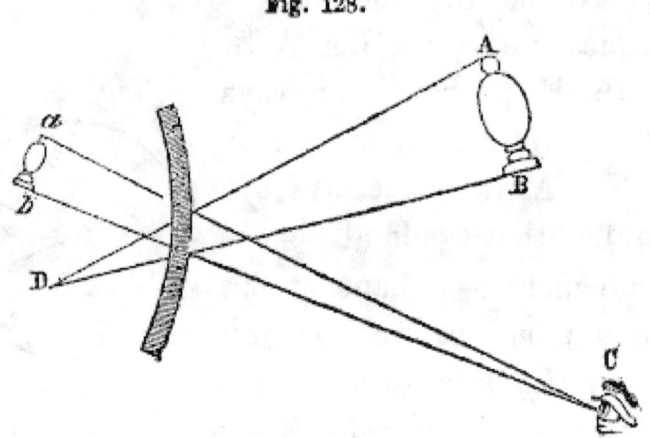

Fig. 128.

What is the true focus of a concave mirror?

827. The true focus of a concave mirror is a point equally distant from the centre and the surface of the sphere of which the mirror is a portion.

When will the image reflected from a concave be upright, and when inverted?

828. When an object is further from the concave surface mirror than its focus, the image will be inverted; but when the object is between the mirror and its focus, the image will be upright, and grow larger in proportion as the object is placed nearer to the focus.

What peculiar property have concave mirrors?

829. Concave mirrors have the peculiar property of forming images in the air. The mirror and the object being concealed behind a screen, or a wall, and the object being strongly illumi-

nated, the rays from the object fall upon the mirror, and are reflected by it through an opening in the screen or wall, forming an image in the air.

Showmen have availed themselves of this property of concave mirrors, in producing the appearance of apparitions, which have terrified the young and the ignorant. These images have been presented with great distinctness and beauty, by raising a fine transparent cloud of blue smoke, by means of a chafing-dish, around the focus of a large concave mirror.

When is the image from a concave mirror larger than the object?

830. The image reflected by a concave mirror is larger than the object when the object is placed between the mirror and its focus.

What is the design of Fig. 129?

831. This is owing to the convergent property of the concave mirror. If the object A B be placed between the concave mirror and its focus f, the rays A and B from its extremities will fall divergent on the mirror, and, on being reflected, become less divergent, as if they proceeded from C. To an

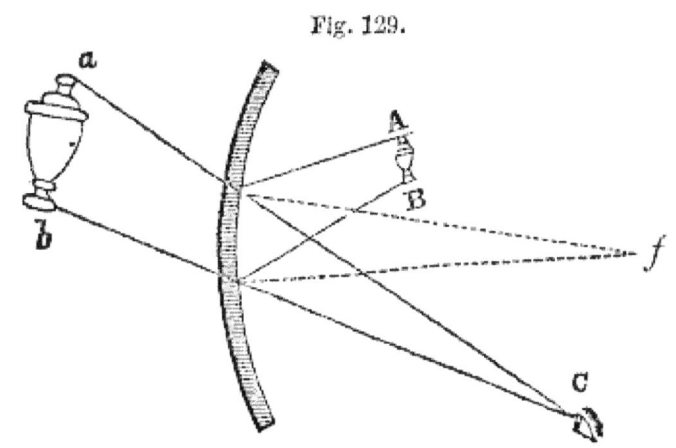

Fig. 129.

eye placed in that situation, namely, at C, the image will appear magnified behind the mirror, at $a\ b$, since it is seen under a larger angle than the object.

832. There are three cases to be considered with regard to the effects of concave mirrors:
1. When the object is placed between the mirror and the principal focus.
2. When it is situated between its centre of concavity and that focus.
3. When it is more remote than the centre of concavity.

1st. In the first case, the rays of light diverging after reflection but in a less degree than before such reflection took place, the image will be larger than the object, and appear at a greater or smaller distance from the surface of the mirror, and behind it. The image in this case will be erect.

2d. When the object is between the principal focus and the centre of the mirror, the apparent image will be in front of the mirror, and beyond the centre, appearing very distant when the object is at or just beyond the focus, and advancing towards it as it recedes towards the centre of concavity, where, as will be stated, the image and the object will coincide. During the retreat of the object the image will still be inverted, because the rays belonging to each visible point will not intersect before they reach the eye. But in this case the image becomes less and less distinct, at the same time that the visual angle is increasing; so that at the centre, or rather a little before, the image becomes confused and imperfect, because at this point the object and the image coincide.

3d. In the cases just considered, the images will appear inverted; and in the case where the object is further from the mirror than its centre of concavity, the image will be inverted. The more distant the object is from the centre, the less will be its image; but the image and object will coincide when the latter is stationed exactly at the centre.

833. The following laws flow from the fundamental law of Catoptrics, namely, that the angles of incidence and reflection are always equal. In estimating these angles, it must be recollected that no line is perpendicular to a convex or concave mirror, which will not, when sufficiently prolonged, pass through the centre of the sphere of which the mirror is a portion. The truth of these statements may be illustrated by simple drawings; always recollecting, in drawing the figures, to make the angles of incidence and reflection equal. The whole may also be shown by the simple experiment of placing the flame of a candle in various positions before both convex and concave mirrors. [*It is recommended that the learner be required to draw a figure to represent each of these laws.*]

834. LAWS OF REFLECTION FROM CONVEX MIRRORS.—(1.) Parallel rays reflected from a CONVEX surface are made to diverge.

(2.) Diverging rays reflected from a CONVEX surface are made more diverging.

(3.) When converging rays tend towards the focus of parallel rays, they will become parallel when reflected from a CONVEX surface.*

(4.) When converging rays tend to a point nearer the surface

---

* For the sake of distinction, the principal focus is called "the focus of parallel rays."—*Peschel*.

than the focus, they will converge less when reflected from a CONVEX surface.

(5.) If converging rays tend to a point between the focus and the centre, they will diverge as from a point on the other side of the centre, further from it than the point towards which they converged.

(6.) If converging rays tend to a point beyond the centre, they will diverge as from a point on the contrary side of the centre, nearer to it than the point towards which they converged.

(7.) If converging rays tend to the centre, when reflected they will proceed in a direction as if from the centre.

835. LAWS OF REFLECTION FROM CONCAVE MIRRORS.— (1.) Parallel rays reflected from a CONCAVE surface are made converging. [*See Note to No.* 837.]

(2.) Converging rays falling upon a CONCAVE surface are made to converge more.

(3.) Diverging rays falling upon a CONCAVE surface, if they diverge from the focus of parallel rays, become parallel.

(4.) If from a point nearer to the surface than that focus, they diverge less than before reflection.

(5.) If from a point between that focus and the centre, they converge, after reflection, to some point on the contrary side of the centre, and further from the centre than the point from which they diverged.

(6.) If from a point beyond the centre, the reflected rays will converge to a point on the contrary side, but nearer to it than the point from which they diverged.

(7.) If from the centre, they will be reflected back to the same point from which they proceeded.

How are objects seen from a convex mirror?

836. As a necessary consequence of the laws which have now been recited, it may be stated, *First*, in regard to CONVEX MIRRORS, the images of objects invariably appear beyond the mirror; in other words, they are virtual images. *Secondly*, they are seen in

their natural position, and, *Thirdly*, they are smaller than the objects themselves; the further the object is from the mirror, and the less the radius of the mirror, the smaller the image will be. If the object be very remote, its image will be in the virtual focus of the mirror.

837. *Secondly*, in regard to CONCAVE MIRRORS.

(1.) The image of an object very remote from a concave mirror, as that of the sun, will be in the focus of the mirror, and the image will be extremely small.*

(2.) Every object which is at a distance from the mirror greater than its centre produces an image between this point and the focus smaller than the object itself, and in an inverted position.

(3.) If the object be at a distance from the mirror equal to the length of its radius, then the image will be at an equal distance from the mirror, and the dimensions of the image will be the same as those of the object, but its position will be inverted.

(4.) If the object be between the focus and the centre of curvature, the image will be inverted, and its size will much exceed that of the object.

These four varieties of inverted images, produced by the reflection of the rays of light from concave mirrors, are sometimes called "*physical spectra.*"

* This is the manner in which concave mirrors become burning-glasses The rays of the sun fall upon them parallel [*see No.* 835], and they are all reflected into one point, called the focus, where the light and heat are as much greater than the ordinary light and heat of the sun as the area of the mirror is greater than the area of the focus. It is related of Archimedes, that he employed burning-mirrors, two hundred years before the Christian era, to destroy the besieging navy of Marcellus, the Roman consul. His mirror was, probably, constructed from large numbers of flat pieces. M. de Vilette constructed a burning-mirror in which the area of the mirror was seventeen thousand times greater than the area of the focus. The heat of the sun was thus increased seventeen thousand times. M. Dufay made a concave mirror of plaster of Paris, gilt and burnished, twenty inches in diameter, with which he set fire to tinder at the distance of fifty feet. But the most remarkable thing of the kind on record is the compound mirror constructed by Buffon. He arranged one hundred and sixty-eight small plane mirrors in such a manner as to reflect radiant light and heat to the same focus, like one large concave mirror. With this apparatus he was able to set wood on fire at the distance of two hundred and nine feet, to melt lead at a hundred feet, and silver at fifty feet.

The existence and position of these spectra may easily be shown experimentally thus:

Experiment.— Hold a candle opposite to a concave mirror, at the distances named in the last four paragraphs respectively. The spectrum can, in each case, be received on a white screen, which must be placed at the prescribed distance from the mirror.

Different optical instruments, especially reflecting telescopes, exhibit the application of these spectra.

(5.) If a luminous body, as, for instance, the flame of an argand lamp, or a burning coal, be placed in the focus of a concave mirror, no image will be produced, but the whole surface of the mirror will be illuminated, because it reflects in parallel lines all the rays of light that fall upon it. This may be made the subject of an experiment so simple as not to require further explanation.

The reflectors of compound microscopes, magic lanterns and lighthouses, by means of which the light given by the luminous body is increased and transmitted in some particular direction that may be desired, are illustrations of the practical application of this principle.

(6.) Lastly, place the object between the mirror and the focus, and the image of the object will appear behind the mirror. It will not be inverted, but its proportions will be enlarged according to the proximity of the object to the focus. It is this circumstance that gives to concave mirrors their magnifying powers, and, because by collecting the sun's rays into a focus they produce a strong heat, they are called burning-mirrors.

What is a Medium in Optics? 838. MEDIA, OR MEDIUMS, AND REFRACTION.— A Medium,* in Optics, is any substance, solid or fluid, through which light can pass.

What is refraction? 839. When light passes *in an oblique direction* from one medium into another, it is turned or bent from its course, and this is called *refrac-*

---

* The proper plural of this word is *media*, although *mediums* is frequently used.

tion. The property which causes it is called *refrangibility.*

What is Dioptrics?

840. DIOPTRICS.—That part of the science of Optics which treats of refracted light is called Dioptrics.

What is meant by a denser and rarer medium in Optics?

841. A medium, in Optics, is called dense or rare according to its refractive power, and not according to its specific gravity. Thus, alcohol, and many of the essential oils, although of less specific gravity than water, have a greater refracting power, and are, therefore, called denser media than water. In the following list, the various substances are enumerated in the order of their refractive power, or, in other words, in the order of their density as media, the last-mentioned being the densest, and the first the rarest, namely: air, ether, ice, water, alcohol, alum, olive oil, oil of turpentine, amber, quartz, glass, melted sulphur, diamond.

What are the fundamental laws of Dioptrics?

842. There are three fundamental laws of Dioptrics, on which all its phenomena depend, namely:

(1.) When light passes from one medium to another in a direction perpendicular to the surface, it continues on in a straight line, without altering its course.

(2.) When light passes in an oblique direction, from a *rarer* to a *denser* medium, it will be turned from its course, and proceed through the denser medium *less* obliquely, and in a line nearer to a perpendicular to its surface.

(3.) When light passes from a denser to a rarer medium in an oblique direction, it passes through the rarer medium in a more oblique direction, and in a line further from a perpendicular to the surface of the denser medium.

Explain Fig. 130.

843. In Fig. 130, the line A B represents a ray of light passing from air into water, in a perpendicular direction. According to the first

OPTICS. 231

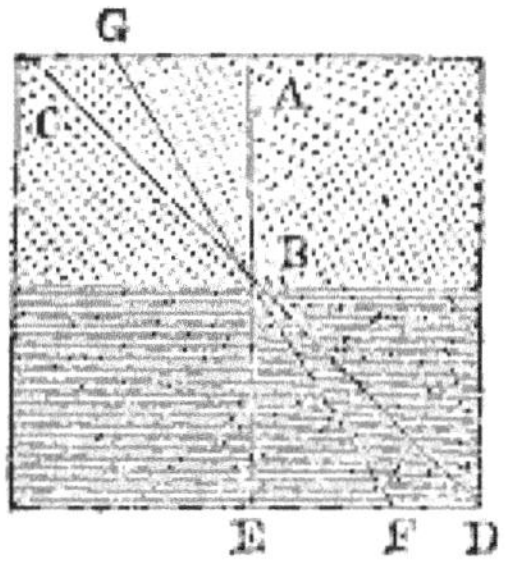

law stated above, it will continue on in the same line through the denser medium to E. If the ray were to pass upward through the denser medium, the water, in the same perpendicular direction to the air, by the same law it would also continue on in the same straight line to A.

Fig. 130.

But, if the ray proceed from a rarer to a denser medium, in an oblique direction, as from C to B, when it enters the denser medium it will not continue on in the same straight line to D, but, by the second law, stated above, it will be refracted or bent out of its course and proceed in a less oblique direction to F, which is nearer the perpendicular A B E than D is.

Again, if the ray proceed from the denser medium, the water, to the rarer medium, the air, namely, from F to B, instead of pursuing its straight course to G, it will be refracted according to the third law above stated, and proceed in a more oblique direction to C, which is further from the perpendicular E B A than G is. The refraction is more or less in all cases in proportion as the rays fall more or less obliquely on the refracting surface.

In what proportion is refraction in all cases?

When are we in danger of mistaking the depth of water, and why?

844. From what has now been stated with regard to refraction, it will be seen that many interesting facts may be explained. Thus, an oar, or a stick, when partly immersed in water, appears bent, because we see one part in one medium, and the other in another medium: the part which is in the water appears higher than it really is, on account of the refraction of the denser medium. For the same reason, when we look *obliquely* upon a body of water it appears more shallow than it really is. But, when we look *perpendicularly* downwards, we are liable to no such deception, because there will be no refraction.

845. Let a piece of money be put into a cup or a bowl, and the cup and the eye be placed in such a position that the side of the cup will just hide the money from the sight; then, keeping the eye

directed to the same spot, let the cup be filled with water,— the money will become distinctly visible.

Why do we not see the sun, moon and stars, in their true places?
846. The refraction of light prevents our seeing the heavenly bodies in their real situation.

The light which they send to us is refracted in passing through the atmosphere, and we see the sun, the stars, &c., in the direction of the refracted ray. In consequence of this atmospheric refraction, the sun sheds his light upon us earlier in the morning, and later in the evening, than we should otherwise perceive it. And, when the sun is actually below the horizon, those rays which would otherwise be dissipated through space are refracted by the atmosphere towards the surface of the earth, causing twilight. The greater the density of the air, the higher is its refractive power, and, consequently, the longer the duration of twilight.

It is proper, however, here to mention that there is another reason, why we do not see the heavenly bodies in their true situation. Light, though it moves with great velocity, is about eight and a half minutes in its passage from the sun to the earth, so that when the rays reach us the sun has quitted the spot he occupied on their departure; yet we see him in the direction of those rays, and, consequently, in a situation which he abandoned eight minutes and a half before. The refraction of light does not affect the appearance of the heavenly bodies when they are vertical, that is, directly over our heads, because the rays then pass vertically, a direction incompatible with refraction.

What effect is produced when light suffers two equal refractions?
847. When a ray of light passes from one medium to another, and through that into the first again, if the two refractions be equal, and in opposite directions, no sensible effect will be produced.

This explains the reason why the refractive power of flat window-glass produces no effect on objects seen through it. The rays suffer two refractions, which, being in contrary directions, produce the same effect as if no refraction had taken place.

What is a Lens?
848. LENSES.— A Lens is a glass, which, owing to its peculiar form, causes the rays

of light to converge to a focus, or disperses them, according to the laws of refraction.

Explain the different kinds of lenses.

849. There are various kinds of lenses, named according to their focus; but they are all to be considered as portions of the internal or external surface of a sphere.

850. A single convex lens has one side flat and the other convex; as A, in Fig. 131.

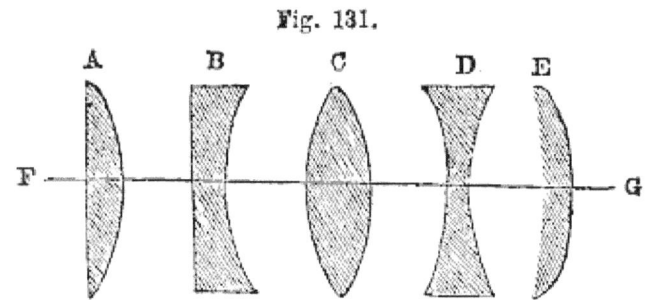

Fig. 131.

851. A single concave lens is flat on one side and concave on the other, as B in Fig. 131.

852. A double convex lens is convex on both sides, as C, Fig. 131.

A double concave lens is concave on both sides, as D, Fig. 131.

A meniscus is convex on one side and concave on the other, as E, Fig. 131.

What is the meaning of a Meniscus?

853. The word *meniscus* is derived from the Greek language, and means literally *a little moon*. This term is applied to a *concavo-convex* lens, from its similarity to a moon in its early appearance. To this kind of lens the term *periscopic* has recently been applied, from the Greek language, meaning literally *viewing on all sides*. When the concave and convex sides of periscopic glasses are even, or parallel, they act as plane glasses; but when the sides are unequal, or not parallel, they will act as concave or convex lenses, according as the concavity or the convexity is the greater.

What is the axis of a lens?

854. The axis of a lens is a line passing through the centre: thus F G, Fig. 131, is the axis of all the five lenses.

855. The peculiar form of the various kinds of lenses causes the light which passes through them to be refracted from its course according to the laws of Dioptrics.

What effect have lenses?

It will be remembered that, according to these laws, light, in passing from a rarer to a denser medium, is *refracted* towards the perpendicular; and, on the contrary, that in passing from a denser to a rarer medium it is refracted further from the perpendicular. In order to estimate the effect of a lens, we must consider the situation of the perpendicular with respect to the surface of the lens. Now, a perpendicular, to any convex or concave surface, must always, when prolonged, pass through the centre of sphericity; that is, in a lens, the centre of the sphere of which the lens is a portion. By an attentive observation, therefore, of the laws above stated, and of the situation of the perpendicular on *each* side of the lens, it will be found, *in general*,—

How must we estimate the effect of a lens?

(1.) *That convex lenses collect the rays into a focus, and magnify objects at a certain distance.*

(2.) *That concave lenses disperse the rays, and diminish objects seen through them.*

What effect have convex and concave lenses respectively?

856. The focal distance of a lens is the distance from the middle of the glass to the focus. This, in a single convex lens, is equal to the diameter of the sphere of which the lens is a portion, and in a double convex lens is equal to the radius of a sphere of which the lens is a portion.

What is the focal distance of a lens?

857. When parallel rays* fall on a convex lens, those only which fall in the direction of the axis of the lens are perpendicular to its surface, and those only will continue

What rays will pass through a lens without refraction?

---

* The rays of the sun are considered parallel at the surface of the earth. They are not so in reality, but, on account of the great distance of that luminary, their divergency is so small that it is altogether inappreciable.

on in a straight line through the lens. The other rays, falling obliquely, are refracted towards the axis, and will meet in a focus.

On what principle are sun-glasses, or burning-glasses, constructed? 858. It is this property of a convex lens which gives it its power as a burning-glass, or sun-glass. All the parallel rays of the sun which pass through the glass are collected together in the focus; and, consequently, *the heat at the focus is to the common heat of the sun as the area of the glass is to the area of the focus.* Thus, if a lens, four inches in diameter, collect the sun's rays into a focus at the distance of twelve inches, the image will not be more than one-tenth of an inch in diameter; the surface of this little circle is 1600 times less than the surface of the lens, and consequently the heat will be 1600 times greater at the focus than at the lens.

859. The following effects were produced by a large lens, or burning-glass, two feet in diameter, made at Leipsic in 1691. Pieces of lead and tin were instantly melted; a plate of iron was soon rendered red-hot, and afterwards fused, or melted; and a burnt brick was converted into yellow glass. A double convex lens, three feet in diameter, and weighing two hundred and twelve pounds, made by Mr. Parker, in England, melted the most refractory substances. Cornelian was fused in seventy-five seconds, a crystal pebble in six seconds, and a piece of white agate in thirty seconds. This lens was presented by the King of England to the Emperor of China.

What is a Multiplying-glass? 860. If a convex lens have its sides ground down into several flat surfaces, it will present as many images of an object to the eye as it has flat surfaces. It is then called a Multiplying-glass. Thus, if one lighted candle be viewed through a lens having twelve flat surfaces, twelve candles will be seen through the lens. The principle of the multiplying-glass is the same with that of a convex or concave lens.

861. The following effects result from the laws of refraction
FACTS WITH REGARD TO CONVEX SURFACES. — (1.) Parallel rays passing out of a rarer into a denser medium, through a CONVEX surface, will become converging.
(2.) Diverging rays will be made to diverge less, to become par-

allel, or to converge, according to the degree of divergency before refraction, or the convexity of the surface.

(3.) Converging rays towards the centre of convexity will suffer no refraction.

(4.) Rays converging to a point beyond the centre of convexity will be made more converging.

(5.) Converging rays towards a point nearer the surface than the centre of convexity will be made less converging by refraction.

[When the rays proceed out of a *denser* into a *rarer* medium, the reverse occurs in each case.]

862. FACTS WITH REGARD TO CONCAVE SURFACES.—(1.) Parallel rays proceeding out of a rarer into a denser medium, through a CONCAVE surface, are made to diverge.

(2.) Diverging rays are made to diverge more, to suffer no refraction, or to diverge less, according as they proceed from a point beyond the centre, from the centre, or between the centre and the surface.

(3.) Converging rays are made less converging, parallel, or diverging, according to their degree of convergency before refraction.

863. The above eight principles are all the necessary consequence of the operation of the three laws mentioned as the fundamental laws of Dioptrics. The reason that so many different principles are produced by the operation of those laws is, that the perpendiculars to a convex or concave surface are constantly varying, so that no two are parallel. But in flat surfaces the perpendiculars are parallel; and one invariable result is produced by the rays when passing from a rarer to a denser, or from a denser to a rarer medium, having a flat surface.

[When the rays proceed out of a denser into a rarer medium, the reverse takes place in each case.]

What kinds of glasses are used in spectacles, and for what purpose?
What kinds of glasses are generally worn by old persons?
What kind by young?

864. Double convex, and double concave glasses, or lenses, are used in spectacles, to remedy the defects of the eye: the former, when by age it becomes too flat, or loses a portion of its roundness: the latter, when by any other cause it assumes too round a form, as in the case of short-sighted (or, as they are sometimes called, near-sighted) persons. Convex glasses are used when the eye is too flat, and concave glasses when it is too round.

These lenses or glasses are generally numbered, by opticians, according to their degree of convexity or concavity; so that, by knowing the number that fits the eye, the purchaser can generally be accommodated without the trouble of trying many glasses.

OPTICS.

865. THE EYE. — The eyes of all animals are constructed on the same principles, with such modifications as are necessary to adapt them to the habits of the animal. The knowledge, therefore, of the construction of the eye of an animal will give an insight of the construction of the eyes of all.

Of what is the eye composed?

866. The eye is composed of a number of coats, or coverings, within which are enclosed a lens, and certain humors, in the shape and performing the office of convex lenses.*

What are the different parts of the eye?

867. The different parts of the eye are:

(1.) The Cornea.
(2.) The Iris.
(3.) The Pupil.
(4.) The Aqueous Humor.
(5.) The Crystalline Lens.
(6.) The Vitreous Humor.
(7.) The Retina.
(8.) The Choroid.
(9.) The Sclerotica.
(10.) The Optic Nerve.

Explain Fig. 132.

868. Fig. 132 represents a front view of the eye, in which *a a* represents the Cornea, or, as it is commonly called, the white of the eye; *e e* is the Iris, having a circular opening in the centre, called the pupil, *p*, which contracts in a strong light, and expands in a faint light, and thus regulates the quantity which is admitted to the tender parts in the interior of the eye.

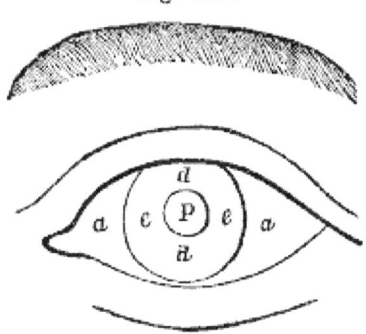

Fig. 132.

Explain Fig. 133.

869. Fig. 133 represents a side view of the eye, laid open, in which *b b* represents the cornea, *e e* the iris, *d d* the pupil, *f f* the aqueous humor, *g g* the crystalline lens, *h h*

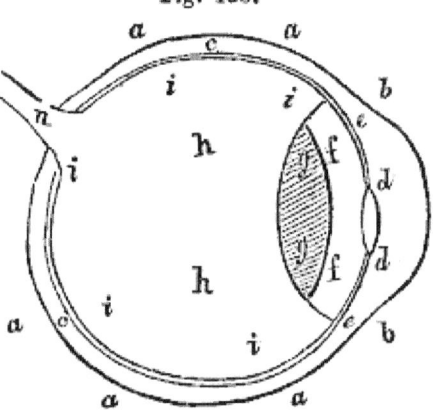

Fig. 133.

---

* The following description of the eye is taken principally from Paxton's Introduction to the Study of Anatomy, edited by Dr. Winslow Lewis, of this city.

the vitreous humor, *i i i i* the retina, *c c* the choroid, *a a a a a* the sclerotica, and *n* the optic nerve.

Describe the Cornea. 870. The Cornea forms the anterior portion the eye. It is set in the sclerotica in the same manner as the crystal of a watch is set in the case. Its degree of convexity varies in different individuals, and in different periods of life. As it covers the pupil and the iris, it protects them from injury. Its principal office is to cause the light which reaches the eye to converge to the axis. Part of the light, however, is reflected by its finely-polished surface, and causes the brilliancy of the eye.

Describe the Iris. 871. The Iris is so named from its being of different colors. It is a kind of circular curtain, placed in the front of the eye, to regulate the quantity of light passing to the back part of the eye. It has a circular opening in the centre, which it involuntarily enlarges or diminishes.

What causes a person's eyes to be black, blue or gray, &c.? 872. It is on the color of the iris that the color of the eye depends. Thus a person is said to have black, blue, or hazel eyes, according as the iris reflects those colors respectively.

What is the Pupil? 873. The Pupil is merely the opening in the iris, through which the light passes to the lens behind. It is always circular in the human eye, but in quadrupeds it is of different shape. When the pupil is expanded to its utmost extent, it is capable of admitting ten times the quantity of light that it does when most contracted.

How can some animals see in the dark? 874. In cats, and other animals which are said to see in the dark, the power of dilatation and contraction is much greater; it is computed that their pupils may receive one hundred times more light at one time than at another. That light only which passes the pupil can be of use in vision; that which falls on the iris, being reflected, returns through the cornea, and exhibits the color of the iris.

When we come from a dark place into a strong light, our eyes suffer pain, because the pupil, being expanded, admits a larger quantity of light to rush in, before it has had time to contract. And, when we go from a strong light into a faint one, we at first imagine ourselves in darkness, because the pupil is then contracted, and does not *instantly* expand.

Describe the Aqueous Humor. 875. The Aqueous Humor is a fluid as clear as the purest water. In shape it resembles a meniscus, and, being situated between the cornea and the crystalline lens, it assists in collecting and transmitting the rays of light from external objects to that lens.

What is the Crystalline Lens? 876. The Crystalline Lens is a transparent body, in the form of a double convex lens, placed between the aqueous and the vitreous humors. Its office is not only to collect the rays to a focus on the retina, but also to increase the intensity of the light which is directed to the back part of the eye.

What is the Vitreous Humor? 877. The Vitreous Humor (so called from its resemblance to melted glass) is a perfectly transparent mass, occupying the globe of the eye. Its shape is like a meniscus, the convexity of which greatly exceeds the concavity.

878. In Fig. 134 the shape of the aqueous and vitreous humors and the crystalline lens is presented. A is the aqueous humor, which is a meniscus, B the crystalline lens, which is a double convex lens, and C the vitreous humor, which is also a meniscus, whose concavity has a smaller radius than its convexity.

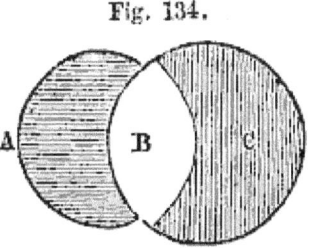

Fig. 134.

What is the Retina?

879. The Retina is the seat of vision. The rays of light, being refracted in their passage by the other parts of the eye, are brought to a focus in the retina, where an inverted image of the object is represented.

What is the Choroid?

880. The Choroid is the inner coat or covering of the eye. Its outer and inner surface is covered with a substance called the *pigmentum nigrum* (or black paint). Its office is, apparently, to absorb the rays of light immediately after they have fallen on the retina. It is the opinion of some philosophers that it is the choroid, and not the retina, which conveys the sensation produced by rays of light to the brain.

Describe the Sclerotica.

881. The Sclerotica is the outer coat of the eye. It derives its name from its hardness. Its office is to preserve the globular figure of the eye, and defend its more delicate internal structure. To the sclerotica are attached the muscles which move the eye. It receives the cornea, which is inserted in it somewhat like a watch-glass in its case. It is pierced by the optic nerve, which, passing through it, expands over the inner surface of the choroid, and thus forms the retina.

What is the Optic Nerve?

882. The Optic Nerve is the organ which carries the impressions made by the rays of light (whether by the medium of the retina, or the choroid) to the brain, and thus produces the sensation of sight.

What optical instrument does the eye resemble?

883. The eye is a natural *camera obscura* [see No. 805], and the images of all objects seen by the eye are represented on the retina in the same manner as the forms of external objects are delineated in that instrument.

Explain Fig. 135.

884. Fig. 135 represents only those parts of the eye which are most essential for the explanation of the

OPTICS. 241

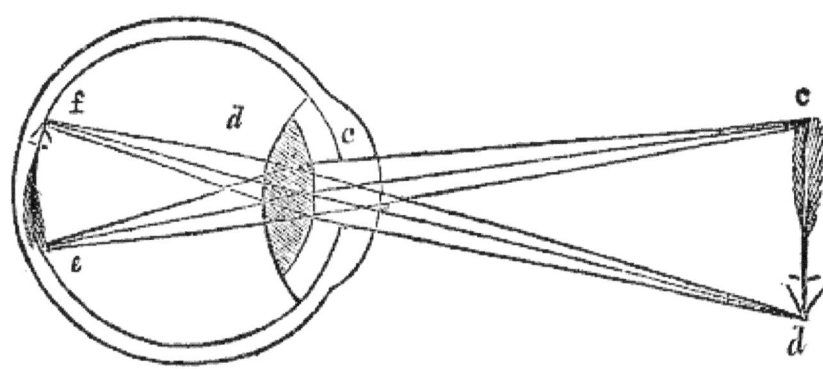

Fig. 135.

phenomenon of vision. The image is formed thus: The rays from the object *c d*, diverging towards the eye, enter the cornea *c*, and cross one another in their passage through the crystalline lens *d*, by which they are made to converge on the retina, where they form the inverted image *f e*.

How is the convexity of the crystalline lens altered, and for what purpose?
885. The convexity of the crystalline humor is increased or diminished by means of two muscles, to which it is attached. By this means, the focus of the rays which pass through it constantly falls on the retina; and an equally distinct image is formed, both of distant objects and those which are near.

How can you account for the apparent position of objects?
886. Although the image is inverted on the retina, we see objects *erect*, because all the images formed on the retina have the same relative position which the objects themselves have; and, as the rays all cross each other, the eye is directed upwards to receive the rays which proceed from the upper part of an object, and downwards to receive those which proceed from the lower part.

Why do we not see double with two eyes?
887. A distinct image is also formed on the retina of each eye; but, as the optic nerves of the two eyes unite, or cross each other, before they reach the brain, the impressions received by the two nerves are united, so that only one idea is excited, and objects are seen single. Although an object may be distinctly seen with only one eye, it has been calculated that the use of *both* eyes makes a difference of about one-twelfth. From the description now

given of the eye, it may be seen what are the defects which are remedied by the use of concave and convex lenses, and how the use of these lenses remedies them.

What defects of the eye are spectacles designed to remedy?

888. When the crystalline humor of the eye is too round, the rays of light which enter the eye converge to a focus *before* they reach the retina, and, therefore, the image will not be distinct; and when the crystalline humor is too *flat* (as is often the case with old persons), the rays will not converge on the retina, but tend to a point beyond it. A convex glass, by assisting the convergency of the crystalline lens, brings the rays to a focus on the retina, and produces distinct vision.

For what defects of the eye is there no remedy?

889. The eye is also subject to imperfection by reason of the humors losing their transparency, either by age or disease. For these imperfections no glasses offer a remedy, without the aid of surgical skill. The operation of couching and removing cataracts from the eye consists in making a puncture or incision through which the diseased part may escape. Its office is then supplied by a lens. If, however, the operator, by accident or want of skill, permit the vitreous humor to escape, the globe of the eye immediately diminishes in size, and total blindness is the inevitable result.

What is a single microscope?

890. A single microscope consists simply of a convex lens, commonly called a magnifying-glass; in the focus of which the object is placed, and through which it is viewed.

891. By means of a microscope the rays of light from an object are caused to diverge less; so that when they enter the pupil of the eye they fall parallel on the crystalline lens, by which they are refracted to a focus on the retina.

Explain Fig. 136.

892. Fig. 136 represents a convex lens, or single microscope, C P. The diverging rays from the object A B are refracted in their passage through the lens C P, and

made to fall parallel on the crystalline lens, by which they are refracted to a focus on the retina R R; and the image is thus magnified, because the divergent rays are collected by the lens and carried to the retina.

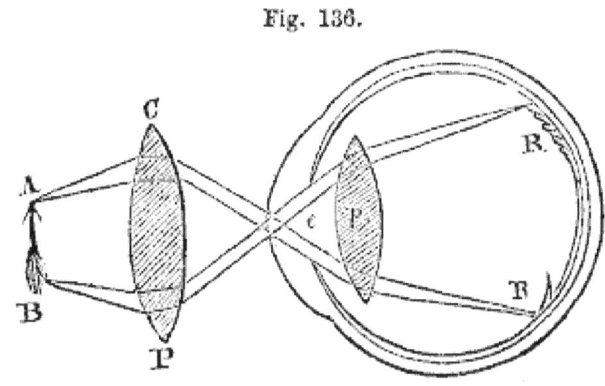

Fig. 136.

What glasses have the greatest magnifying powers? 893. Those lenses or microscopes which have the shortest focus have the greatest magnifying power; and those which are the most bulging or convex have the shortest focus. Lenses are made small because a reduction in size is necessary to an increase of curvature.

What is a double microscope? 894. A double microscope consists of two convex lenses, by one of which a magnified image is formed, and by the other this image is carried to the retina of the eye.

Explain Fig. 137. 895. Fig. 137 represents the effect produced by the lenses of a double microscope. The rays which diverge from the object A B are collected by the lens L M (called the object-glass, because it is nearest to the object), and form an

Fig. 137.

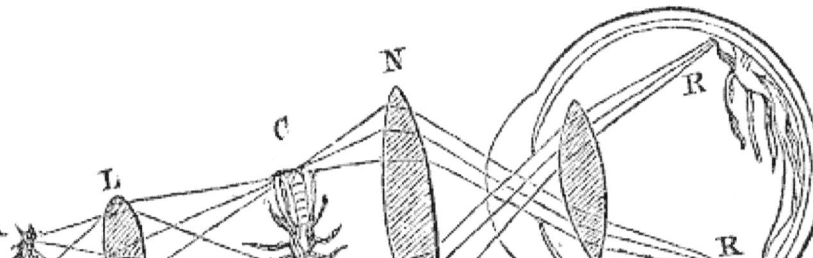

inverted magnified image at C D. The rays which diverge from this image are collected by the lens N O (called the eye-glass, because it is nearest to the eye), which acts on the principle of the single microscope, and forms a still more magnified image on the retina R R.

What is the solar microscope?   896. The solar microscope is a microscope with a mirror attached to it, upon a movable joint, which can be so adjusted as to receive the sun's rays and reflect them upon the object. It consists of a tube, a mirror or looking-glass, and two convex lenses. The sun's rays are reflected by the mirror through the tube upon the object, the image of which is thrown upon a white screen, placed at a distance to receive it.

897. The microscope, as above described, is used for viewing transparent objects only. When opaque objects are to be viewed, a mirror is used to reflect the light on the side of the object; the image is then formed by light reflected *from* the object, instead of being transmitted through it.

How is the magnifying power of single and double microscopes ascertained?   898. The magnifying power of a single microscope is ascertained by dividing the least distance at which an object can be distinctly seen by the naked eye by the focal distance of the lens. This, in common eyes, is about seven inches. Thus, if the focal distance of a lens be only $\frac{1}{4}$ of an inch, then the *diameter* of an object will be magnified 28 times (because 7 divided by $\frac{1}{4}$ is the same as 7 multiplied by 4), and the *surface* will be magnified 784 times.

The magnifying power of the compound microscope is found in a similar manner, by ascertaining the magnifying power, first of one lens, and then of the other.

The magnifying power of the solar microscope is in proportion as the distance of the image from the object-glass is greater than that of the object itself from it. Thus, if the distance of the object from the object-glass be $\frac{1}{4}$ of an inch, and the distance of the image, or picture, on the screen, be ten feet, or 120 inches, the object will be magnified in length 480 times, or in surface 230,000 times.

A lens may be caused to magnify or to diminish an object. If the object be placed at a distance from the focus of a lens, and the image be formed in or near the focus, the image will be diminished; but, if the object be placed near the focus, the image will be magnified.

What is the Magic Lantern? The Magic Lantern is an instrument constructed on the principle of the solar microscope, but the light is supplied by a lamp instead of the sun.

899. The objects to be viewed by the magic lantern are generally painted with transparent colors, on glass slides, which are

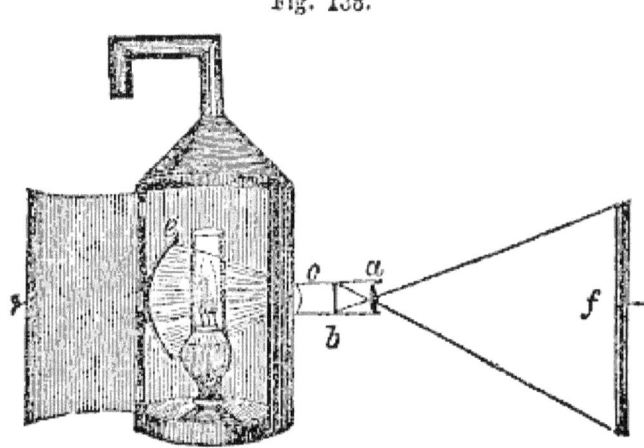

Fig. 138.

received into an opening in the front of the lantern. The light from the lamp in the lantern passes through them, and carries the pictures painted on the slides through the lenses, by means of which a magnified image is thrown upon the wall, on a white surface prepared to receive it.

Describe Fig. 138. Fig. 138 represents the magic lantern. The rays of light from the lamp are received upon the concave mirror *e*, and reflected to the convex lens *c*, which is called the condensing lens, because it concentrates a large quantity of light upon the object painted on the slide, inserted at *b*. The rays from the illuminated object at *b* are carried divergent through the lens *a*, forming an image on the screen at *f*. The image will increase or diminish in size, in proportion to the distance of the screen from the lens *a*.

246   NATURAL PHILOSOPHY.

How are "Dissolving Views" represented?

900. DISSOLVING VIEWS. — The exhibition called "*Dissolving Views*" is made by means of two magic lanterns of equal power, so as to throw pictures of the same magnitude in the same position on the screen. By the proper adjustment of sliding tubes and shutters, one picture on the screen is made brighter while the other becomes fainter, so that the one seems to dissolve into the other. In the hands of a skilful artist * this is an exhibition of the most pleasing kind.

What is a Telescope?

901. TELESCOPES. — A Telescope is an instrument for viewing distant objects, and causing them to appear nearer to the eye.

How are telescopes constructed?

902. Telescopes are constructed by placing lenses of different kinds within tubes that slide within each other, thus affording opportunity of adjusting the distances between the lenses within.

903. They are also constructed with mirrors, in addition to the lenses, so that, instead of looking directly at an object, the eye is directed to a magnified image of the object, reflected from a concave mirror. This has given rise to the two distinctions in the kinds of telescopes in common use, called respectively the Refracting and the Reflecting Telescope.

How many kinds of telescopes are there?

How is the Refracting Telescope constructed?

904. The Refracting Telescope is constructed with lenses alone, and the eye is directed toward the object itself.

How does a Reflecting Tele-

905. The Reflecting Telescope is constructed with one or more mirrors, in addi-

---

* Mr. John A. Whipple, of this city, has given several exhibitions of this kind, with great success. A summer scene seemed to dissolve into the same scene in mid-winter; a daylight view was gradually made to faint successively into twilight and moonshine; and many changes of a most interesting nature showed how pleasing an exhibition might be made by a skilful combination of science and art

scope differ from a Refracting? tion to the lenses; and the image of the object, reflected from a concave mirror, is seen, instead of the object itself.

906. Each of these kinds of telescope has its respective advantages, but refracting telescopes have been so much improved that they have in some degree superseded the reflecting telescopes.

What is an Achromatic Telescope? 907. Among the improvements which have been made in the telescope, may be mentioned, as the most important, that peculiar construction of the lenses by which they are made to give a pencil of white light, entirely colorless. Lenses are generally faulty in causing the object to be partly tinged with some color, which is imperfectly refracted. The fault has been corrected by employing a double object-glass, composed of two lenses of different refracting power, which will naturally correct each other. The telescopes in which these are used are called Achromatic. Common telescopes have a defect arising from the convexity of the object-glass, which, as it is increased, has a tendency to tinge the edges of the images. To remedy this defect, *achromatic* lenses were formed by the union of a convex lens of crown glass with a concave lens of flint glass. Owing to the difference of the refracting power of these two kinds of glass, the images became *free from color* and more distinct; and hence the glasses which produce them were called *Achromatic*, that is, *free from color*.

Lenses are also subject to another imperfection, called *spherical aberration*, arising from the different degrees of thickness in the centre and edges, which causes the rays that are refracted through them respectively, to come to different focuses, on account of the greater or less refracting power of these parts, consequent on their difference in thickness. To correct this defect, lenses have been constructed of gems and crystals, &c., which have a higher refractive power than glass, and require less sphericity to produce equal effects.

What is the simplest form of the telescope? 908. The simplest form of the telescope consists of two convex lenses, so combined as to increase the angle of vision under which the

248 NATURAL PHILOSOPHY.

object is seen. The lenses are so placed that the distance between them may be equal to the sum of their focal distances.

Which is the Object-glass, and which the Eye-glass, of a telescope?

909. The lens nearest to the eye is called the Eye-glass, and that at the other extremity is called the Object-glass.

How are objects seen through telescopes of the simplest construction?

910. Objects seen through telescopes of this construction (namely, with two glasses only) are always inverted, and for this reason this kind of instrument is principally used for astronomical purposes, in which the inversion of the object is immaterial. Hence, this is also called the Night-glass.

What is the difference between a day and a night telescope?

911. The common day telescope, or spy-glass, is an instrument of the same sort, with the addition of two, or even three or four glasses, for the purpose of presenting the object upright, increasing the field of vision, and diminishing the aberration caused by the dissipation of the rays.

Explain Fig. 139.

912. Fig. 139 represents a night-glass, or astronomical telescope. It consists of a tube A B C D, containing two glasses, or lenses. The lens A B, having a longer focus, forms the object-glass; the other lens D C is the eye-glass. The rays from a very

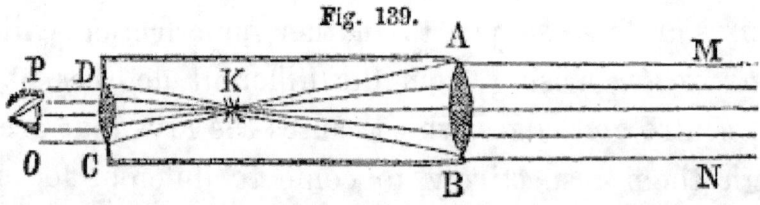

Fig. 139.

distant body, as a star, and which may be considered parallel to each other, are refracted by the object-glass A B to a focus at K. The image is then seen through the eye-glass D C, magnified as many times as the focal length of the eye-glass is contained in the focal length of the object-glass. Thus, if the focal length of the eye-glass D C be contained 100 times in that of the object-glass A B, the star will be seen magnified 100 times. It will be seen, by the figure, that the image is inverted; for the ray M A, after refraction, will be seen in the direction C O, and the ray N B in the direction D P.

Explain Fig. 140.
913. Fig. 140 represents a day-glass, or terrestrial telescope, commonly called a spy-glass.

This, likewise, consists of a tube A B H G, containing four lenses, or glasses, namely, A B, C D, E F, and G H. The lens A B is the object-glass, and G H the eye-glass. The two additional eye-glasses, E F and C D, are of the same size and shape, and placed at equal distances from each other

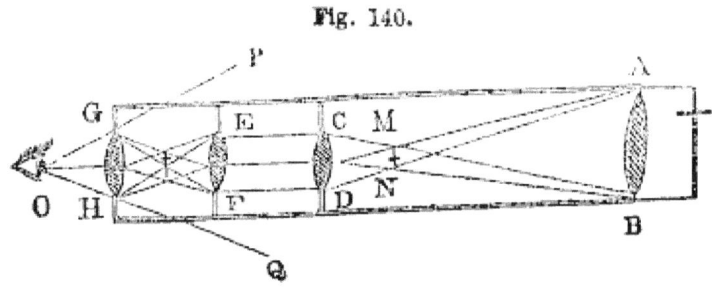
Fig. 140.

in such a manner that the focus of the one meets that of the next lens. These two eye-glasses E F and C D are introduced for the purpose of collecting the rays proceeding from the inverted image M N, into a new upright image, between G H and E F; and the image is then seen through the last eye-glass G H, under the angle of vision P O Q.

What are Opera Glasses?
Opera Glasses are constructed on the principle of the refracting telescope. They are, in fact, nothing more than two small telescopes, united in such a manner that the eye-glasses of each may be moved together, so as to be adjusted to the eyes of different persons.

Of what does the Reflecting Telescope consist?
914. THE REFLECTING TELESCOPE.—The Reflecting Telescope, in its simplest form, consisted of a concave mirror and a convex eye-glass. The mirror throws an image of the object, and the eye-glass views that image under a larger angle of vision.

This instrument was subsequently improved by Newton, and since him by Cassegrain, Gregory, Hadley, Short, and the Herschels.

Explain Fig. 141.
915. Fig. 141 represents the Gregorian Telescope. It consists of a large tube, containing two concave metallic mirrors, and two plano-convex eye-glasses. The rays from a distant object are received through the open end of the tube, and proceed from $r\ r$

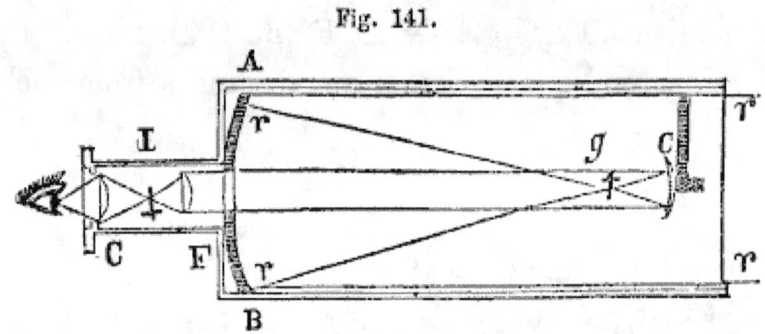

Fig. 141.

to $r\ r$, at the large mirror A B, which reflects them to a focus at g, whence they diverge to the small mirror C, which reflects them parallel to the eye-glass F, through a circular aperture in the middle of the mirror A B. The eye-glass F collects those reflected rays into a new image at I, and this image is seen magnified through the second eye-glass G.

It is thus seen that the mirrors bring the object near to the eye, and the eye-glasses magnify it. Reflecting telescopes are attended with the advantage that they have greater magnifying power, and do not so readily decompose the light. It has already been stated that the improvements in refractors have given them the greater advantage.

How does the Cassegrainian telescope differ from the Gregorian?
916. The Cassegrainian telescope differs from that which has been described, in having the smaller mirror convex. This construction is attended with two advantages; first, it is superior in distinctness of its images, and, second, it dispenses with the necessity of so long a tube.

917. The telescopes of Herschel and of Lord Rosse dispense with the smaller mirror. This is done by a slight inclination of the large mirror, so as to throw the image on one side, where it is viewed by the eye-glass. The observer sits with his back towards the object to be viewed. Herschel's gigantic telescope was erected at Slough, near Windsor, in 1789. The diameter of the speculum or mirror was four feet, and the mirror weighed 2118 pounds; its focal distance was forty feet.

What peculiarities are there in the telescopes of Herschel and the Earl of Rosse?

918. The telescope of Lord Rosse is the largest that has ever been constructed. The diameter of the speculum is six feet, and its focal distance fifty-six feet. The diameter of the tube is seven feet, and the tube and speculum weigh more than fourteen tons. The cost of the instrument was about sixty thousand dollars.

The telescope lately imported for Harvard University is a refractor. It is considered one of the best instruments ever constructed.

What is Chromatics?

919. CHROMATICS.—That part of the science of Optics which relates to colors is called Chromatics.

Of what is light composed?

920. Light is not a simple thing in its nature, but is composed of rays of different colors, each of which has different degrees of refrangibility, and has also certain peculiarities with regard to reflection.

Of what color are bodies composed?

921. Some substances reflect some of the rays that fall upon them and absorb the others, some appear to reflect all of them and absorb none, while others again absorb all and reflect none. Hence, bodies in general have no color of themselves, independent of light, but every substance appears of that color which it reflects.

What are white and black?

922. White is a due mixture of all colors in nice and exact proportion. When a body reflects all the rays that fall upon it, it will appear white, and the purity of the whiteness depends on the perfectness of the reflection.

923. Black is the deprivation of all color, and, when a body reflects none of the rays that fall upon it, it will appear black.

924. Some bodies reflect two or more colors either partially or perfectly, and they therefore present the varied hues which we perceive, formed from the mixture of rays of different colors.*

What are the colors of light? 925. The colors which enter into the composition of light, and which possess different degrees of refrangibility, are seven in number, namely, red, orange, yellow, green, blue, indigo, and violet.

What is a Prism? 926. A Prism is a solid, triangular piece of highly-polished glass.

927. A prism which will answer the same purpose as a solid one may be made of three pieces of plate glass, about six or eight inches long and two or three broad, joined together at their edges, and made water-tight by putty. The ends may be fitted to a triangular piece of wood, in one of which an aperture is made by which to fill

---

* When the eye has become fatigued by gazing intently on any object, of a red or of any other color, the retina loses, to some extent, its sensitiveness to that color, somewhat in the same manner that the ear is deafened for a moment by an overpowering sound. If that object be removed and another be presented to the eye, of a different color, into the composition of which red enters, the eye, insensible to the red, will perceive the other colors, or the compound color which they would form by the omission of the red, and the object thus presented would appear of that color. The truth of this remark may be easily tested. Fix the eye intently for some time on a red wafer on a sheet of white paper. On removing the wafer, the white disk beneath it will transmit all the colors of white light; but the eye, insensible to the red, will perceive the blue or green colors at the other end of the spectrum, and the other spot where the red wafer was will appear of a bluish-green, until the retina recovers its sensibility for red light. The colors thus substituted by the fatigued eye are called the accidental color.

The accidental colors of the seven prismatic colors, together with black and white, are as follows:

	Accidental Color.
Red	Bluish Green.
Orange	Blue.
Yellow	Indigo.
Green	Violet reddish.
Indigo	Orange red.
Violet	Orange yellow.
Black	White.
White	Black.

OPTICS. 253

it with water, and thus to give it the appearance and the refractive power of a solid prism.

What effect has a prism on the light that passes through it?
928. When light is made to pass through a prism, the different-colored rays are refracted or separated, and form an image on a screen or wall, in which the colors will be arranged in the order just mentioned.

Explain Fig. 142.
929. Fig. 142 represents rays of light passing from the aperture, in a window-shutter A B, through the prism P. Instead of continuing in a straight course to E, and there forming an image, they will be refracted, in their passage through the prism, and form an image on the screen C D. But,

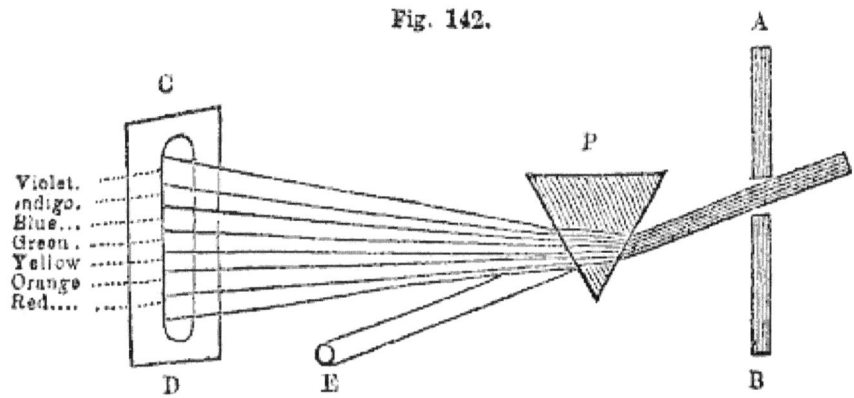

Fig. 142.

as the different-colored rays have different degrees of refrangibility, those which are refracted the least will fall upon the lowest part of the screen, and those which are refracted the most will fall upon the highest part. The red rays, therefore, suffering the smallest degree of refraction, fall on the lowest part of the screen, and the remaining colors are arranged in the order of their refraction.

930. It is supposed that the red rays are refracted the least, on account of their greater momentum; and that the blue, indigo and violet, are refracted the most, because they have the least momentum. The same reason, it is supposed, will account for the red appearance of the sun through a fog, or at rising and setting. The increased quantity of the atmosphere which the oblique rays must traverse, and its being loaded with mists and vapors, which are usually formed at those times, prevents the other rays from reaching us.

A similar reason will account for the blue appearance of the sky.

As these rays have less momentum, they cannot traverse the atmosphere so readily as the other rays, and they are, therefore, reflected back to our eyes by the atmosphere. If the atmosphere did not reflect any rays, the skies would appear perfectly black.

How can the rays refracted by a prism be reünited? 931. If the colored rays which have been separated by a prism fall upon a convex lens, they will converge to a focus, and appear white. Hence it appears that white is not a simple color, but is produced by the union of several colors.

932. The spectrum formed by a glass prism being divided into 360 parts, it is found that the red occupies 45 of those parts, the orange 27, the yellow 48, the green 60, the blue 60, the indigo 40, and the violet 80. By mixing the seven primitive colors in these proportions, a white is obtained; but, on account of the impurity of all colors, it will be of a dingy hue. If the colors were more clearly and accurately defined, the white thus obtained would appear more pure also. An experiment to prove what has just been said may be thus performed: Take a circular piece of board, or card, and divide it into parts by lines drawn from the centre to the circumference. Then, having painted the seven colors in the proportions above named, cause the board to revolve rapidly around a pin or wire at the centre. The board will then appear of a white color. From this it is inferred that the whiteness of the sun's light arises from a due mixture of all the primary colors.

933. The colors of all bodies are either the simple colors, as refracted by the prism, or such compound colors as arise from a mixture of two or more of them.

What are the three simple colors? 934. From the experiment of Dr. Wollaston, it appears that the seven colors formed by the prism may be reduced to four, namely, red, green, blue, and violet; and that the other colors are produced by combinations of these, but violet is merely a mixture of blue and red, and green is a mixture of blue and yellow. A better division of the simple colors is blue, yellow, and red.

935. Light is found to possess both heat and chemical action.

The prismatic spectrum presents some remarkable phenomena with regard to these qualities: for, while the red rays appear to be the seat of the maximum of heat, the violet, on the contrary, are the apparent seat of the maximum of chemical action.

936. Light, from whatever source it proceeds, is of the same nature, composed of the various-colored rays; and although some substances appear differently by candle-light from what they appear by day, this result may be supposed to arise from the weakness or want of purity in artificial light.

937. *There can be no light without colors, and there can be no colors without light.*

938. That the above remarks in relation to the colors of bodies are true, may be proved by the following simple experiment. Place a colored body in a dark room, in a ray of light that has been refracted by a prism; the body, of whatever color it naturally is, will appear of the color of the ray in which it is placed; for, since it receives no other colored rays, it can reflect no others.

939. Although bodies, from the arrangement of their particles, have a tendency to absorb some rays and reflect others, they are not so uniform in their arrangement as to reflect only pure rays of one color, and perfectly absorb all others; it is found, on the contrary, that a body reflects in great abundance the rays which determine its color, and the others in a greater or less degree in proportion as they are nearer or further from its color, in the order of refrangibility. Thus, the green leaves of a rose will reflect a few of the red rays, which will give them a brown tinge. Deepness of color proceeds from a deficiency rather than an abundance of reflected rays. Thus, if a body reflect only a few of the green rays, it will appear of a dark green. The brightness and intensity of a color shows that a great quantity of rays are reflected. That bodies sometimes change their color, is owing to some chemical change which takes place in the internal arrangement of their parts, whereby they lose their tendency to reflect certain colors, and acquire the power of reflecting others.

How is a rainbow produced? 940. The rainbow is produced by the refraction of the sun's rays in their passage through a shower of rain; each drop of which acts as a prism in separating the colored rays as they pass through it.

941. This is proved by the following considerations: *First*, a rainbow is never seen except when rain is falling and the sun shining at the same time; and that the sun and the bow are always in opposite parts of the heavens; and, *secondly*, that the same appearance may be produced artificially, by means of water thrown into the air, when the spectator is placed in a proper

position, with his back to the sun; and, *thirdly*, that a similar bow is generally produced by the spray which arises from large cataracts or waterfalls. The Falls of Niagara afford a beautiful exemplification of the truth of this observation. A bow is always seen there when the sun is clear and the spectator's back is towards the sun.

942. As the rainbow is produced by the refraction of the sun's rays, and every change of position is attended by a corresponding change in the rays that reach the eye, it follows that no two persons can see exactly the same rainbow, or, rather, the same appearance from the same bow.

943. POLARIZATION OF LIGHT. — The Polarization of Light is a change produced on light by the action of certain media, by which it exhibits the appearance of having polarity, or poles possessing different properties. This property of light was first discovered by Huygens in his investigations of the cause of double refraction, as seen in the Iceland crystal. The attention of the scientific world was more particularly directed to it by the discoveries of Malus, in 1810. The knowledge of this singular property of light has afforded an explanation of several very intricate phenomena in Optics, and has afforded corroborating evidence in favor of the undulatory theory; but the limits of this volume will not allow an extended notice of this singular property.

944. OF THE THERMAL, CHEMICAL, AND OTHER NON-OPTICAL EFFECTS OF LIGHT. — The science of Optics treats particularly of light as the medium of vision. But there are other effects of this agent, which, although more immediately connected with the science of chemistry, deserve to be noticed in this connexion.

945. The thermal effects of light, that is, its agency in the excitation of heat when it proceeds directly from the sun, are well known. But it is not generally known that these effects are extremely unequal in the differently colored rays, as they are refracted by the prism. It has already been stated that the red rays appear to possess the thermal properties in the greatest degree, and that in the other rays in the spectrum there is a decrease of thermal power towards the violet, where it ceases altogether. But, on the contrary, that the chemical agency is the most powerful in the violet, from which it constantly decreases towards the red, where it ceases altogether. Whether these thermal and chemical powers exist in all light, from whatever source it is derived, remains yet to be ascertained. The chromatic intensity of the colored spectrum is greatest in the yellow, from whence it decreases both ways, terminating almost abruptly in the red, and decreasing by almost imperceptible shades towards the violet, where it becomes faint, and then wholly indistinct. Thus it appears that the greatest heating power resides where the chemical power is feeblest, and the greatest chemical

power where the heating power is feeblest, and that the optical power is the strongest between the other two.

946. The chemical properties of light are shown in this, that the light of the sun, and in an inferior degree that of day when the sun is hidden from view, is a means of accelerating chemical combinations and decompositions. The following experiment exhibits the chemical effects of light:

Place a mixture of equal parts (by measure) of chlorine and hydrogen gas in a glass vessel, and no change will happen so long as the vessel be kept in the dark and at an ordinary temperature; but, on exposing it to the daylight, the elements will slowly combine and form hydrochloric acid; if the glass be set in the sun's rays, the union will be accompanied with an instantaneous detonation. The report may also be produced by transmitting ordinary daylight through violet or blue glass to the mixture, but by interposing a red glass between the vessel and the light all combination of the elements is prevented.

What is meant by Photography, or Heliography? 947. The chemical effects of light have recently been employed to render permanent the images obtained by means of convex lenses. The art of thus fixing them is termed Photography, or Heliography. These words are Greek derivatives; the former meaning *"writing or drawing by means of light,"* the latter, *"writing or drawing by the aid of the sun."*

Who is the author of Photography? 948. The mode in which the process is performed is essentially as follows: The picture, formed by a camera obscura, is received on a plate, the surface of which has been previously prepared so as to make it as susceptible as possible of the chemical influence of light. After the lapse of a longer or shorter time, the light will have so acted on the plate that the various objects the images of which were projected upon it will appear, with all their gradations of light and shade, most exactly depicted in black and white, no color being present. This is the process commonly known by the name of Daguerreotype, from M. Daguerre, the author of the discovery. Since his original discovery, he has ascertained that by isolating and electrifying the plate it acquires such a sensibility to the chemical influence of light that one-tenth of a second is a sufficient time to obtain the requisite luminous impression for the formation of the picture.

949. The chemical effects of light are seen in the varied colors of the vegetable world. Vegetables which grow in dark places are either white or of a palish-yellow. The sunny side of fruits is of a richer tinge than that which grows in the shade. Persons whose daily employment keeps them much within doors are pale, and more or less sickly, in consequence of such confinement.

From what has now been detailed with regard to the nature, the effects, and the importance of light, we may see with what reason the great epic poet of our language has apostrophized it in the words:

> "Hail, holy Light! offspring of Heaven, first born,
> Bright effluence of bright essence increate;"

and why the author of the "Seasons" has in a similar manner addressed it in the terms:

> "Prime cheerer, Light!
> Of all material beings first and best!
> Efflux divine! Nature's resplendent robe!
> Without whose vesting beauty all were wrapt
> In unessential gloom; and thou, O Sun!
> Soul of surrounding worlds, in whom best seen
> Shines out thy Maker! may I sing of thee?"

What is Electricity?

950. ELECTRICITY.— Electricity is the name given to an imponderable agent which pervades the material world, and which is visible only in its effects.

What are its simplest effects?

951. It is exceedingly elastic, susceptible of high degrees of intensity, with a tendency to equilibrium unlike that of any other known agent. Its simplest exhibition is seen in the form of attraction and repulsion.

952. If a piece of amber, sealing-wax, or smooth glass, perfectly clean and dry, be briskly rubbed with a dry woollen cloth, and immediately afterwards held over small and light bodies, such as pieces of paper, thread, cork, straw, feathers, or fragments of gold-leaf, strewed upon a table, these bodies will be attracted, and fly towards the surface that has been rubbed, and adhere to it for a certain time.

953. The surfaces that have acquired this power of attraction are said to be *excited;* and the substances thus susceptible of being excited are called *electrics*, while those which cannot be excited in a similar manner are called *non-electrics*.

What are the electrical divisions of all substances?

954. The science of Electricity, therefore, divides all substances into two kinds, namely, *Electrics*, or those substances which can be excited, and *Non-electrics*, or those substances which cannot be excited.

955. The word Electricity is derived from a Greek word, which signifies amber, because this substance was supposed to possess, in a remarkable degree, the property of producing the fluid, when excited or rubbed. The property itself was first discovered by Thales of Miletus, one of the seven wise men of Greece. The word is now used to express both the fluid itself and the science which treats of it.

What are the prevailing theories of electricity?

956. The nature of electricity is entirely unknown. Some philosophers consider it a fluid; others consider it as two fluids of opposite qualities; and others again deny its materiality, and deem it, like attraction, a mere property of matter. The theory of Dr. Franklin was, that it is a *single* fluid, disposed to diffuse itself equally among all substances, and exhibiting its peculiar effects only when a body by any means becomes possessed of more or less than its proper share. That when any substance has more than its natural share it is *positively* electrified, and that when it has less than its natural share it is *negatively* electrified; that *positive* electricity implies a redundancy, and *negative* electricity a deficiency, of the fluid. The prevalent theory at the present day is that it consists of two fluids, bearing the names of positive and negative.

957. Professor Faraday has proposed a nomenclature of electricity, which has been adopted in some scientific treatises. From the Greek words ἤλεκτρον, (electricity, or *amber*, from which it was first produced), and ἰδὸς (a way or path), he formed the word *electrodes*, that is, ways or paths of electricity. The course of positive electricity he called *the anode* (from the Greek ἄνοδος, an ascending or entering way), and the course of the negative electricity *the cathode* (from the Greek καθοδος, a descending way, or path of exit). The terms positive and negative are, however, more frequently employed to designate the extremities of the channels through which electricity passes. Positive electricity is sometimes expressed by the term *plus*, or its character $+$; and negative electricity by the term *minus*, or its character $-$.

How may electricity be excited?

958. Electricity may be excited by several modes — as, 1st, *by friction*, whence it is called *Frictional Electricity*; 2dly, *by chemical action*, called, from its discoverers, *Galvanic*, or *Voltaic Electricity*; 3dly, *by the action of heat*, whence it is called

Thermo-Electricity; 4thly, *by Magnetism.* Frictional Electricity forms the subject of that branch of Electricity usually treated under the head of Natural Philosophy; *Electricity* excited by chemical action forms the subject of *Galvanism;* and Electricity produced by the agency of heat, or by Magnetism, is usually considered in connection with the subject of Electro-Magnetism. The intimate connection between these several subjects shows how close are the links of the chain by which all the departments of physical science are united.

What is meant by a Conductor and a Non-conductor of electricity?
959. The electric fluid is readily communicated from one substance to another. Some substances, however, will not allow it to pass through or over them, while others give it a free passage. Those substances through which it passes without obstruction are called *Conductors*, while those through which it cannot readily pass are called *Non-conductors;* and it is found, by experiment, that all *electrics** are *non-conductors*, and all *non-electrics* are good *conductors* of electricity.

960. The following substances are *electrics*, or non-conductors of electricity; namely,

Atmospheric air (when dry),	Feathers,
Glass,	Amber,
Diamond,	Sulphur,
All precious stones,	Silk,
All gums and resins,	Wool,
The oxides of all metals,	Hair,
Beeswax,	Paper,
Sealing-wax,	Cotton.

All these substances must be dry, or they will become more or less conductors.

* The terms "electrics" and "non-electrics" have fallen into disuse.

961. The following substances are non-electrics, or conductors of electricity; namely,

All metals,	Living animals,
Charcoal,	Vapor, or steam.

962. The following are imperfect conductors (that is, they conduct the electric fluid, but not so readily as the substances above mentioned); namely,

Water,	Common wood,
Green vegetables,	Dead animals,
Damp air,	Bone,
Wet wood,	Horn, &c.

All substances containing moisture.

When is a conductor said to be insulated? 963. When a conductor is surrounded on all sides by non-conducting substances, it is said to be *insulated*.

964. As glass is a non-conducting substance, any conducting substance surrounded with glass, or standing on a table or stool with glass legs, will be *insulated*.

965. As the air is a non-conductor when dry, a substance which rests on any non-conducting substance will be insulated, unless it communicate with the ground, the floor, a table, &c.

How is a conductor charged? 966. When a communication is made between a conductor and an excited surface, the electricity from the excited surface is immediately conveyed by the conductor to the ground; but, if the conductor be insulated, its whole surface will become electrified, and it is said to be charged.

What is the grand reservoir of electricity? 967. The earth may be considered as the principal reservoir of electricity; and when a communication exists, by means of any conducting substance, between a body containing more than its natural share of the fluid and the earth, the body will immediately lose its redundant quantity, and the fluid will escape to the earth. Thus, when a person holds a metallic tube to an excited surface, the electricity escapes from the surface to the tube, and passes from the tube through the person to the floor; and the floor being connected with the earth by conducting substances, such as the timbers, &c., which support the building, the electricity will finally pass off, by a regular succession of conducting substances, from the excited surface to the earth. But, if the chain of conducting substances be interrupted,—that is, if any non-conducting substance occur between the excited surface and the course which the fluid takes in its progress to the earth,—the conducting substances will be insulated, and become charged with electricity. Thus, if an excited surface be connected by a long chain to a metallic tube, and the metallic tube be held by a person who is standing on a stool with glass legs, or on a cake of sealing-wax, resin, or any other non-conducting substance, the electricity cannot pass to the ground, and the person, the chain and the tube, will all become electrified.

What is the simplest mode of exciting electricity?

968. The simplest mode of exciting electricity is by friction.

Thus, if a thick cylinder of sealing-wax, or sulphur, or a glass tube, be rubbed with a silk handkerchief, a piece of clean flannel, or the fur of a quadruped, the electric fluid will be excited, and may be communicated to other substances from the electric thus excited.

Whatever substance is used, it must be perfectly dry. If, therefore, a glass tube be used, it should previously be held to the fire, and gently warmed, in order to remove all moisture from its surface.

What is meant by Vitreous and Resinous electricity?

969. The electricity excited in glass is called the *Vitreous* or *positive* electricity; and that obtained from sealing-wax, or other resinous substances, is called *Resinous*, or *negative* electricity.

What are the effects when a body is charged with either kind of electricity?

970. The vitreous and resinous, or, in other words, the positive and negative electricities, always accompany each other; for, if any surface become positive, the surface with which it is rubbed will become negative, and if any surface be made positive, the *nearest* conducting surface will become negative; and, if positive electricity be communicated to one side of an *electric*, (as a pane of glass, or a glass vial), the opposite side will become negatively electrified, and the plate or the glass is then said to be charged.

971. When one side of a metallic, or other conductor, receives the electric fluid, its whole surface is instantly pervaded; but when an electric is presented to an electrified body, it becomes electrified in a small spot only.

What is the effect when two bodies oppositely electrified are united?

972. When two surfaces oppositely electrified are united, their powers are destroyed; and, if their union be made through the human body, it produces an affection of the nerves, called an electric shock.

What is the law of electrical attraction and repulsion?

973. Similar states of electricity repel each other; and dissimilar states attract each other.

Thus, if two pith-balls, suspended by a silk thread, are both positively or both negatively electrified, they will repel each other; but if one be positively and the other negatively electrified, they will attract each other.

What is the Leyden jar?

974. The Leyden jar is a glass vessel used for the purpose of accumulating the electric fluid, procured from excited surfaces.

Explain Fig. 143.

975. Fig. 143 represents a Leyden jar. It is a glass jar, coated both on the inside and the outside with tin-foil, with a cork, or wooden stopper, through

which a metallic rod passes, terminating upwards in a brass knob, and connected by means of a wire, at the other end, with the inside coating of the jar. The coating extends both on the inside and outside only to within two or three inches of the top of the jar. Thus prepared, when an excited surface is applied to the brass knob, or connected with it by any conducting surface, it parts with its electricity, the fluid enters the jar, and the jar is said to be charged.

Fig. 143.

When a jar is charged, where is the electricity? 976. When the Leyden jar is charged, the fluid is contained on the surface of the glass. The coating serves only as a conductor to the fluid; and, as this conductor within the glass is insulated, the fluid will remain in the jar until a communication be made, by means of some conducting substance, between the inside and the outside coating of the jar. If then a person apply one hand or finger to the brass knob, and the other to the outside coating of the jar, a communication will be formed by means of the brass knob with the inside and outside of the jar, and the jar will be discharged. *A vial or jar that is insulated cannot be charged.*

What is an Electrical Battery? 977. An electrical battery is composed of a number of Leyden jars connected together.

The inner coatings of the jars are connected together by chains or metallic bars attached to the brass knobs of each jar; and the outer coatings have a similar connection established by placing the vials on a sheet of tin-foil. The whole battery may then be charged like a single jar. For the sake of convenience in discharging the battery, a knob connected with the tin-foil on which the jars stand projects from the bottom of the box which contains the jars.

What is the jointed discharger? 978. The *jointed* discharger is an instrument used to discharge a jar or battery.

Explain Fig. 144. Fig. 144 represents the jointed discharger. It consists of two rods, generally of brass, terminating at one end in brass balls, and connected together at the other end by a joint, like that of a pair of tongs, allowing them to be opened or closed. It is furnished with a glass handle, to secure the person who holds it from the effects of a shock.

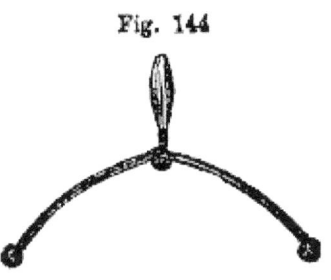

Fig. 144

When opened, one of the balls is made to touch the outside coating of the jar, or the knob connected with the bottom of the battery, and the other is applied to the knob of the jar or jars. A communication being thus formed between the inside and the outside of the jar, a discharge of the fluid will be produced.

Where must a body be placed, in order to receive a charge of electricity?

979. When a charge of electricity is to be sent through any particular substance, the substance must form a part of the *circuit of electricity*; that is, it must be placed in such a manner that the fluid cannot pass from the inside to the outside surface of the jar, or battery, without passing through the substance in its passage.

What effect have sharp metallic points?

980. Metallic rods, with sharp points, silently attract the electric fluid.

If the balls be removed from the jointed discharger, and the two rods terminate in sharp points, the electricity will pass off silently, and produce but little effect.

How may a Leyden jar or battery be silently discharged?

981. A Leyden jar, or a battery, may be silently discharged by presenting a metallic point, even that of the finest needle, to the knob; *but the point must be brought slowly towards the jar.*

On what principle are lightning-rods constructed?

982. It is on this principle that lightning-rods are constructed. The electric fluid is silently drawn from the cloud by the sharp points on the rods, and is thus prevented from suddenly exploding on high buildings.

What is meant by

983. Electricity of one kind or the other is generally *induced* in surrounding bodies by the *vicin-*

266   NATURAL PHILOSOPHY.

Electricity by Induction?   *ity* of a highly-excited electric. This mode of communicating electricity *by approach* is styled *induction*.

984. A body, on approaching another body powerfully electrified, will be thrown into a contrary state of electricity. Thus, a feather, brought near to a *glass tube* excited by friction, will be attracted to it; and, therefore, previously to its touching the tube, negative electricity must have been induced in it. On the contrary, if a feather be brought near to excited *sealing-wax*, it will be attracted, and, consequently, positive electricity must have been induced in it before contact.

What is Electricity by Transfer?   985. When electricity is communicated from one body to another *in contact* with it, it is called electricity *by transfer*.

What is an Electrical Machine, and on what principle is it constructed?   986. The electrical machine is a machine constructed for the purpose of accumulating or collecting electricity, and transferring it to other substances.

987. Electrical Machines are made in various forms, but all on the same principle, namely, the attraction of metallic points. The electricity is excited by the friction of silk on a glass surface, assisted by a mixture or preparation called an amalgam, composed of mercury, tin, and zinc. That recommended by Singer is made by melting together one ounce of tin and two ounces of zinc, which are to be mixed, while fluid, with six ounces of mercury, and agitated in an iron or thick wooden box, until cold. It is then to be reduced to a very fine powder in a mortar, and mixed with a sufficient quantity of lard to form it into a paste.

The glass surface is made either in the form of a cylinder or a circular plate, and the machine is called a cylinder or a plate machine, according as it is made with a cylinder or with a plate.

Explain Fig. 145.   988. Fig. 145 represents a plate electrical machine. A D is the stand of the machine, L L L L are the four glass legs, or posts, which support and insulate the parts of the machine. P is the glass plate (which in some machines is a hollow cylinder) from which the electricity is excited. and H is the handle by which the plate (or cylinder) is turned. R is a leather cushion, or rubber, held closely to both sides of the glass plate by a brass clasp, supported by the post G L, which is called the rubber-post. S is a silk bag, embraced by the same clasp that holds the leather cushion or rubber; and it is connected by strings S S S attached to its three other corners, and to the legs L L and the fork F of the prime conductor. C is the prime conductor, terminating at one end with a movable

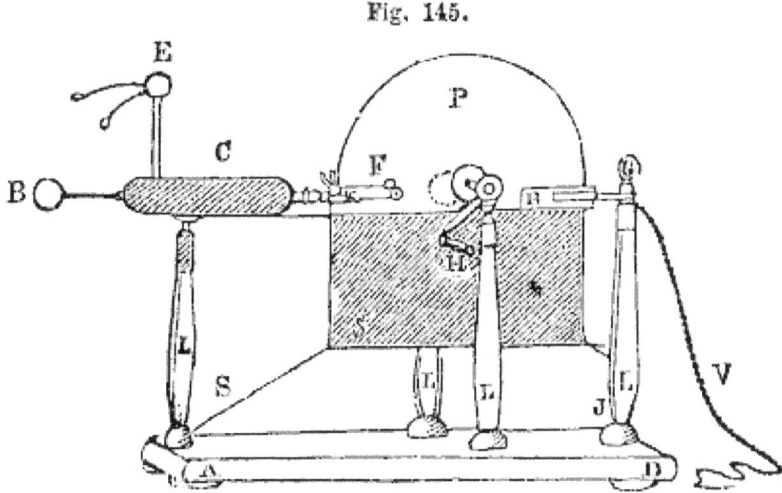

Fig. 145.

brass ball, B, and at the other by the fork F, which has one prong on each side of the glass plate. On each prong of the fork there are several sharp points projecting towards the plate, to collect the electricity as it is generated by the friction of the plate against the rubber. V is a chain or wire, attached to the brass ball on the rubber-post, and resting on the table or the floor, designed to convey the fluid from the ground to the plate. When negative electricity is to be obtained, this chain is removed from the rubber-post and attached to the prime conductor, and the electricity is to be gathered from the ball on the rubber-post.

Explain the operation of

989. OPERATION OF THE MACHINE. — By turning the handle H, the glass plate is pressed by the rub-

the Electrical Machine. ber. The friction of the rubber against the glass plate (or cylinder) produces a transfer of the electric fluid from the rubber to the plate; that is, the cushion becomes negatively and the glass positively electrified. The fluid which thus adheres to the glass, is carried round by the revolution of the cylinder; and, its escape being prevented by the silk bag, or flap, which covers the plate (or cylinder) until it comes to the immediate vicinity of the metallic points on the fork F, it is attracted by the points, and carried by them to the prime conductor. Positive electricity is thus accumulated on the prime conductor, while the conductor on the rubber-post, being deprived of this electricity, is negatively electrified. The fluid may then be collected by a Leyden jar from the prime conductor, or conveyed, by means of a chain attached to the prime conductor, to any substance which is to be electrified. If both of the conductors be insulated, but a small portion of the electric fluid can be excited; for this reason, *the chain must in all cases be attached to the rubber-post, when positive electricity is required, and to the prime conductor when negative electricity is wanted.*

What is an Electrometer, and on what principle is it constructed? 990. On the prime conductor is placed an Electrometer, or *measurer of electricity.* It is made in various forms, but always on the principle that similar states of electricity repel each other.

It sometimes consists of a single pith-ball, attached to a light rod in the manner of a pendulum, and behind is a graduated arc, or circle, to measure the repulsive force by degrees. Sometimes it is more simply made (as in the figure), consisting of a wooden ball mounted on a metallic stick, or wire, having two pith-balls, suspended by silk, hair, or linen threads. When the machine is worked, the pith-balls, being both similarly electrified, repel each other; and this causes them to fly apart, as is represented in the figure; and they will continue elevated until the electricity is drawn off. But, if an uninsulated conducting substance touch the prime conductor, the pith-balls will fall. The height to which the balls rise, and the quickness with which they are elevated, afford some test of the power of the machine. This simple apparatus may be attached to any body the electricity of which we wish to measure.

The balls of the electrometer, when elevated, are attracted by any resinous substance, and repelled by any vitreous substance that has been previously excited by friction.

991. If an electric, or a non-conductor, be presented to the prime conductor, when charged, it will produce no effect on the balls; but if a non-electric, or any conducting substance, be presented to the conductor, the balls of the electrometer will fall. This shows that the conductor has parted with its electricity, and that the fluid has passed off to the earth through the substance, and the hand of the person presenting it.

Describe Bennett's Electroscope. 992. An *Electroscope* is an instrument, of more delicate construction, to detect the presence of electricity. The most sensitive of this kind of apparatus is that called Bennett's Gold-leaf Electroscope, improved by Singer. It consists of two strips of gold-leaf suspended under a glass covering, which completely insulates them. Strips of tin-foil are attached to the sides of the glass, opposite the gold-leaf, and when the strips of gold-leaf diverge, they will touch the tin-foil, and be discharged. A pointed wire surmounts the instrument, by which the electricity of the atmosphere may be observed.

993. An Electrophorus is a simple apparatus by which small portions of electricity may be generated by induction. It consists of a disc, or circular cake of resinous substance,* on which is laid a smaller circular disc of metal, with a glass handle. Rub the resinous disc with hair or the fur of some animal, and the metallic disc, being pressed down on the resin by the finger, may then be raised by the glass handle. It will contain a small portion of electricity, which may be communicated to the Leyden jar, and thus the jar may *slowly* be charged.

* A mixture of Shell-lac, resin and Venice-turpentine, cast in a tin mould.

994. EXPERIMENTS WITH THE ELECTRICAL MACHINE. — In peforming experiments with the Electrical Machine, great care must be taken that all its parts be perfectly dry and clean. Moisture and dust, by carrying off the electricity as fast as it is generated, prevent successful action. Clear and cold weather should be chosen, if possible, as the machine will always perform its work better then.

995. When the machine is turned, if a person touch the prime conductor, the fluid passes off through the person to the floor without his feeling it. But if he present his finger, his knuckle, or any part of the body, *near* to the conductor, without touching it, a spark will pass from the conductor to the knuckle, which will produce a sensation similar to the pricking of a pin or needle.

996. If a person stand on a stool with glass legs, or any other non-conductor, he will be *insulated*. If in this situation he touch the prime conductor, or a chain connected with it, when the machine is worked, sparks may be drawn from any part of the body in the same manner as from the prime conductor. While the person remains insulated, he experiences no sensation from being filled with electricity; or, if a metallic point be presented to any part of his body, the fluid may be drawn off silently, without being perceived. But if he touch a blunt piece of metal, or any other conducting substance, or if he step from the stool to the floor, he will feel the electric shock; and the shock will vary in force according to the quantity of fluid with which he is charged.

997. THE TISSUE FIGURE. Fig. 146 is a figure with a dress of fancy paper cut into narrow strips. When placed on the prime conductor, or, being insulated, is connected with it, the strips being all electrified will recede and form a sphere around the head. On presenting a metallic point to the electrified strips, very singular combinations will take place. If the electrometer be

Fig. 146.

removed from the prime conductor, and a tuft of feathers, or hair, fastened to a stick or wire, be put in its place, on turning the machine the feathers or hair will become electrified, and the separate hairs will rise and repel each other. A toy is in this way constructed, representing a person under excessive fright. On touching the head with the hand, or any conducting substance not insulated, the hair will fall.

How is the Leyden jar charged? 998. The Leyden jar may be charged by presenting it to the prime conductor when the machine is worked. If the ball of the jar touch the prime conductor it will receive the fluid silently; but, if the ball of the jar be held at a small distance from the prime conductor, the sparks will be seen darting from the prime conductor to the jar with considerable noise.

999. The jar may in like manner be filled with negative electricity by applying it to the ball on the rubber-post, and connecting the chain with the prime conductor.

1000. If the Leyden jar be charged from the prime conductor (that is, with positive electricity), and presented to the pith-balls of the electrometer, they will be repelled; but if the jar be charged from the brass ball of the rubber-post (that is, with negative electricity), they will be attracted.

1001. If the ball of the prime conductor be removed, and a pointed wire be put in its place, the current of electricity flowing from the point when the machine is turned may be perceived by placing a lighted lamp before it; the flame will be blown *from* the point; and this will be the case in what part soever of the machine the point is placed, whether on the prime conductor or the rubber; or if the point be held in the hand, and the flame placed between it and the machine, thus showing that in all cases the fluid is blown *from* the point. Delicate apparatus may be put in motion by the electric fluid when issuing from a point. In this way electrical orreries, mills, &c., are constructed.

1002. If the electrometer be removed from the prime con-

272   NATURAL PHILOSOPHY.

ductor, and a pointed wire be substituted for it, a wire with sharp points bent in the form of an S, balanced on it, will be made to revolve rapidly. In a similar manner the motion of the sun and the earth around their common centre of gravity, together with the motion of the earth and the moon, may be represented. This apparatus is sometimes called an Electrical Tellurium. It may rest on the prime conductor or upon an insulated stand.

Describe Fig. 147. 1003. A chime of small bells on a stand, Fig. 147, may also be rung by means of brass balls suspended from the revolving wires. The principle of this revolution is similar to that mentioned in connection with the revolving jet, Fig. 98, which is founded on the law that action and reaction are equal and in opposite directions.

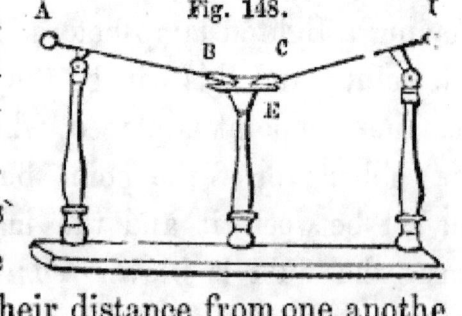

Fig. 147.

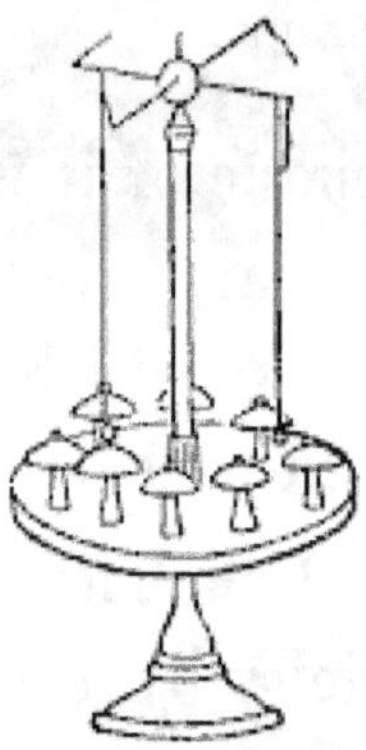

1004. If powdered resin be scattered over dry cotton-wool, loosely wrapped on one end of the jointed discharger, it may be inflamed by the discharge of the battery or a Leyden jar. Gunpowder may be substituted for the resin.

1005. The *universal discharger* is an instrument for directing a charge of electricity through any substance, with certainty and precision.

Explain Fig. 148. 1006. It consists of two sliding rods, A B and C D, terminating at the extremities, A and B, with brass balls, and at the other ends which rest upon the ivory table or stand E, having a fork, to which any small substance may be attached. The whole is insulated by glass legs, or pillars. The rods slide through collars, by which means their distance from one another may be adjusted.

Fig. 148.

1007. In using the universal discharger one of the rods or slides must be connected by a chain, or otherwise, with the outside, and the other with the inside coating of the jar or battery. By this means the substance through which the charge is to be sent is placed within the electric circuit.

1008. By means of the universal discharger, any small metallic substance may be burnt. The substance must be placed in the forks of the slides, and the slides placed within the electric circuit, in the manner described in the last paragraph. In the same manner, by bringing the forks on the slides into contact with a substance placed upon the ivory stand of the discharger, such as an egg, a piece of a potato, water, &c., it may be illuminated.

1009. Ether or alcohol may be inflamed by a spark communicated from a person, in the following manner: The person standing on the insulating stool receives the electric fluid from the prime conductor by touching the conductor or any conducting substance in contact with it; he then inserts the knuckles of his hand in a small quantity of sulphuric ether, or alcohol, held in a shallow metallic cup, by another person, who is not insulated, and the ether or alcohol immediately inflames. In this case the fluid passes from the conductor to the person who is insulated, and he becomes charged with electricity. As soon as he touches the liquid in the cup, the electric fluid, passing from him to the spirit, sets it on fire.

1010. The electrical bells are designed to show the effects of electrical attraction and repulsion.

1011. In some sets of instruments, the bells are insulated on a separate stand; but the mode here described is a convenient mode of connecting them with the prime conductor.

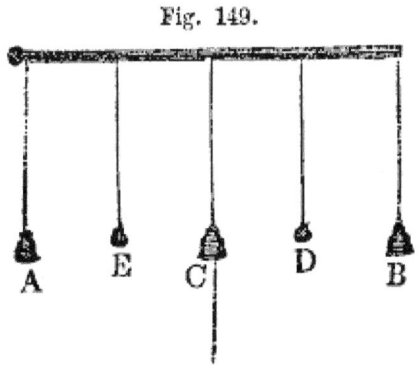

Fig. 149.

Explain Fig. 149

1012. They are thus to be applied: The ball B of the prime conductor, with its rod, is to be unscrewed, and the rod on which the bells are suspended is to be screwed in its place. The middle bell is to be connected by a chain with the table or the floor. When the machine is turned, the balls suspended between the bells will be alternately attracted and repelled by the bells, and cause a constant ringing. If the battery be charged, and connected with the prime conductor, the bells will continue to ring until all the fluid from the battery has escaped.

It may be observed, that the fluid from the prime conductor passes readily from the two outer bells, which are suspended by chains; they, therefore, attract the two balls towards them. The balls, becoming electrified by contact with the outer bells, are repelled by them, and driven to the middle bell, to which they communicate their electricity; having parted with their electricity, they are repelled by the middle bell, and again attracted by the outer ones, and thus a constant ringing is maintained. The fluid which is communicated to the middle bell, is conducted to the earth by the chain attached to it.

Explain what Fig. 150 represents.

1013. SPIRAL TUBE.—The passage of the electric fluid from one conducting substance to another, is beautifully exhibited by means of a glass tube, having a brass ball at each end, and coated in

Fig. 150.

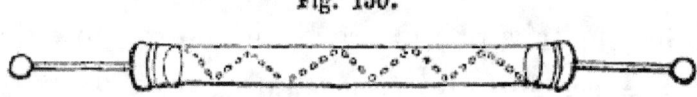

the inside with small pieces of tin-foil, placed at small distances from each other in a spiral direction, as represented in Fig. 150.

1014. In the same manner various figures, letters and words, may be represented, by arranging similar pieces of tin-foil between two pieces of flat glass. These experiments appear more brilliant in a darkened room.

Explain Fig. 151.

1015. THE HYDROGEN PISTOL.—The hydrogen pistol is made in a variety of forms, sometimes in the exact form of a pistol, and sometimes in

ELECTRICITY. 275

the form of a piece of ordnance. The form in Fig. 151 is a simple and cheap contrivance, and is sufficient to explain the manner in which the instrument is to be used in any of its forms. It is to be filled with hydrogen gas, and a cork inserted, fitting tightly. When thus prepared, if the insulated knob K be presented to the prime conductor, it will immediately explode.

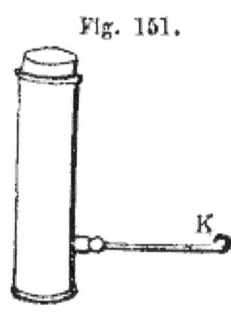
Fig. 151.

Explain Fig. 152.

1016. A very convenient and economical way of procuring hydrogen gas for this and other experiments, is by means of *the hydrogen gas generator*, as represented in Fig. 152. It consists of a glass vessel, with a brass cover, in the centre of which is a stop-cock; from the inside of the cover another glass vessel is suspended, with its open end downwards. Within this a piece of zinc is suspended by a wire. The outer vessel contains a mixture of sulphuric acid and water, about nine parts of water to one of acid. When the cover, to which the inner glass is firmly fixed, is placed upon the vessel, the acid, acting upon the zinc, causes the metal to absorb the oxygen of the water, and the hydrogen, the other constituent part of the water, being thus disengaged, rises in the inner glass, from which it expels the water; and when the stop-cock is turned the hydrogen gas may be collected in the hydrogen pistol, or any other vessel. In the use of hydrogen gas for explosion, it will be necessary to dilute the gas with an equal portion of atmospheric air.

Fig. 152.

Describe the Electrical Sportsman.

1017. ELECTRICAL SPORTSMAN.— Fig. 153 represents the Electrical Sportsman. From the larger ball of a Leyden jar two birds, made of pith (a substance procured in large quantities from the cornstalk, the whole of which, except the outside, is composed of *pith*), are suspended by a linen thread, silk, or hair. When the jar is charged, the birds will rise, as represented in the figure,

276 NATURAL PHILOSOPHY.

on account of the repulsion of the fluid in the jar.

1018. If the jar be then placed on the tin-foil of the stand, and the smaller ball placed within a half inch of the end of the gun, a discharge will be produced, and the birds will fall.

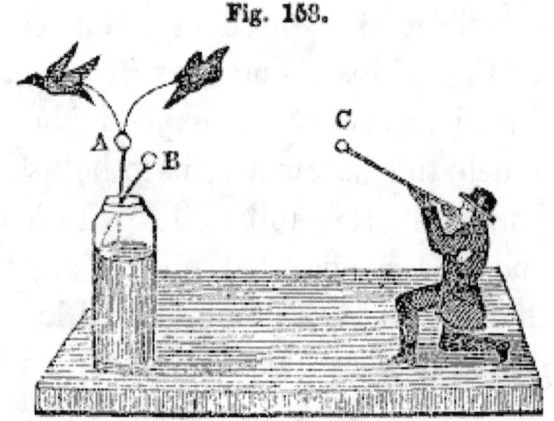

Fig. 153.

Explain Fig. 154.

1019. If images, made of pith, or small pieces of paper, are placed under the insulated stool, and a connection be made between the prime conductor and the top of the stool, the images will be alternately attracted and repelled; or, in other words, they will first rise to the electrified top of the stool, and thus becoming themselves electrified, will be repelled, and fall to the ground, the floor, or the table; where, parting with their electricity, they will again be attracted by the stool, thus rising and falling with considerable rapidity. In order to conduct this experiment successfully, the images, &c., must be placed within a short distance of the bottom of the stool.

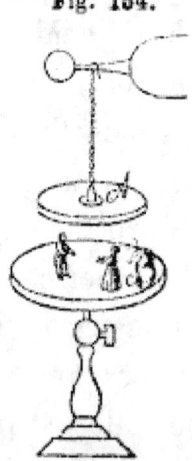

Fig. 154.

1020. On the same principle light figures may be made to dance when placed between two discs, the lower one being placed upon a sliding stand with a screw to adjust the distance, and the upper one being suspended from the prime conductor, as in Fig. 154.

1021. A hole may be perforated through a quire of paper by charging the battery, resting the paper upon the brass ball of the battery, and making a communication, by means of the jointed discharger, between the ball of one of the jars, and the brass ball of the box. The paper, in this case, will be between the ball of the battery and the end of the discharger.

1022. Gold-leaf may be forced into the pores of glass by placing it between two slips of window-glass, pressing the slips of glass firmly together, and sending a shock from a battery through them.

If gold-leaf be placed between two cards, and a strong charge be passed through them, it will be completely fused.

1023. When electricity enters at a point, it appears in the form of a star; but when it goes out from a point, it puts on the appearance of a brush.

Describe Fig. 155.
1024. The thunder-house, Fig. 155, is designed to show the security afforded by lightning-rods when lightning strikes a building. This is done by placing a highly-combustible material in the inside of the house, and passing a charge of electricity through it. On the floor of the house is a surface of tin-foil. The hydrogen pistol, being filled with hydrogen gas from the gasometer, must be placed on the floor of the thunder-house, and connected with the wire on the opposite side.

Fig. 155.

The house being then put together, a chain must be connected with the wire on the side opposite to the lightning-rod, and the other end placed in contact either with a single Leyden jar or with the battery. When the jar, thus situated, is charged, if a connection be formed between the jar and the points of the lightning-rod, the fluid will pass off silently, and produce no effect. But, if a small brass ball be placed on the points of the rod, and a charge of electricity be sent to it from the jar or the battery, the gas in the pistol will explode, and throw the parts of the house asunder with a loud noise.

1025. The success of this experiment depends upon the proper connection of the jar with the lightning-rod and the electrical pistol On the side of the house opposite to the lightning-rod there is a wire, passing through the side, and terminating on the outside in a hook. When the house is put together, this wire, in the inside, must touch the tin-foil on the floor of the house. The hydrogen pistol must stand on the tin-foil, and its insulated knob, or wire, projecting from its side, must be connected with the lower end of the lightning-rod, extending into the inside of the house. A communication must then be made between the hook on the outside of the house and the outside of the jar, or battery. This is conveniently done by attaching one end of a chain to the hook, and holding the other end in the hand against the side of a charged jar. By presenting the knob of the jar to the points of the lightning-rod no effect is produced; but if a brass ball be placed on the points at P, and the knob of the jar be presented to the ball, the explosion will take place. If the charged jar be *very suddenly* presented to the points, the explosion *may* take place; and the jar *may* be silently discharged if it be brought very slowly to the ball. The thunder-house is sometimes put together with magnets.

What is lightning and thunder?
1026. The phenomena of lightning are caused by the rapid motion of vast quantities of electric matter. Thunder is the noise which accompanies the passage of electricity through the air.

What is supposed to be the cause of the northern lights?
1027. The *aurora borealis* (or northern lights) is supposed to be caused by the electric fluid passing through highly-rarefied air; and most of the great convulsions of nature, such as earthquakes, whirlwinds, hurricanes, water-spouts, &c., are generally accompanied by electricity, and often depend upon it.

1028. The electricity which a body manifests by being brought near to an excited body, without receiving a spark from it, is said to be acquired by *induction*. When an insulated but unelectrified conductor is brought near an insulated charged conductor, the end near to the excited conductor assumes a state of opposite electricity, while the farther end assumes the same kind of electricity,— that is, if the conductor be electrified positively, the unelectrified conductor will be negative at the nearer end, and positive at the further end, while the middle point evinces neither positive nor negative electricity. [See No. 993.

1029. The experiments which have now been described exemplify all the elementary principles of the science of electricity. These experiments may be varied, multiplied, and extended in innumerable forms, by an ingenious practical electrician. Among other things with which the subject may be made interesting, may be mentioned the following facts, &c.

1030. A number of feathers, suspended by strings from an insulated conducting substance, will rise and present the appearance of a flight of birds. As soon as the substance is discharged, the feathers will fall. The experiment may be varied by placing the sportsman on the prime conductor, without the use of the Leyden jar, to which the birds are attached.

1031. Instead of the Leyden jar, a plate of common glass (a pane of window-glass, for instance) may be coated on both sides with tin-foil, leaving the edges bare. A bent wire balanced on the edge of the glass, to the ends of which balls may be attached, with an image at each end, may be made to represent two persons tilting, on the same principle by which the electrical bells are made to ring.

1032. Miniature machinery has been constructed, in which the power was a wheel, with balls at the ends of the spokes, situated within the attractive influence of two larger balls, differently electrified. As the balls on the spokes were attracted by one of the larger balls, they changed their electrical state, and were attracted by the other, which, in its return, repelled them, and thus the motion being given to the wheel was communicated by cranks at the end of the axle to the saws above.

1033. When the hand is presented to the prime conductor, a spark is communicated, attended with a slightly painful sensation. But, if a pin or a needle be held in the hand with the point towards the conductor, neither spark nor pain will be perceived, owing to the attracting (or, perhaps, more properly speaking, the *receiving*) *power* of the point.

1034. That square rods are better than round ones to conduct electricity silently to the ground, and thus to protect buildings, may be proved by causing each kind of rod to approach the prime conductor when charged. It will thus be perceived that, while little effect is produced on the pith-balls of the electrometer by the near approach of the round rod, on the approach of the square one the balls will immediately fall. The round rod, also, will produce an explosion and a spark from the ball of the prime conductor, while the square one will draw off the fluid silently.

1035. The effects of pointed conductors upon clouds charged with electricity may be familiarly exemplified by suspending a small fleece of cotton-wool from the prime conductor, and other smaller fleeces from the upper one, by small filaments. On presenting a point to them they will be repelled, and all drawn together; but, if a blunt conductor approach them, they will be attracted.

1036. From a great variety of facts, it has been ascertained, that lightning-rods afford but little security to any part of a building beyond twenty feet from them; and that when a rod is *painted* it loses its conducting power.

What are the best kinds of lightning-rods? 1037. The lightning-rods of the most approved construction, and in strictest accordance with philosophical principles, are composed of *small* square rods, similar to nail-rods. They run over the building, and down each of the corners, presenting many elevated points in their course. At each of the corners, and on the chimneys, the rods should be elevated several feet above the building. If the rods are twisted, it will be an improvement, as thereby the sharp surfaces presented to collect the fluid will point in more varied directions.

1038. The removal of silk and woollen garments, worn during the day in cold weather, is often accompanied by a slight noise, resembling that of sparks issuing from a fire. A similar effect is produced on passing the hand softly over the back of a cat. These effects are produced by electricity.

1039. It may here be remarked, that the terms positive and negative, are merely relative terms, as applied to the subject of electricity. Thus, a body which is possessed of its natural share of electricity, is positive in respect to one that has less, and negative in respect to one that has more than its natural share of the fluid. So, also, one that has more than its natural share is positive with regard to one that has only its natural share, or less than its natural share, and negative in respect to one having a larger share than itself.

1040. The experiments with the spiral tube connected with Fig. 150 may be beautifully varied by having a collection of such tubes placed on a stand; and a jar coated with small strips, resembling a brick wall, presents, when it is charged, a beautiful appearance in the dark.

1041. The electric fluid occupies no perceptible space of time in its passage through its circuit. The rapidity of its motion has been estimated as high as 288,000 miles in a second of time. It always seems to prefer the shortest passage, when the conductors are equally good. Thus, if two, ten, a hundred, or a thousand or more persons, join hands, and be made part of the circuit of the fluid in passing from the inside to the outside of a Leyden jar, they will all feel the shock at the same moment of time. But, in its passage, the fluid always prefers the best conductors. Thus, if two clouds, differently electrified, approach one another, the fluid, in its passage from one cloud to the other, will sometimes take the earth in its course, because the air is a bad conductor.

1042. In thunder-storms the electric fluid sometimes passes from the clouds to the earth, and sometimes from the earth to the clouds, and sometimes, as has just been stated, from one cloud to the earth, and from the earth to another cloud.*

What are comparatively safe and unsafe positions during a thunder-storm? 1043. It is not safe, during a thunder-storm, to take shelter under a tree, because the tree attracts the fluid, and, the human body being a better conductor than the tree, the fluid will leave the tree and pass into the body.

It is also unsafe to hold in the hand edge-tools, or any sharp point which will attract the fluid.

The safest position that can be chosen during a thunder-storm is a recumbent posture on a feather bed; and in all situations a recumbent is safer than an erect position. No danger is to be apprehended from lightning when the interval between the flash and the noise of the explosion is as much as three or four seconds. This space of time may be conveniently measured by the beatings of the pulse, if no time-piece be at hand.

1044. Lightning-rods were first proposed by Dr. Franklin, to whom is also ascribed the honor of the discovery that thunder and lightning are the effects of electricity. He raised a kite, constructed of a silk handkerchief adjusted to two light strips of cedar, with a pointed wire fixed to it; and, fastening the end of the twine to a key, and the key, by means of a piece of silk lace, to a post (the silk lace serving to insulate the whole apparatus), on the approach of a

* Among the common effects of lightning one of the most familiar is its effect on milk. The reason that milk frequently turns sour during the prevalence of a thunder-storm, or when the air is surcharged with electricity, may be thus explained: The air consists of two gases, called oxygen and nitrogen, *mixed* together, but not chemically *combined*. Oxygen *combined* with nitrogen produces five deadly poisons; namely, nitrous oxide, nitric oxide, hyponitrous acid, nitrous acid, and nitric acid, according to the proportion of each gas which enters into the combination. The electric fluid causes these gases, which are merely *mixed* in the air, chemically to *combine*, and form an acid, which causes the milk to turn sour.

thunder-cloud, he was able to collect sparks from the key, to charge Leyden jars, and to set fire to spirits. This experiment established the identity of lightning and electricity. The experiment was a dangerous one, as was proved in the case of Professor Richman, of St. Petersburgh, who fell a sacrifice to his zeal for electrical science by a stroke of lightning from his apparatus.

What are the Electrical Animals? 1045. Among the most remarkable facts connected with the science of electricity, may be mentioned the power possessed by certain species of fishes of giving shocks, similar to those produced by the Leyden jar. There are three animals possessed of this power, namely, the Torpedo, the Gymnotus Electricus (or Surinam Eel), and the Silurus Electricus. But, although it has been ascertained that the Torpedo is capable of giving shocks to the animal system, similar to those of the Leyden jar, yet he has never been made to afford a spark, nor to produce the least effect upon the most delicate electrometer. The Gymnotus gives a small but perceptible spark. The electrical powers of the Silurus are inferior to those of the Torpedo or the Gymnotus, but still sufficient to give a distinct shock to the human system. This power seems to have been bestowed upon these animals to enable them to secure their prey, and to resist the attacks of their enemies. Small fishes, when put into the water where the Gymnotus is kept, are generally killed or stunned by the shock, and swallowed by the animal when he is hungry. The Gymnotus seems to be possessed of a new kind of sense, by which he perceives whether the bodies presented to him are conductors or not. The consideration of the electricity developed by the organs of these animals of the aquatic order, belongs to that department called *Animal Electricity*.

1046. It will be recollected that the phenomena which have now been described, with the exception of what has just been stated as belonging to animal electricity, belong to the subject of *frictional electricity*. But there are other forms in which this subtle agent presents itself, which are yet to be described, which show that its operations are not confined to beautiful experiments, such as have already been presented, nor to the terrific and tremendous effects that we witness in the storm and the thunder-gust. Its powerful agency works unseen on the intimate relations of the parts and properties of bodies of every description, effecting changes in their constitution and character so wonderfully minute, thorough and universal, that it may almost be considered as the chief agent of nature, the prime minister of Omnipotence, the vicegerent of creative power.

What is Galvanism? 1047. GALVANISM, OR VOLTAIC ELECTRICITY. — Galvanism, or Voltaic Electricity, is a branch of electricity which derives its name from Galvani, who first discovered the principles which form its basis.

1048. Dr. Aloysius Galvani was a Professor of Anatomy in Bologna, and made his discoveries about the year 1790. His wife, being consumptive, was advised to take, as a nutritive article of diet, some soup made of the flesh of frogs. Several of these animals, recently skinned for that purpose, were lying on a table in his laboratory, near an electrical machine, with which a pupil of the professor was amusing himself in trying experiments. While the machine was in action, he chanced to touch the bare nerve of the leg of one of the frogs with the blade of a knife that he held in his hand, when suddenly the whole limb was thrown into violent convulsions. Galvani, being informed of the fact, repeated the experiment, and examined minutely all the circumstances connected with it. In this way he was led to the discovery of the principles which form the basis of this science. The science was subsequently extended by the discoveries of Professor Volta, of Pavia, who first constructed the galvanic or voltaic pile, in the beginning of the present century.

To produce electricity mechanically (as has been stated under the head of frictional electricity), it is necessary to excite an electric or non-conducting substance by friction. But galvanic action is produced by the *contact* of different conducting substances having a chemical action on one another.

How does galvanism differ from frictional electricity? 1049. Frictional electricity is produced by the mechanical action of bodies on one another; but galvanism, or galvanic electricity, is produced by their chemical action.

What is the difference in the effects of frictional and 1050. The motion of the electric fluid, excited by galvanic power, differs from that explained under the head of frictional electricity in its in-

chemical electricity? tensity and duration; for, while the latter exhibits itself in sudden and intermitted shocks and explosions, the former continues in a constant and uninterrupted current so long as the chemical action continues, and is interrupted only by the separation of the substances by which it is produced.*

What is most sensitive to the galvanic fluid? 1051. The nerves and muscles of animals are most easily affected by the galvanic fluid; and the voltaic or galvanic battery possesses the most surprising powers of chemical decomposition.

How is the galvanic fluid excited? 1052. The galvanic fluid, or influence, is excited by the contact of pieces of different metal, and sometimes by different pieces of the same metal.

1053. If a living frog, or a fish, having a slip of tin-foil on its back, be placed upon a piece of zinc, spasms of the muscles will be excited whenever a communication is made between the zinc and the tin-foil.

1054. If a person place a piece of one metal, as a half-dollar, above his tongue, and a piece of some other metal (as zinc) below the tongue, he will perceive a peculiar taste; and, in the dark, will

---

* The different action of gravity on the particles of water while in the liquid state, and the same particles in the solid state in the form of ice, has been explained in the early pages of this volume. In the one case each particle gravitates independently, while in the form of ice they gravitate in one mass. The fall of a body of ice would therefore produce more serious injury than the fall of the same quantity of water in the liquid form. There is a kind of analogy (which, though not sufficient for a philosophical explanation, may serve to give an insight into the difference between the effects produced by frictional electricity and that obtained by chemical means,) between the gravitation of water and ice, respectively, and the motion of frictional and chemical electricity. If the water be dropped in an infinitely narrow stream, its effects, although mechanically equal, would be so gradual as to be imperceptible. So, also, if a given portion of electricity be set in motion as it were in one mass, and an equal quantity move in an infinitely narrow current, there will be a corresponding difference in its apparent results. The difference in *intensity* may perhaps be partially understood by this illustration, although a strict analogy may fail to have been made out, owing in part to the nature of an imponderable agent. A strict analogy cannot exist between the operations of two agents, one of which is ponderable and the other imponderable. But, that there is something like an analogy existing in the cases cited, will appear from statements which have been made on good authority, namely, that there is a greater quantity of electricity developed by the action of a single drop of acid on a very minute portion of zinc, than is usually brought into action in the darkest cloud that shrouds the horizon.

see a flash of light whenever the outer edges of the metals are in contact.

1055. A faint flash may be made to appear before the eyes by putting a slip of tin-foil upon the bulb of one of the eyes, a piece of silver in the mouth, and making a communication between them. In these experiments no effect is produced so long as the metals are kept apart; but, on bringing them into contact, the effects above described are produced.

What is essential to produce galvanic action? 1056. It is essential in all cases to have three elements to produce galvanic action. In the experiments which have already been mentioned in the case of the frogs, the fish, the mouth and the eye, the *moisture* of the animal, or of the mouth, supplies the place of the acid, so that the *three* constituent parts of the circle are completed.

How are the conductors of galvanism divided? 1057. The conductors of the galvanic fluid, like those of frictional electricity, are divided into the *perfect* and the *imperfect*. Metallic substances, plumbago and charcoal, the mineral acids and saline solutions, are *perfect* conductors. Water, oxydated fluids, as the acids, and all the substances that contain these fluids, alcohol, ether, sulphur, oils, resins and metallic oxydes, are *imperfect* conductors.

What kind of acid must be employed in galvanism? 1058. The acid employed in the galvanic circuit must always be one that has a strong affinity for one of the metals in the circuit. When zinc is employed, sulphuric acid may form one of the three elements, because that acid has a strong affinity for zinc.

What is a law of chemical action? 1059. *A certain quantity of electricity is always developed whenever chemical action takes place between a fluid and a solid body.* This is a general law of chemical action; and, indeed, it has been ascertained that there is so intimate a connection between electrical and chemical changes, that the chemical action can proceed only to a certain extent, unless the electrical equilibrium, which has been disturbed, be again restored. Hence, we find that in the simple, as well as in the compound galvanic circle, the oxydation of the zinc proceeds with activity whenever the galvanic circle is completed; and that it ceases, or at least takes place very slowly, whenever the circuit is interrupted.

What is necessary in order to excite galvanic action? 1060. To produce any galvanic action it is necessary to form what is called a galvanic circle; that is, a certain order or succession of substances capable of exciting electricity.

Of what is the simplest galvanic circle composed? 1061. The simplest galvanic circle is composed of three conductors, one of which must be solid, and one fluid; the third may be either solid or fluid.

What is the usual process for obtaining galvanic electricity? 1062. The process usually adopted for obtaining galvanic electricity is, to place between two plates of different kinds of metal a fluid capable of exerting some chemical action on one of the plates, while it has no action, or a different action, on the other. A communication is then formed between the two plates.

Explain Fig. 156. 1063. Fig. 156 represents a simple galvanic circle. It consists of a vessel containing a portion of diluted sulphuric acid, with a plate of zinc, Z, and of copper, C, immersed in it. The plates are separated at the bottom, and the circle is completed by connecting the two plates on the outside of the vessel by means of wires. The same effect will be produced, if, instead of using the wires, the metallic plates come into direct contact.

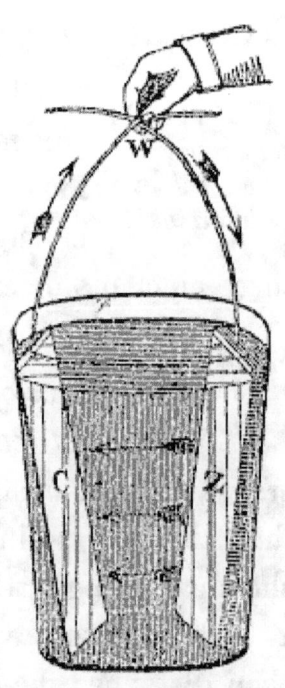

Fig. 156.

What are the essential parts of a galvanic circle? 1064. In the above arrangement, there are three elements or essential parts, namely, the zinc, the copper, and the acid. The acid, acting chemically upon the zinc, produces an alteration in the electrical state of the metal. The zinc, communicating its natural share of the electrical fluid to the acid, becomes *negatively* electrified. The copper, attracting the same fluid from the acid, becomes *positively* electrified. Any conducting substance, therefore, placed within the line of communication between the positive and negative points, will receive the charge thus to be obtained. The arrows in Fig. 156 show the direction of the current of positive electricity, namely, from the zinc to the fluid, from the fluid to the copper, from the copper back through the wires to the zinc, passing from zinc to copper *in* the acid, and from copper to zinc *out of* the acid. The substance submitted to the action of the electric current must be placed in the line of communication between the copper and the zinc. The wire connected with the copper is called the *positive pole*, and that connected with the zinc the *negative pole*, and in all cases *the substance submitted to galvanic action must be placed between the positive and negative poles.*

Where must a substance be placed to be affected by galvanic action?

1065. The electrical effects of a simple galvanic circle, such as has now been described, are, in general, too feeble to be perceived, except by very delicate tests. The muscles of animals, especially those of cold-blooded animals, such as frogs, &c., the tongue, the eye, and other sensitive parts of the body, being very easily affected, afford examples of the operation of simple galvanic circles. In these, although the quantity of electricity set in motion is exceedingly small, it is yet sufficient to produce very considerable effects; but it produces little or no effect on the most delicate electrometer.

How may galvanic action be increased?

1066. The galvanic effects of a simple circle may be increased to any degree, by a repetition of the same simple combination. Such repetitions constitute compound galvanic circles, and are called galvanic piles, or galvanic batteries, according to the mode in which they are constructed.

1067. It appears at first view to be a singular fact, that, in a *simple* galvanic circle, composed of zinc, acid and copper, the zinc end will always be negative, and the copper end positive; while, in all *compound* galvanic circles composed of the same elements, the zinc will be positive, and the copper *negative*. This apparent difference arises from the compound circle being usually terminated by two superfluous plates.

What is the Voltaic pile? 1068. The voltaic pile consists of alternate plates of two different kinds of metal, separated by woollen cloth, card, or some similar substance.

Explain Fig. 157. 1069. Fig. 157 represents a voltaic pile. A voltaic pile may be constructed in the following manner: Take a number of plates of silver, and the same number of zinc, and also of woollen cloth,— the cloth having been soaked in a solution of sal ammoniac in water.

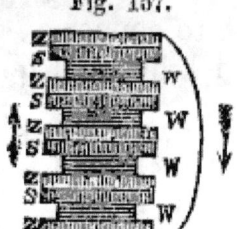

Fig. 157.

With these a pile is to be formed, in the following order, namely: a piece of silver, a piece of zinc, a piece of cloth, and thus repeated. These are to be supported by three glass rods, placed perpendicularly, with pieces of wood at the top and bottom, and the pile will then be complete, and will afford a constant current of electric fluid through any conducting substance. Thus, if one hand be applied to the lower plate, and the other to the upper one, a shock will be felt, which will be repeated as often as the contact is renewed.

Instead of silver, copper plates, or plates of other metal, may be used in the above arrangement. The arrows in the figure show the course of the current of electricity in the arrangement of silver, zinc, &c.

1070. Voltaic piles have been constructed of layers of gold and silver paper. The effect of such piles remains undisturbed for years. With the assistance of two such piles, an approximation to *perpetual motion*, in a self-moving clock, has been invented by an Italian philosopher. The motion is produced by the attraction and repulsion of the piles exerted on a pith-ball, on the principle of the electrical bells. The top of one of the piles was positive, and the bottom negative. The other pile was in an opposite state; namely, the top negative, and the bottom positive.

What is the galvanic battery? 1071. The voltaic, or galvanic battery, is a combination of metallic plates, immersed in pairs in a fluid which exerts a chemical action on one of each pair of the plates, and no action, or, at least, a different action, on the other.

What is the direction of the current in the galvanic battery? 1072. The electricity excited by the battery proceeds *from* the solid *to* the fluid, which acts upon it chemically. Thus, in a battery composed of zinc, diluted sulphuric acid and copper, the acid acts upon the zinc, and not on the copper. The galvanic fluid proceeds, therefore, from the zinc to the acid, from the acid to the copper, &c. Instead of using two different metals to form the galvanic circuit, one metal, in different states, may be employed; — the essential principle being, that one of the elements shall be more powerfully affected by some chemical agent than the other. Thus, if a galvanic pair be made of the same metal, one part must be softer than the other (as is the case with cast and rolled zinc); or a greater amount of surface must be exposed to corrosion on one side than on the other; or a more powerful chemical agent be used on one side, so that a current will be sent from the part most corroded, through the liquid, to the part least corroded, whenever the poles are united, and the circuit thereby completed.

Explain Fig. 158. 1073. Fig. 158 represents a voltaic battery. It consists of a trough made of baked wood, wedgewood-ware, or some other non-conducting substance. It is divided into grooves, or partitions, for the reception of the acid, or a saline solution, and the plates of zinc or copper (or other metal) are immersed by pairs in the grooves. These

Fig. 158.

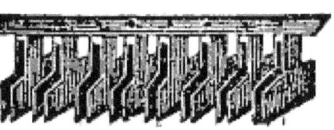

pairs of plates are united by a slip of metal passing from the one and soldered to the other; each pair being placed so as to enclose a partition between them, and each cell or groove in the trough containing a plate of zinc, connected with the copper plate of the succeeding cell, and a copper plate joined with the zinc plate of the preceding cell. These pairs must commence with copper and terminate with zinc, or commence with zinc and terminate with copper. The communication between the first and last plates is made by wires, which thus complete the galvanic circuit. The substance to be submitted to galvanic action is placed between the points of the two wires.

How can a compound battery of great power be obtained?
1074. A compound battery of great power is obtained by uniting a number of these troughs. In a similar manner, a battery may be produced by uniting several piles, making a metallic communication between the last plate of the one and the first plate of the next, and so on, taking care that the order of succession of the plates in the circuit be preserved inviolate.

Describe the Couronne des tasses.
1075. The *Couronne des tasses*, represented in Fig. 159, is another form of the galvanic battery. It consists of a number of cups, bowls, or glasses, with the zinc and copper plates immersed in them, in the order represented in the figure; Z indicating the zinc, and C the copper plates; the arrows denoting the course of the electric fluid.

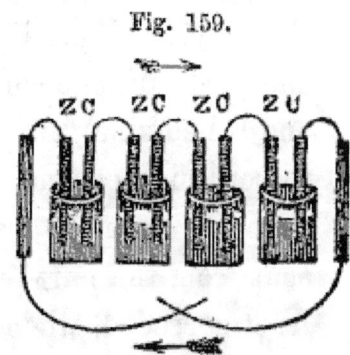

Fig. 159.

1076. The electric shock from the voltaic battery may be received by any number of persons, by joining hands, having previously wetted them.

Describe Smee's Battery.
1077. SMEE'S GALVANIC BATTERY is represented in Fig. 160, and affords an instance of a battery in its simplest form. It consists of a glass vessel (as a tumbler) on which rests the frame that supports the apparatus within

Two screw-cups rise from the frame, to which wires may be attached for the conveyance of the electric current in any direction. One of the screw-cups communicates with a thin strip of platinum, or platinum-foil, which is suspended within the glass vessel between two plates of zinc, thus presenting each surface of the platinum to a surface of zinc; and the galvanic action is in proportion to the extent of the opposite surfaces of the two metals, and their nearness to each other. The other screw-cup is connected with the two zinc plates. The screw-cup connected with the platinum is insulated from the metallic frame which supports it, by rosewood, and a thumb-screw confines the zinc plates, so that they can be renewed when necessary. The liquid employed for this battery is sulphuric acid, or oil of vitriol, diluted with ten parts of water by measure. To prevent the action of the acid upon the zinc plates, their surfaces are commonly amalgamated, or combined with mercury which prevents any chemical action of the acid with the zinc until the galvanic circuit is established, when the zinc is immediately attacked by the acid.

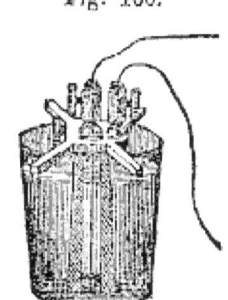

Fig. 160.

Explain Fig. 161. 1078. Fig. 161 represents a series of three pairs of this battery, in which it will be observed that the

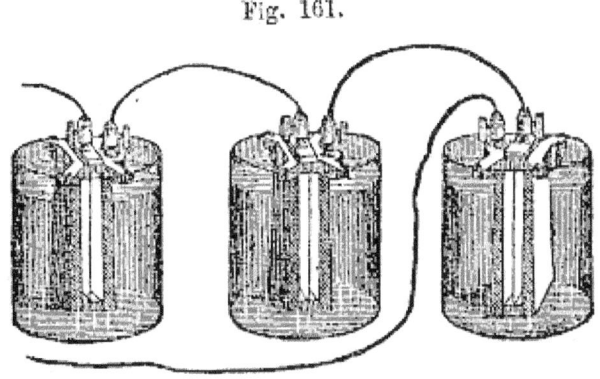

Fig. 161.

platinum of one is connected with the zinc of the next, and that the terminal wires proceed, consequently, one from a platinum plate, and the other from a zinc plate, as in a single pair.

292   NATURAL PHILOSOPHY.

Describe the sulphate of copper battery by Figures 162 and 163.

1078. SULPHATE OF COPPER BATTERY. — Fig. 162 represents a sulphate of copper battery, and Fig. 163 a vertical section of the same battery. It consists of a double cylinder of copper, C C, Fig. 163, with a bottom of the same metal, which serves the double purpose of a galvanic plate and a vessel to contain the exciting solution. The solution is contained in the space between the two copper cylinders. A movable cylinder of zinc, Z, is let down into the solution whenever the battery is to be used. It rests on three arms of wood or ivory at the top, by means of which it is insulated. Thus suspended in the solution, the surfaces of zinc and copper, respectively, face each other. A screw-cup, N, is attached to the zinc, and another, P, to the copper cylinder, to receive the wires. When a communication is made between the two cups, electricity is excited. The liquid employed in this battery is a solution of sulphate of copper (common blue vitriol) in water. A saturated solution is first made, and to this solution as much more water is added.

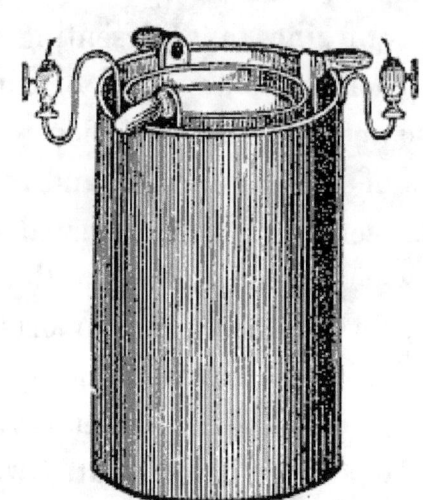
Fig. 162.

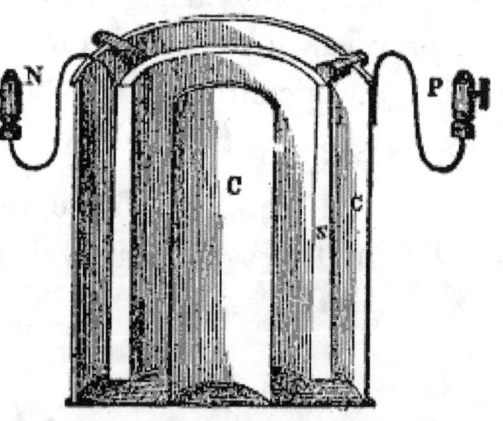
Fig. 163.

1079. A pint of water will dissolve about a quarter of a pound of blue vitriol. The solution described above will therefore contain about two ounces of the salt to the pint. The addition of alcohol in small quantities increases the permanency of the action of the solution. The zinc cylinder should always be taken out of the solution when the battery is not in use; but the solution may remain in the battery. The battery will keep in good action for twenty or thirty minutes at a time.

1080. The sulphate of copper battery, although not so energetic as Smee's, is found very convenient in a large class of experiments, and is particularly recommended to those who are inexpert in the use of acids; because the sulphate of copper, being entirely neutral, will not injure the color nor the texture of organic substances.

Describe the protected sulphate of copper battery.

1081. There is another form of the sulphate of copper battery, called the *Protected Sulphate of Copper Battery*, which differs from the one described in having a porous cell of earthenware, or leather, interposed between the zinc and the copper, thus forming two cells, in the outer of which sulphate of copper may be used, and in the inner one a solution of sulphate of soda (Glauber salt), or chloride of sodium (common salt), or even dilute sulphuric acid. This battery will continue in use for several days, and it is therefore of great use in the electrotype process.

Describe Grove's battery.

1082. GROVE'S BATTERY.—This is the most energetic battery yet known, and is the one most generally used for the magnetic telegraph. The metals employed are platinum and zinc, and the solutions are strong nitric acid in contact with the platinum, and sulphuric acid diluted with ten or twelve parts of water in contact with the zinc. This battery must be used with great care, on account of the strength of the acids used for the solutions, which send out injurious fumes, and which are destructive to organic substances. Fig. 164 represents Grove's battery. The containing vessel is glass; within this is a thick cylinder of amalgamated zinc, standing on short legs, and divided by a longitudinal opening on one side, in order to allow the acid to circulate freely. Inside of this is a porous cell of unglazed porcelain, containing the nitric acid, and strip of platinum. The platinum is supported by a strip of brass fixed by a thumb-screw and an insulating piece of ivory to the

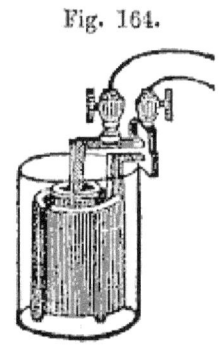

Fig. 164.

arm proceeding from the zinc cylinder. The amalgamated zinc is not acted upon by the diluted sulphuric acid until the circuit of the battery is completed. But, as the nitric acid will filter through the porous cell, and act upon the zinc, it is advisable to remove the zinc from the acid when the battery is to remain inactive. The action of Grove's battery may be considered as three times greater than that of the sulphate of copper battery.

What are the effects of a powerful voltaic battery?

1083. The spark from a powerful voltaic battery acts upon and inflames gunpowder, charcoal, cotton, and other inflammable bodies, fuses all metals, burns up or disperses diamonds and other substances on which heat in other forms produces little or no effect.

1084. The most striking effects of Galvanism on the human frame, after death, were exhibited at Glasgow, a few years ago. The subject on which the experiments were made was the body of the murderer Clydesdale, who was hanged at that city. He had been suspended an hour, and the first experiment was made in about ten minutes after he was cut down. The galvanic battery employed consisted of 270 pairs of four-inch plates. On the application of the battery to different parts of the body, every muscle was thrown into violent agitation; the leg was thrown out with great violence, breathing commenced, the face exhibited extraordinary grimaces, and the finger seemed to point out the spectators. Many persons were obliged to leave the room from terror or sickness; one gentleman fainted, and some thought that the body had really come to life.

How are the hands protected when using a battery?

1085. The wires, by which the circuit of the battery is completed, are generally covered with glass tubes, in order that they may be held or directed to any substance.

In what respects does the electricity produced by the galvanic battery differ from that obtained by the machine?

1086. There are three principal circumstances in which the electricity produced by the galvanic or voltaic battery differs from that obtained by the ordinary electrical machine; namely,

(1.) The very low degree of *intensity* of that produced by the galvanic battery, compared with that obtained by the machine.

1087. By *intensity* is here. meant something analogous to what is implied by density as applied to matter; but in the one case it is a ponderable agent, in the other an imponderable, so that a strict analogy cannot be made out between them. The term density cannot be applied to any of the imponderable agents, light, sound, heat or electricity. We speak of the *intensity* of light, an *intensity* of heat, &c. Hence, the word intensity is properly applied to electricity, and we speak of its *tension*, instead of its density.

Which will develop the greater quantity of electricity, the galvanic battery or the machine? The *quantity* of electricity obtained by galvanic action is much greater than can be obtained by the machine; but it flows, as it were, in narrow streams.

The action of the electrical machine may be compared to a mighty torrent, dashing and *exhausting itself* in one leap from a precipitous height. The galvanic action may be compared to a steady stream, supplied by an inexhaustible fountain. In other words, the *momentum* of the electricity excited by galvanism is less than that from the electrical machine; but the *quantity*, as has been stated, is greater.

(2.) The very large quantity of electricity which is set in motion by the voltaic battery; and,

(3.) The continuity of the current of voltaic electricity, and its perpetual reproduction, even while this current is tending to restore the equilibrium.

1088. Whenever an electrical battery is charged, how great soever may be the quantity that it contains, the whole of the power is *at once* expended, as soon as the circuit is completed. Its action may be sufficiently energetic while it lasts, but it is exerted only for an instant, and, like the destructive operation of lightning, can effect during its momentary passage only sudden and violent changes, which it is beyond human power to regulate or control. On the contrary, the voltaic battery continues, for an indefinite time, to develop and supply vast quantities of electricity, which, far from being lost by returning to the source, circulate in a perpetual stream, and with undiminished force. The effects of this continued current on the bodies subjected to its action will therefore be more definite, and will be constantly accumulating; and their amount, in process of time, will be incomparably greater than even those of the ordinary electrical explosion. It is therefore found that changes in the composition of bodies are effected by galvanism which can be accomplished by no other means. The science of galvanism therefore, has extended the field and multiplied the means of investigation in the kindred sciences, especially that of Chemistry.

How are attraction and repulsion manifested in the galvanic battery?

1089. A common electrical battery may be charged from a voltaic battery of sufficient size; but a battery constructed of a small number of pairs, even though the plates are large, furnishes no indication of attraction or repulsion equal to that which is given by the feeblest degree of excitation to a piece of sealing-wax. A galvanic battery consisting of fifty pairs of plates will affect a delicate gold-leaf electrometer; and, with a series of one thousand pairs, even pith balls are made to diverge.

On what does the effect of the voltaic battery depend?

1090. The effect of the voltaic pile on the animal body depends chiefly on the *number* of plates that are employed; but the intensity of the spark and its chemical agencies increase more with the *size* of the plates than with their number.

Mention some of the familiar effects of galvanism.

1091. Galvanism explains many facts in common life.

Porter, ale, or strong beer, is said to have a peculiar taste when drunk from a pewter vessel. The peculiarity of taste is caused by the galvanic circle formed by the pewter, the beer, &c., and the moisture of the under lip.

Works of metals the parts of which are soldered together soon tarnish in the places where the metals are joined.

Ancient coins composed of a mixture of metal have crumbled to pieces, while those composed of pure metal have been uninjured.

The nails and the copper in sheathing of ships are soon corroded about the place of contact. These are all the effects of galvanism.

There are persons who profess to be able to find out seams in brass and copper vessels by the tongue which the eye cannot discover; and, by the same means, to distinguish the base mixtures which abound in gold and silver trinkets.

1092. From what has now been stated, it will be seen that the effects of galvanic action depend on two circumstances; namely, 1st, the size of the plates employed in the circuit; and, 2dly, the number of the pairs constituting a battery. But there is a remarkable circumstance to be noticed in this connexion; namely, that there is one class of facts dependent on the extension of the size of the plates, and another on the increase of their number. *The power to develop heat and magnetism is dependent on the size of the plates*, that is, on the extent of the surface acted upon by the chemical agent; while the power to decompose chemical compounds, and to affect the animal system, is affected in a greater ratio by the increase of the number of the pairs.

On what does the power of a battery to produce heat and to affect the animal system respectively depend?

1093. The name *Calorimotor* (that is, *the mover of heat*) was applied by Dr. Hare, of Philadelphia, to a very powerful apparatus which he constructed, with large plates, and which he found possessed of a very remarkable power in producing heat. Batteries constructed for this purpose usually consist of from one to eight pairs of plates. They are made in various forms; sometimes the sheets of copper and zinc are coiled in concentric spirals, sometimes placed side by side; and they may be divided into a great number of small plates, *provided that all the zinc plates are connected together, and all the copper plates together, and*

What is a Calorimotor?

then that the experiments are performed in a channel of communication, opened between the SETS OF PLATES, *and not between* PAIRS, *as in the common battery;* for it is immaterial whether one large surface be used, or many small ones electrically connected together. The effect of all these arrangements, by which the metallic surface of a single pair is augmented, is to increase the quantity produced.

1094. The galvanic or voltaic battery is one of the most valuable acquisitions of modern science. It has proved in many instances the key by which science has entered into the innermost recesses of nature, and discovered the secret of many of her operations. It has, in great measure, lifted the hitherto impenetrable veil that has concealed the mysterious workings in the material world, and has opened a field for investigation and discovery as inviting as it is boundless. It has strengthened the sight and enlarged the view of the philosopher and the man of science, and given a degree of certainty to scientific inquiry hitherto known to be unreached, and supposed to be unattainable; and, if it has not yet satisfied the hopes of the alchemist, nor emulated the gold-converting touch of Midas, it has shown, almost to demonstration, that science may yet achieve wonders beyond the stories of mythology, and realize the familiar adage that "*truth is stranger than fiction.*"

What is Magnetism?
1095. MAGNETISM.— Magnetism treats of the properties and effects of the magnet, or loadstone.

1096. The term *loadstone*, or, more properly, leadstone, was applied to an ore of iron in the lowest state of oxidation, from its attractive properties towards iron, and its power of communicating its power to other masses of iron. It received the name of Magnet from Magnesia, in Asia Minor (now called Guzelhizar), about fifteen miles from Ephesus, where its properties were first well known. The term magnet is now applied to those substances which, naturally or artificially, are endowed either permanently or temporarily with the same attractive power.

1097. Certain ores of iron are found to be naturally possessed of magnetic properties, and are therefore called natural or native magnets, or loadstones. Besides iron and some of the compounds, *nickel*, and, perhaps, *cobalt*, also possess magnetic properties. But all conductors of electricity are capable of exerting the magnetic properties of attraction and repulsion while conveying a current of electricity, as will be shown under the head of Electro-Magnetism.

1098. That part of science which relates to the development of magnetism by means of a current of electricity will be noticed under the head of Electro-Magnetism, in which connexion will also be mentioned the development of electricity by magnetism, to which the term Magneto-Electricity has been applied.

What are the two kinds of magnets? 1099. There are two kinds of magnets, namely, the native or natural magnet, and the artificial.

1100. The native magnet, or loadstone, is an ore of iron, found in iron mines, and has the property of attracting iron, and other substances which contain it.

What is a permanent magnet? 1101. A permanent artificial magnet is a piece of iron to which permanent magnetic properties have been communicated.

Which is the more useful, the permanent or the artificial magnet? 1102. For all purposes of accurate experiment, the artificial is to be preferred to the native magnet.

1103. If a straight bar of soft iron be held in a vertical position (or, still better, in a position slightly inclined to the perpendicular, the lower end deviating to the north), and struck several smart blows with a hammer, it will be found to have acquired, by this process, all the properties of a magnet; or, in other words, it will become an artificial magnet.

What are the properties of a magnet? 1104. The properties of a magnet are,—polarity; attraction of unmagnetic iron; attraction *and repulsion* of *magnetic* iron; the power of communicating magnetism to other iron. Besides these properties, the magnet has recently been discovered to be possessed of electrical properties. These will be considered in another connexion.

What is the polarity of a magnet? 1105. By the *polarity* of a magnet is meant the property of pointing or turning to the north and south poles. The end which points to the north is called the north pole of the magnet, and the other the south pole.

1106. The attractive power of a magnet is generally stated to be greatest *at the poles;* but the actual poles, or points of greatest magnetic intensity, in a steel magnet, are not exactly at the ends, but a little within them.

How will a magnet move when freely suspended?
1107. When a magnet is supported in such a manner as to move freely, it will spontaneously assume a position directed *nearly* north and south.

What are the magnetic poles?
1108. The points to which the poles of a magnet turn are the *magnetic poles*. These do not exactly coincide with the astronomical poles of the earth; but, although the value of the magnetic needle has been predicated on the supposition that its polarity is a tendency to point exactly to the north and south poles of the earth, the recent discovery of the magnetic poles, as the points of attraction, has not depreciated the value of the compass, because the variation is known, and proper allowances can be made for such variation.

How are magnets supported?
1109. There are several ways of supporting a magnet, so as to enable it to manifest its polarity. *First*, by suspending it, accurately balanced, from a string. *Secondly*, by poising it on a sharp point. *Thirdly*, by attaching it to some buoyant substance, and allowing it to float freely on water.

What is the law of magnetic attraction and repulsion?
1110. Different poles of magnets attract, and similar poles repel each other.

There is here a close analogy between the attractive and repulsive powers of the positive and the negative forms of electricity, and the northern and southern polarities of the magnet. The same law obtains with regard to both; namely, *between like powers there is repulsion, between unlike there is attraction.*

1111. A magnet, whether native or artificial, attracts iron or steel which has no magnetic properties; but it both *attracts* and *repels* those substances when they are magnetic; that is the north pole of one magnet will attract the south pole of another, and the south pole of one will attract the north of another; but the north pole of the one *repels* the north pole of the other, and the south pole of one repels the south pole of another.

1112. If either pole of a magnet be brought near any small piece of soft iron, it will attract it. Iron filings will also adhere in clusters to either pole.

To what bodies are the magnetic properties most easily communicated?

1113. A magnet may communicate its properties to other unmagnetized bodies. But these properties can be generally conveyed to no other substances than iron, nickel or cobalt, without the aid of electricity.

Coulomb has discovered that "*all solid bodies are susceptible of magnetic influence.*" But the "*influence*" is perceptible only by the nicest tests, and under peculiar circumstances.

What are permanent magnets?

1114. All permanent natural and artificial magnets, as well as the bodies on which they act, are either iron in its pure state, or such compounds as contain it.

What effect has the use of a magnet on its power?

1115. The powers of a magnet are increased by action, and are impaired and even lost by long disuse.

What is a horse-shoe or U magnet?

1116. When the two poles of a magnet are brought together, so that the magnet resembles in shape a horse-shoe, or the capital letter U, it is called a horse-shoe magnet, or a U magnet; and it may be made to sustain a considerable weight, by suspending substances from a small iron bar, extending from one pole to the other. This bar is called the keeper. A small addition may be made to the weight every day.

1117. Soft iron acquires the magnetic power very readily, and also loses it as readily; hardened iron or steel acquires the property with difficulty, but retains it permanently.

What follows when a magnet is divided? 1118. When a magnet is broken or divided, each part becomes a perfect magnet, having both a north and south pole.

This is a remarkable circumstance, since the central part of a magnet appears to possess but little of the magnetic power; but, when a magnet is divided *in the centre*, this very part assumes the magnetic power, and becomes possessed in the one part of the north, and in the other of the south polarity.

1119. The magnetic power of iron or steel appears to reside wholly on the surface, and is independent of its mass.

In what do magnetism and electricity resemble each other? 1120. In this respect there is a strong resemblance between magnetism and electricity. Electricity, as has already been stated, is wholly confined to the surface of bodies. In a few words, magnetism and electricity may be said to resemble each other in the following particulars:

(1.) Each consists of two species, namely, the vitreous and the resinous (or, the positive and negative) electricities; and the northern or southern (sometimes called the *Boreal* and the *Austral*) polarity.

(2.) In both magnetism and electricity, those of the same name repel, and those of different names attract each other.

(3.) The laws of induction in both are similar.

(4.) The influence, in both cases (as has just been stated), resides at the *surface*, and is wholly independent of their *mass*.

What effect has heat on a magnet? 1121. Heat weakens, and a great degree of heat destroys the power of a magnet; but the magnetic attraction is undiminished by the interposition of any bodies, except iron, steel, &c.

What other causes will affect the polarity of a magnet?

1122. Electricity frequently changes the poles of a magnet; and the explosion of a small quantity of gunpowder, on one of the poles, produces the same effect. Electricity, also, sometimes renders iron and steel magnetic, which were not so before the charge was received.

What is the effect of a double magnet?

1123. The effect produced by two magnets, used together, is much more than double that of either one used alone.

What is meant by "the dipping" of a magnet, and how is it corrected?

1124. When a magnet is suspended freely from its centre, the two poles will not lie in the same horizontal direction. This is called the inclination or the *dipping* of the magnet.

1125. The tendency of a magnetic needle to dip is corrected, in the mariner's and surveyor's compasses, by making the south ends of the needles intended for use in northern latitudes somewhat heavier than the north ends. Compass-needles, intended to be employed on long voyages, where great variations of latitude may be expected, are furnished with a small sliding-weight, by the adjusting of which the tendency *to dip* may be counteracted. The cause of the dipping of the needle is the superior attraction caused by the closer proximity of the pole of the magnet to the magnetic pole of the earth. In north latitude, the north pole of the needle dips; in south latitude, the south pole.

In what direction does a magnet point when freely suspended?

1126. The magnet, when suspended, does not invariably point *exactly* to the north and south points, but varies a little towards the east or the west. This variation differs at different places, at different seasons, and at different times in the day.

1127. The variation of the magnetic needle from what has been supposed its true polarity was a phenomenon that for centuries had baffled the science of the philosopher to explain. Recent discoveries have given a satisfactory explanation of this apparent anomaly.* The earth has, in fact, four magnetic poles, two of which are strong and two are weak. The strongest north pole is in America, — the weakest, in Asia. The earth itself is considered as a magnet, or, rather, as composed in part of magnetic substances, so that its action at the surface is irregular. The variation of the needle from the true geographical meridian is therefore subject to changes more or less irregular. †

What gift has the science of Magnetism bestowed on navigation? What is the Mariner's Compass?

1128. The science of Magnetism has rendered immense advantages to commerce and navigation, by means of the mariner's compass. The Mariner's Compass consists of a magnetized bar of steel, called a *needle;* having at its centre a cap fitted to it, which is supported on a sharp-pointed pivot

---

* The following statement has been made in the *National Intelligencer*, on the authority of its London correspondent:

Mr. Faraday, in a late lecture before the Royal Institution upon the Magnetic Forces, made the following important announcement.

" A German astronomer has for many years been watching the spots on the sun, and daily recording the result. From year to year the groups of spots vary. They are sometimes very numerous, sometimes they are few. After a while it became evident that the variation in number followed a descending scale through five years, and then an ascending scale through five subsequent years, — so that the periodicity of the variations became a visible fact.

" While our German friend was busy with his groups of sun-spots, an Englishman was busy with the variations of the magnetic needle. He, too, was a patient recorder of patient observation. On comparing his tabular results with those of the German astronomer, he found that the variations of the magnetic needle corresponded with the variations of the sun-spots, — that the years when the groups were at their maximum, the variations of the needle were at their maximum, and so on through their series. This relation may be coincident merely, or derivative; if the latter, then do we connect astral and terrestrial magnetism, and new reaches of science are open to us."

† This subject is very ably treated in " Davis' Manual of Magnetism" (edition of 1847), to which the student is referred, as probably the best elementary treatise on the subject that has been published. Mr. Davis is one of those scientific and skilful mechanics (of whom there are not a few among us) who have, as it were, forced their way into the temple of science amid discouragements and difficulties, but have deposited richer gifts on the altar than most of those whose contributions were expected. He has originated many improvements in this department of science; and his devotion to the subject has probably rendered him as familiar with all the peculiar phenomena relating to it as any one in or out of the country.

Mr. Davis has been succeeded by two intelligent and skilful young men, Palmer and Hall, Magnetical Instrument Makers, 526 Washington-street, Boston. They are also the agents for the sale of his works, " The Book of the Telegraph," and " Medical Electricity."

fixed in the base of the instrument. A circular plate, or card, the circumference of which is divided into degrees, is attached to the needle, and turns with it. On an inner circle of the card the thirty-two points of the mariner's compass are inscribed.

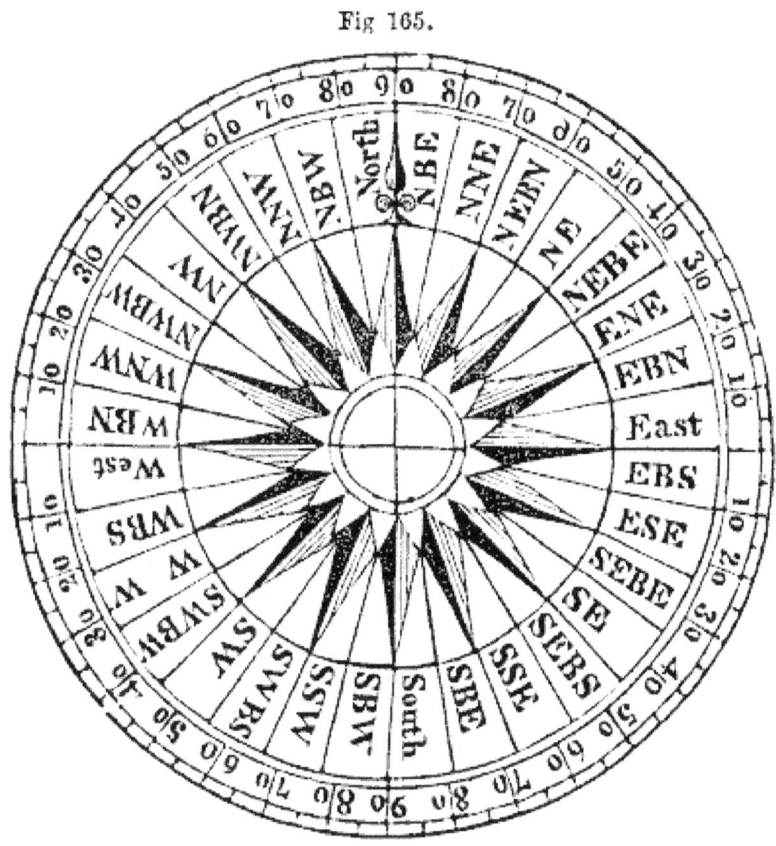

Fig. 165.

1129. The needle is generally placed *under* the card of a mariner's compass, so that it is out of sight; but small needles, used on land, are placed above the card, not attached to it, and the card is permanently fixed to the box.

1130. The compass is generally fitted by two sets of axes to an outer box, so that it always retains a horizontal position, even when the vessel rolls. When the artificial magnet or *needle* is kept thus freely suspended, so that it may turn north or south, the pilot, by looking at its position, can ascertain in what direction his vessel is proceeding; and, although the needle varies a little from a correct polarity, yet this variation is neither so great, nor so irregular, as seriously to impair its use as a guide to the vessel in its course over the pathless deep.

1131. The invention of the mariner's compass is usually ascribed to Flavio de Melfi, or Flavio Gioia, a Neapolitan, about the year 1302. Some authorities, however, assert that it was brought from China by Marco Paolo, a Venetian, in 1260. The invention is also claimed both by the French and English.

Of what use has the mariner's compass been?
1132. The value of this discovery may be estimated from the consideration that, before the use of the compass, mariners seldom trusted themselves out of sight of land; they were unable to make long or distant voyages, as they had no means to find their way back. This discovery enabled them to find a way where all is trackless; to conduct their vessels through the mighty ocean, out of the sight of land; and to prosecute those discoveries, and perform those gallant deeds, which have immortalized the names of Cook, of La Perouse, Vancouver, Sir Francis Drake, Nelson, Parry, Franklin and others.

Which pole of a magnet is the more powerful?
1133. The north pole of a magnet is more powerful in the northern hemisphere, or north of the equator, and the south pole in the southern parts of the world.

1134. When a piece of iron is brought sufficiently near to a magnet, it becomes itself a magnet; and bars of iron that have stood long in a perpendicular situation are generally found to be magnetical.

How are artificial magnets made?
1135. Artificial magnets are made by applying one or more powerful magnets to pieces of soft iron. The end which is touched by the north pole becomes the south pole of the new magnet, and that touched by the south pole becomes the north pole. The magnet which is employed in magnetizing a steel bar loses none of its power by being thus employed; and, as the effect is increased when two or more magnets are used, with one magnet a number of bars may be magnetized, and then combined together; by which means their power may be

indefinitely increased. Such an apparatus is called a *magnetic magazine*.

1136. There are several methods of making artificial magnets. One of the most simple and effectual consists in passing a strong horse-shoe magnet over bars of soft iron.

In making bar (or straight) magnets, the bars must be laid lengthwise, on a flat table, with the marked end of one bar against the unmarked end of the next; and in making horse-shoe magnets, the pieces of steel, previously bent into their proper form, must be laid with their ends in contact, so as to form a figure like two capital U's, with their tops joined together, thus, ⊂⊃; observing that the marked ends come opposite to those which are not marked; and then, in either case, a strong horse-shoe magnet is to be passed, with moderate pressure, over the bars, taking care to let the marked end of this magnet precede and its unmarked end follow it, and to move it constantly over the steel bars, so as to enter or commence the process at a mark, and then to proceed to an unmarked end, and enter the next bar at its marked end, and so proceed.

After having thus passed over the bars ten or a dozen times on each side, and in the same direction as to the marks, they will be converted into tolerably strong and permanent magnets. But if, after having continued the process for some time, the exciting magnet be moved over the bars in a contrary direction, or if its south pole should be permitted to precede after the north pole has been first used, the previously-excited magnetism will disappear, and the bars will be found in their original state.

This mode of making artificial magnets is likely to be wholly superseded by the new mode by electrical aid, which will be noticed in connexion with Electro-magnetism.

How is a magnetic magazine constructed? 1137. A magnetic magazine may be made by taking several horse-shoe magnets of equal size, and, after having magnetized them, uniting them together by means of screws.

1138. A magnetic needle is made by fastening the steel on a piece of board, and drawing magnets over it from the centre outwards.

How should a horse-shoe magnet be kept? 1139. A horse-shoe magnet should be kept *armed*, by a small bar of iron or steel, connecting the two poles. The bar is called "*the keeper*."

Interesting experiments may be made by a magnet, even of no great power, with steel or iron filings, small needles, pieces of ferruginous substances, and black sand which contains iron. Such substances may be made to assume a variety of amusing forms and positions by moving the magnet *under* the card, paper or table, on which they are placed. Toys, representing fishes, frogs, aquatic birds, &c., which are made to appear to bite at a hook, birds floating on the water, &c., are constructed on magnetic principles, and sold in the shops.

What is Electro-magnetism? 1140. Electro-magnetism relates to magnetism which is induced by the agency of electricity.

1141. The passage of the two kinds of electricity (namely, the positive and the negative) through their circuit is called the electric currents; and the science of Electro-magnetism explains the phenomena attending those currents. It has already been stated that from the connecting wires of the galvanic circle, or battery, there is a constant current of electricity passing from the zinc to the copper, and from the copper to the zinc plates. In the single circle these currents will be negative from the zinc, and positive from the copper; but in the compound circles, or the battery, the current of positive electricity will flow from the zinc to the copper, and the current of negative electricity from the copper to the zinc. From the effect produced by electricity on the magnetic needle, it had been conjectured, by a number of eminent philosophers, that magnetism, or magnetic attraction, is in some manner caused by electricity. In the year 1819, Professor Œrsted, of Copenhagen, made the grand discovery of the power of the electric current to induce magnetism; thus proving the connexion between magnetism and electricity. In a short time after the discovery of Professor Œrsted, Mr. Faraday discovered that an electrical spark could be taken from a magnet; and thus the common source of magnetism and electricity was fully proved. In a paper published a few years ago, this distinguished philosopher has very ably maintained the identity of common electricity, voltaic electricity, magnetic electricity (or electro-magnetism), thermo-electricity, and animal electricity. The phenomena exhibited in all these five kinds of electricity differ merely in degree, and the state of intensity in the action of the fluid. The discovery of Professor Œrsted has been followed out by Ampère, who, by his mathematical and experimental researches, has presented a theory of the science less obnoxious to objections than that proposed by the professor. The discovery of Œrsted was limited to the action of the electric current on needles *previously magnetized;* it was afterwards ascertained by Sir Humphrey Davy and M. Arago that magnetism may be developed in steel not previously possessing it, if the steel be placed in the electric current. Both of these philosophers, independently of each other, ascertained that the uniting wire, becoming a magnet, attracts iron filings and collects sufficient to acquire the diameter of a common quill; but the moment the connexion is broken, all the filings drop off, and the attraction diminishes with the decaying energy of the pile. Filings of brass or copper, or wood-shavings, are not attracted at all.

1142. All the effects of electricity and galvanism that have hitherto been described have been produced on bodies *interposed* between the extremities of conductors, proceeding from the positive and negative poles. It was not known, until the discoveries of Professor Œrsted were made, that any effect could be produced when the electric circuit is *uninterrupted.*

What is the difference between electricity and electro-magnetism? It will presently be seen that this constitutes the great distinction between electricity and electro-magnetism, namely, that one describes the effect of electricity when *interrupted* in its course, and that the other more especially explains the effect of an uninterrupted current of electricity.

What are the principal facts of electro-magnetism? 1143. The principal facts in connexion with the science of electro-magnetism are,—

(1.) That the electrical current, passing *uninter_ ruptedly* through a wire connecting the two ends of a galvanic battery, produces an effect upon the magnetic needle.

(2.) That electricity will induce magnetism.

(3.) That a magnet, or a magnetic magazine, will induce electricity.

(4.) That the combined action of electricity and magnetism, as described in this science, produces a rotatory motion of certain kinds of bodies, in a direction pointed out by certain laws.

(5.) That the periodical variation of the magnetic needle

from the true meridian, or, in other words, the variation of the compass, is caused by the influence of the electric currents.

(6.) That the magnetic influence is not confined to iron, steel, &c., but that most metals, and many other substances, may be converted into temporary magnets by electrical action.

(7.) That the magnetic attraction of iron, steel, &c., may be prodigiously increased by electrical agency.

(8.) That the *direction* of the electric current may, in all cases, be ascertained.

(9.) That magnetism is produced whenever concentrated electricity is passed through space.

(10.) That while in common electrical and magnetic attractions and repulsions those of the same name are mutually repulsive, and those of different names attract each other, in the attractions and repulsions of *electric currents* it is precisely the reverse, the repulsion taking place only when the wires are so situated that the currents are in *opposite direction*.

The consideration of the subject of electricity induced by magnetism properly belongs to the subject of Magneto-electricity, in which connexion it will be particularly noticed.

How is the direction of a current of electricity ascertained? 1144. The direction of the electric current is ascertained by means of the magnetic needle. If a sheet of paper be placed over a horse-shoe magnet, and fine black sand, or steel filings, be dropped loosely on the paper, the particles will be disposed to arrange themselves in a regular order, and in the direction of curve lines. This is, undoubtedly, the effect of some influence, whether that of electricity, or of magnetism alone, is not material at present to decide.

How will a freely-suspended magnet place itself in relation to the electrical current? 1145. A magnet freely suspended tends to assume a position at right angles to the direction of a current of electricity passing near it.

1146. If a wire, which connects the extremities of a voltaic battery, be brought over and parallel with a magnetic needle at rest, or with its poles properly directed north and south, that end of the needle next to the negative pole of the battery will move towards the west, whether the wire be on one side of the needle or the other, provided only that it be parallel with it.

1147. Again, if the connecting wire be lowered on either side of the needle, so as to be in the horizontal plane in which the needle should move, it will not move in that plane, but will have a tendency to revolve in a *vertical* direction; in which, however, it will be prevented from moving, in consequence of the attraction of the earth, and the manner in which it is suspended. When the wire is to the east of the needle, the pole nearest to the negative extremity of the battery will be elevated; and when it is on the west side, that pole will be depressed.

1148. If the connecting wire be placed below the plane in which the needle moves, and parallel with it, the pole of the needle next to the negative end of the wire will move towards the east, and the attractions and repulsions will be the reverse of those observed in the former case.

How does the electro-magnetic current act?

1149. The action of the conducting-wire in these cases exhibits a remarkable peculiarity. All other known forces exerted between two points act in the direction of a straight line connecting these points, and such is the case with electric and magnetic actions, separately considered; but the electric current exerts its magnetic influence laterally, at right angles to its own course. Nor does the magnetic pole move either directly towards or directly from the conducting-wire, but tends to revolve around it without changing its distance. Hence the force must be considered as acting in the direction of a *tangent* to the circle in which the magnetic pole would move.

What effect has a voltaic battery on unmagnetized steel?

1150. The two sides of an unmagnetized steel needle will become endued with the north and south polarity, if the needle be placed parallel with the connecting wire of a voltaic battery, and nearly or quite in contact with it. But, if the needle be placed at right angles with the connecting wire, it will become permanently magnetic; one of its extremities pointing to the north pole and the other to the south, when it is freely suspended and suffered to vibrate undisturbed.

To what may magnetism be communicated by the voltaic battery, and what is the process called?

1151. Magnetism may be communicated to iron and steel by means of electricity from an electrical machine; but the effect can be more conveniently produced by means of the voltaic battery. This phenomenon is called *electro-magnetic induction*.

What is a Helix?

1152. A Helix is a spiral line, or a line wound into the shape of a cork-screw.

What use is made of a helix in connexion with the battery?

1153. If a helix be formed of wire, and a bar of steel be enclosed within the helix, on applying the conducting-wires of the battery to the extremities of the helix, the steel bar will immediately become magnetic. The electricity from a common electrical machine, when passed through the helix, will produce the same effect.

And what must first be done with the wire of the helix?

1154. The wire which forms the helix should be coated with some non-conducting substance, such as silk wound around it; as it may then be formed into close coils, without suffering the electric fluids to pass from surface to surface, which would impair its effect.

1155. If such a helix be so placed that it may move freely, as when made to float on a basin of water, it will be attracted and repelled by the opposite poles of a common magnet.

1156. If a magnetic needle be surrounded by coiled wire, covered with silk, a very minute portion of electricity through the wire will cause the needle to deviate from its proper direction.

What is an Electro-magnetic Multiplier?
1157. A needle thus prepared is called an Electro-magnetic Multiplier. It is, in fact, a very delicate electroscope, or rather *galvanometer*, capable of pointing out the direction of the electric current in all cases.

What is meant by the Electro-magnetic Rotation?
1158. Among the most remarkable of the facts connected with the science of electro-magnetism is what is called the Electro-magnetic Rotation. Any wire through which a current of electricity is passing has a tendency to revolve around a magnetic pole in a plane perpendicular to the current, and that without reference to the axis of the magnet the pole of which is used. In like manner a magnetic pole has a tendency to revolve around such a wire.

1159. Suppose the wire perpendicular, its upper end positive, or attached to the positive pole of the voltaic battery, and its lower end negative; and let the centre of a watch-dial represent the magnetic pole: if it be a north pole, the wire will rotate round it in the direction that the hands move; if it be a south pole, the motion will be in the opposite direction. From these two, the motions which would take place if the wire were inverted, or the pole changed, or made to move, may be readily ascertained, since the relation now pointed out remains constant.

Explain Fig. 166.
1160. Fig. 166 represents the ingenious apparatus, invented by Mr. Faraday, to illustrate the electro-magnetic rotation. The central pillar supports a piece of thick copper wire, which, on the one side, dips into the mercury contained in a small glass cup *a*. To a pin at the bottom of this cup a small cylindrical magnet is attached by a thread, so that one pole shall rise a little above the surface of the mercury, and be at liberty to move around the wire. The bottom of the cup is perforated, and has a cop-

per pin passing through it, which, touching the mercury on the inside, is also in contact with the wire that proceeds outwards, on that side of the instrument. On the other side of the instrument b, the thick copper wire, soon after turning down, terminates, but a thinner piece of wire forms a communication between it and the mercury on the cup beneath. As freedom of motion is re-

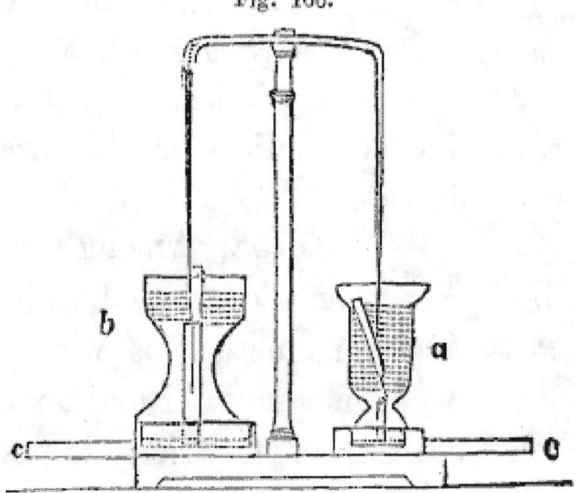

Fig. 166.

garded in the wire, it is made to communicate with the former by a ball and socket-joint, the ball being held in the socket by a thread; or else the ends are bent into hooks, and the one is then hooked to the other. As good metallic contact is required, the parts should be amalgamated, and a small drop of mercury placed between them; the lower ends of the wire should also be amalgamated. Beneath the hanging wire a small circular magnet is fixed in the socket of the cup b, so that one of its poles is a little above the mercury. As in the former cup, a metallic connexion is made through the bottom, from the mercury to the external wire.

If now the poles of a battery be connected with the horizontal external wires $c\ c$, the current of electricity will be through the mercury and the horizontal wire, on the pillar which connects them, and it will now be found that the movable part of the wire will rotate around the magnetic pole in the cup b, and the magnetic pole round the fixed wire in the other cup a, in the direction before mentioned.

By using a very delicate apparatus, the magnetic pole of the earth may be made to put the wire in motion.

Explain Fig. 167.   1161. Fig. 167 represents another ingenious contrivance, invented by M. Ampère, for illustrating the electro-magnetic rotation; and it has the advantage of comprising within itself the voltaic combination which is employed. It consists of a cylinder of copper about two inches high, and a little less than two inches internal diameter, within which is a small cylinder, about one inch in diameter. The two cylinders are connected together by a bottom, having an aperture in its centre the size of the smaller cylinder, leaving a circular cell, which may be filled with acid. A piece of strong copper wire is fastened across the top of the inner cylinder, and from the middle of it rises, at a right angle, a piece of copper wire, supporting a very small metal cup, containing a few globules of mercury. A cylinder of zinc, open at each end, and about an inch and a quarter in diameter, completes the voltaic combination. To the latter cylinder a wire, bent like an inverted U, is soldered at opposite sides; and in the bend of this wire a metallic point is fixed, which, when inserted in the little cup of mercury, suspends the zinc cylinder in the cell, and allows it a free circular motion. An additional point is directed downwards from the central part of the stronger wire, which point is adapted to a small hole at the top of a powerful bar magnet. When the apparatus with one point only is charged with diluted acid, and set on the magnet placed vertically, the zinc cylinder revolves in a direction determined by the magnetic pole which is uppermost. With two points, the copper revolves in one direction and the zinc in a contrary direction.

Fig. 167.

1162. If, instead of a bar magnet, a horse-shoe magnet be employed, with an apparatus on each pole similar to that which has now been described, the cylinders in each will revolve in opposite directions. The small cups of mercury mentioned in the preceding description are sometimes omitted, and the points are inserted in an indentation on the inverted U.*

* The phenomenon of electro-magnetic rotation is beautifully illustrated

316 NATURAL PHILOSOPHY.

How is the magnetizing power of the battery increased?

1163. The magnetizing power of the conducting wires of a battery is very greatly increased by coiling it into a helix, into which the body to be magnetized may be inserted. A single circular turn is more efficient than a straight wire, and each turn adds to the power within a certain limit, whether the whole forms a single layer, or whether each successive turn encloses the previous one.

How is a helix of great power obtained?

1164. When a helix of great power is required, it is composed of several layers of wire. The wire forming the coil must be insulated by being wound with cotton, to prevent any lateral passage of the current.

Explain Fig. 168.

1165. Fig. 168 represents a helix on a stand. A bar of soft iron, N S, being placed within the helix, is connected with the battery by

Fig. 168.

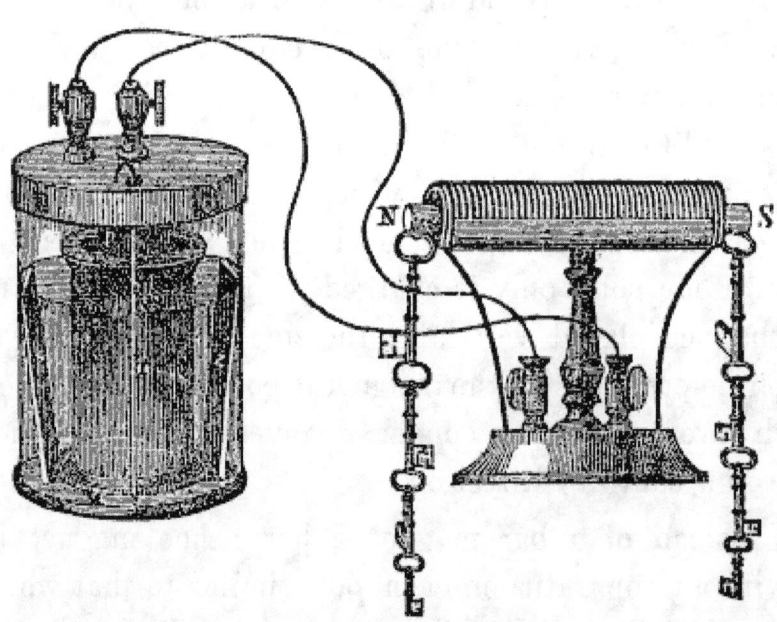

by Mr. Davis, in his treatise on Magnetism, to which reference has already been made. He has invented and prepared a great variety of ingenious contrivances for the illustration of this subject, and his book should be in the hands of all who desire a thorough acquaintance with all that has been discovered in the new department of science, in which magnetism and electricity are combined. The author has been indebted to Mr. Davis' volume for a number of explanations which are incorporated in this work.

means of the screw-cups on the base of the stand. The two extremities of the bar instantly become strongly magnetic, and keys, or pieces of iron, iron filings, nails, &c., will be held up so long as the connexion with the battery is sustained. But, so soon as the connexion is broken, the bar loses its magnetic power, and the suspended articles will fall. The bar can be made alternately to take up and drop such magnetizable articles as are brought near it, as the connexion with the battery is made or broken.

1166. A steel bar placed within the helix acquires the polarity less readily, but retains it after the connexion is broken. Small rods or bars of steel, needles, &c., may be made permanent magnets in this way.

What is an Electro-magnet?

1167. A bar temporarily magnetized by the electric current is called an Electro-magnet.

How can the poles of an electro-magnet be discriminated?

1168. To ascertain the poles of an electro-magnet, it must be observed that *the north pole will be at the furthest end of the helix when the current circulates in the direction of the hands of a watch.*

1169 Magnets of prodigious power have been formed by means of voltaic electricity.

What was the power of the electro-magnets constructed by Prof. Henry and Dr. Ten Eyck?

1170. An electro-magnet was constructed by Professor Henry and Dr. Ten Eyck which was capable of supporting a weight of 750 pounds. They have subsequently constructed another, which will sustain 2063 pounds. It consists of a bar of soft iron, bent into the form of a horse-shoe, and wound with twenty-six strands of copper bell-wire, covered with cotton threads, each thirty-one feet long; about eighteen inches of the ends are left projecting, so that only twenty-eight feet of each actually surround the iron. The aggregate length of the coils is therefore 728 feet. Each strand is wound on a little less than an inch; in the middle of the horse-shoe it forms three thicknesses of wire; and on the ends, or near the poles, it is wound so as to form six thicknesses. Being connected with a battery consisting of plates, containing a little less than forty-eight square feet of surface, the magnet supported the prodigious weight stated above, namely 2063 pounds.

Explain Fig. 169.

1171. HELIACAL RING. Fig. 169 represents a heliacal ring, or ring of wire bent in the form of a helix, with the ends of the wire left free to be inserted in the screw-cups of a battery. Two semicircular pieces of soft unmagnetized iron, furnished with rings,—the upper one for the hand, the lower one for weights,—are prepared to be inserted into the helix, in the manner of the links of a chain. As soon as the ends of the helix are inserted into the screw-cups of the battery, the rings will be held together, with **great force**, by magnetic attraction.

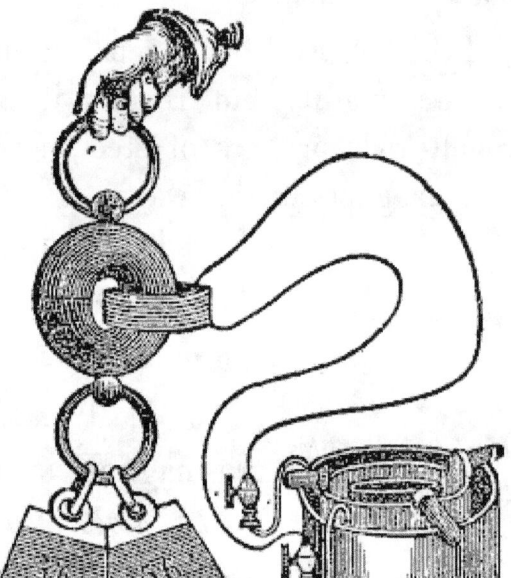

Fig. 169.

1172. That the attraction is caused, or that the magnetism is induced, by the circulation of electricity *around* the coils, may be proved by the following interesting experiment. Hold the heliacal ring *horizontally* over a plate of small nails, and suspend an unmagnetized bar *perpendicularly* on the *outside* of the ring, over the nails, and there will be no attraction. Suspend the bar *perpendicularly through* the helix, and the nails will all attach themselves to it in the form of tangents to the circles formed by the coils of the heliacal ring.

How are horse-shoe magnets

1173. *Communication of Magnetism to Steel by the Electro-magnet.*— Bars of the U form

ELECTRO-MAGNETIC TELEGRAPH.

most readily made?

are most readily magnetized by drawing them from the bend to the extremities, across the poles of the U electro-magnet, in such a way that both halves of the bar may pass at the same time over the poles to which they are applied. This should be repeated several times, recollecting always to draw the bar in the same direction.

Explain Fig. 170.

1174. Fig. 170 represents the U electro-magnet, with the bar to be magnetized. When the bar is thick, both surfaces should be drawn across the electro-magnet, keeping each half applied to the same pole. To remove the magnetism, it is only necessary to

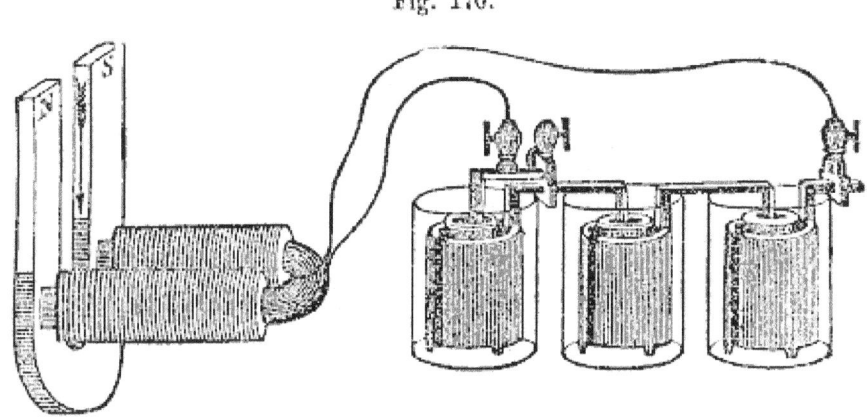

Fig. 170.

reverse the process by which it was magnetized, that is, to draw the bar across the electro-magnet in a contrary direction.

On what fundamental principle is the Electric Telegraph conconstructed?

1175. THE ELECTRO-MAGNETIC TELEGRAPH.* — From the description which has now been given of the electro-magnetic power, it will readily be perceived that a great force can be made to act simply by bringing a wire into contact with another conductor, and that the force can be instantly arrested in its operation by removing the wire from the contact; in other words, that by connecting and disconnecting a helix with a battery, a prodigious power can be made succes-

---

* The word *telegraph* is compounded of two Greek words, τηλε (*tele*), signifying *at a distance*, and γραφω (*grapho*), to write, that is, to signify or to write at a distance. The word *telescope* is another compound of the word τηλε with the word σκοπεω (*scopio*), to see, — an instrument *to see at a distance.*

sively to act and cease to act. Advantage has been taken of this principle in the construction of the American electro-magnetic telegraph, which was matured by Professor Morse, and first put into operation between the cities of Baltimore and Washington, in 1844.* It was not, however, until Professor Henry, of Princeton, New Jersey, had discovered the mode of constructing the powerful electro-magnets which have been noticed, that this form of the telegraph became possible.

Whose discoveries rendered the magnetic telegraph possible?

1176. The principles of its construction may be briefly stated as follows:

Explain the manner in which the electric telegraph performs its work.

An electro-magnet is so arranged with its armature that when the armature is attracted it communicates its motion to a lever, to which a blunt point is attached, which marks a narrow strip of paper, drawn under it by machinery resembling clock-work, whenever the electro-magnet is in action. When the electro-magnet ceases to act, the armature falls, and, communicating its motion to the lever, the blunt point is removed from its contact with the paper. By this means, if one of the wires from the battery is attached to the screw-cup, whenever the other wire is attached to the remaining cup the armature is powerfully attracted by the magnet, and the point on the lever presses the paper into the groove of a roller, thereby making an indentation on the paper, corresponding in length to the time during which the contact with the battery is maintained, the paper being drawn slowly under the roller.

1177. In the construction of the electric telegraph the first object of consideration is the development of the agent. The agent is the electric fluid, which is brought into action by a powerful battery.

What is the agent in the electric telegraph?

* For a particular description of this wonderful invention, the student is referred to Davis' treatise on Magnetism, in which the parts are all described with a minuteness which leaves little more to be desired. The history, also, of the successive steps by which it was brought to its present degree of perfection, is to be found in the same connexion.

ELECTRO-MAGNETIC TELEGRAPH.   321

Explain Fig. 171.   Fig. 171 represents a battery composed of twelve cups, on the principle of Grove's battery, each cup containing a thick cylinder of zinc, with a porous cell, two acids, and a strip of platinum, as described in Fig. 164. The chemical action of the acids on the

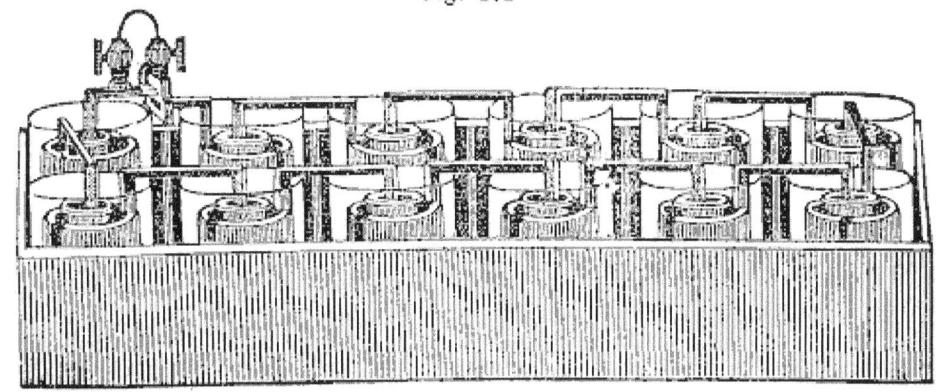

Fig. 171

zinc generates a powerful current of electricity towards the screw-cups at A B.

Explain Fig. 172.   1178. The second step in the construction of the telegraph is represented by Fig. 172. The wires from the battery represented in Fig. 171 are carried to the screw-cups in the apparatus represented by Fig. 172, called the signal-key, A to A and B to B, respectively. It will be observed that the cups of the signal-key are insulated, and that the electric fluid can finish its circuit only when the finger depresses the knob and makes it come in contact with the metallic strip below, thus forming a communication between the screw-cups. The signal-key thus regulates the completion of the circuit, and the flow of the current of electricity, at the will of the operator.

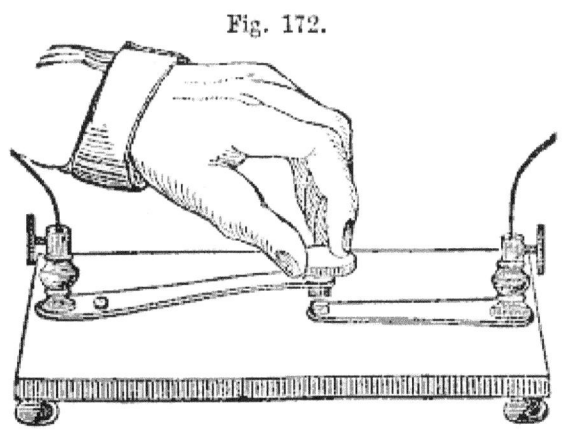

Fig. 172.

Explain Fig. 173.

1179. The signal-key is made in several forms in the different telegraphs, and in Fig. 173 is represented in its more perfect construction. It consists of a lever, mounted on a horizontal axis, with a knob of ivory for the hand at the extremity of the long arm, which is at the left in the cut. This lever is thrown up by a spring, so as to avoid contact with the button on the frame below, except when the lever is depressed for the purpose of completing the circuit.

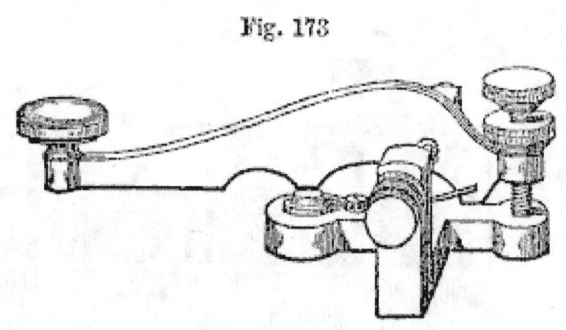

Fig. 173

A regulating screw is seen at the extremity of the short arm of the lever, which graduates precisely the amount of motion of which it is at any time susceptible.

Explain Fig. 174.

1180. The third and last part of the telegraph is the registering apparatus, represented in Fig. 174.

Here are two screw-cups, for the insertion of the wires from a distant battery. An iron in the shape of a U magnet stands

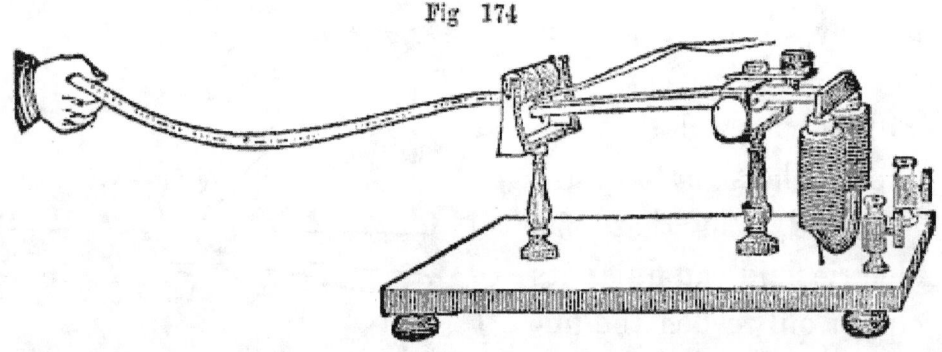

Fig 174

at the left of the screw-cups, each arm of which is surrounded by a helix or coil of wire, the ends of which, passing down through the stand, are connected below with the screw-cups. It will then be seen that when the signal-key is depressed the electric circuit is completed, and that the electricity, passing through the coils of wire, renders the U-shaped iron highly magnetic,

ELECTRO-MAGNETIC TELEGRAPH. 323
Fig. 175.

and it attracts the armature down. The armature is fixed to the shorter arm of a lever, and when the shorter arm is attracted down, the longer arm, with a point affixed, is forced upward and makes an indentation upon a strip of paper. The length of the indentation on the paper will depend on the length of time that the signal-key is depressed. When the signal-key is permitted again to rise, the electric current is broken, the U-shaped iron ceases to be a magnet, and, the armature being no longer attracted, the weight of the longer arm will cause that arm to fall, and no mark is made on the paper.

When the telegraph was first constructed, it was thought necessary to have two wires in order to form the circuit. It has since been found that the earth itself will serve for one-half the circuit, and that one wire will alone be necessary to perform the work of the telegraph.

Explain Fig. 175. 1180. Fig. 175 represents the manner in which the electric telegraph is put into operation. On the left of the figure is seen the operator, with the battery at his feet and his finger on the signal-key. From one screw-cup of the battery extends a wire which traverses the whole distance between two cities, elevated on posts for security. In the distant city the wire reaches another screw-cup to which it is attached, while from another screw-cup at the same station another wire is attached, which extends back to the operator first mentioned. The depression of the signal-key forms a connexion between the two poles of the battery by means of the wire, and the fluid will traverse the whole distance between the two stations in preference to leaping over the space between the two screw-cups. The right of the figure represents the receiver of the information, reading the message which has thus been imprinted by the point.

Explain Fig. 176. 1182. In the preceding figures the mere outlines have been given, in order that they may be distinctly understood. To present the strip of paper so that it may readily receive the impression, additional machinery becomes necessary. The complete registering machine is shown in Fig. 176, in which S represents a large

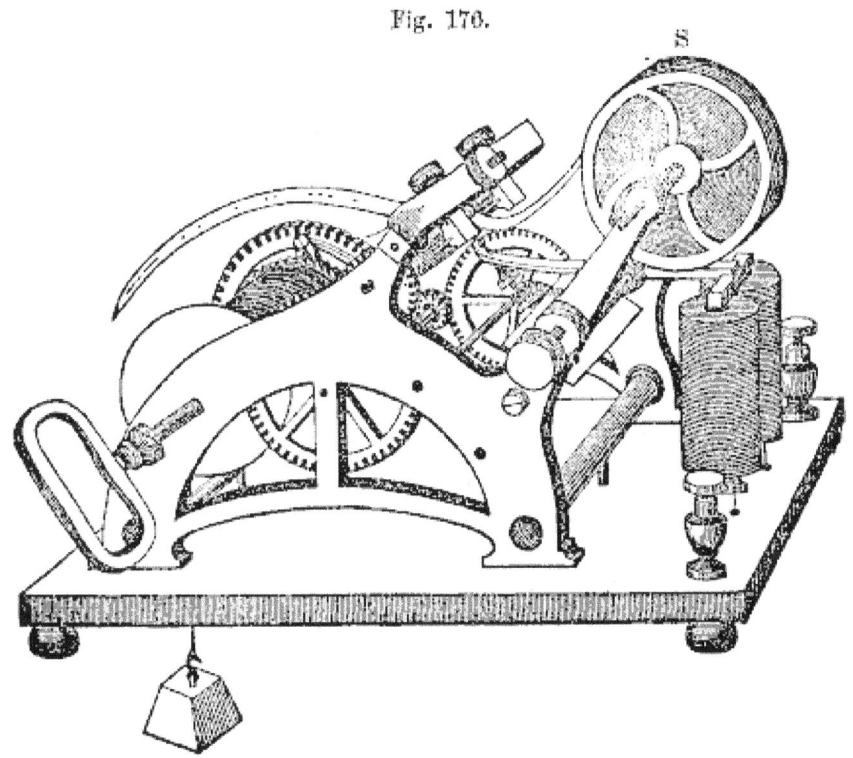

Fig. 176.

spool on which the paper is wound, and clock-work with rollers, to give the paper a steady motion toward the point by which the marks are to be made. A bell is sometimes added, which is struck by a hammer when the lever first begins to move, in order to draw the attention of the operator.

1183. It will be recollected that this form of the magnetic telegraph is familiarly known as Morse's, the machine making nothing but straight marks on the slip of paper. But these straight marks may be made long or short, at the pleasure of the operator. If the key be pressed down and instantly be permitted to rise, it will make a short line, not longer than a hyphen. By means of a conventional alphabet, in which the letters are expressed by the repetition and combination of marks varying in length, any message may be conveniently spelt out, so as to be distinctly understood at the distant station. These are the essential features of Morse's Telegraph.

1184. It is necessary, in long lines of telegraphs, to combine the effects of several batteries to supply the loss of power in traversing long circuits. This is done by local batteries or relays, as they are sometimes called, familiarly known in connexion with Morse's telegraph. The use of the relays may be dispensed with by increasing the power of the battery, or distributing it in groups along the line. It is sometimes divided by arranging one-half at each end of the line. For every twenty miles an addition of one of Grove's pint cups should be made. The expense of acids for each cup for two days does not much exceed one cent. For a line of telegraph extending around the earth, twelve hundred Grove's cups would be required, distributed at equal distances, fifty in a group.

1185. BAIN'S TELEGRAPH.—The telegraph known by the name of Bain's telegraph, the simplest now in use, differs from Morse's principally in its mode of registering. It performs its work by the decomposition of a saline solution. The pen or point is stationary. A circular tablet, moved by clock-work, under the point, receives the point in concentric grooves, and the writing is arranged in spiral lines, occupying but little space.

Explain Fig. 177. Fig. 177 represents Bain's telegraph. The pen-holder is connected with the positive wire of the battery, and the tablet with the negative. The circuit is com-

Fig. 177.

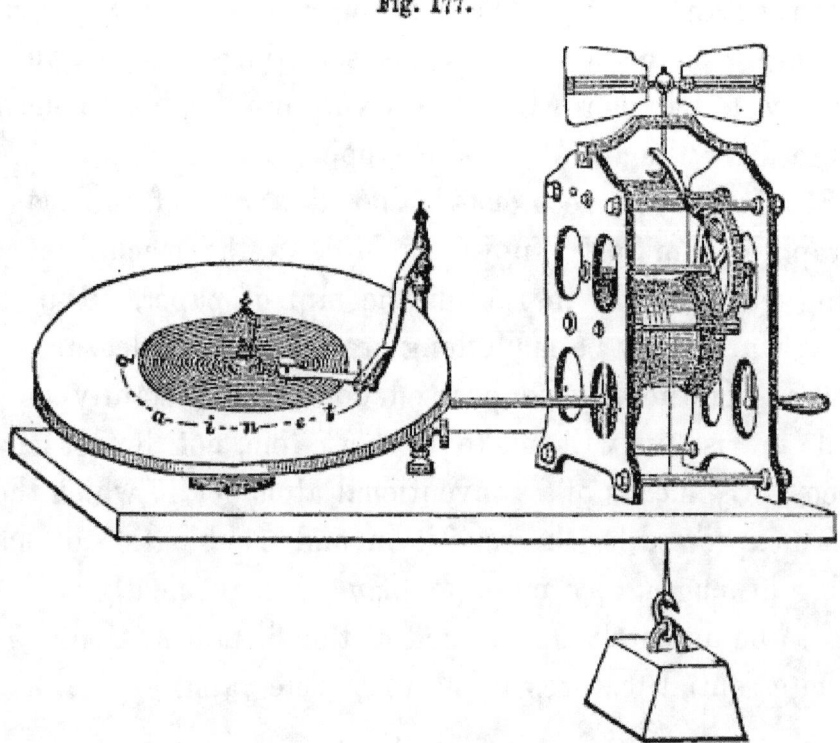

pleted by paper moistened with a solution of the yellow prussiate of potash, acidulated with nitric or sulphuric acid. The pen-wire is of iron. When the circuit is completed, the solution attacks the pen, dissolves a portion of its iron, and forms the color known as Prussian blue, which stains the paper. The alphabet used by this line is the same in principle as that used in the telegraph of Morse. The advantage of this telegraph consists in the rapidity with which the disks at both ends are made to revolve, by which a message may be communicated at the rate of a thousand letters in a minute.

Explain Fig. 178. 1186. The call commonly used in connexion with Bain's telegraph is represented in Fig. 178. It consists of a U magnet, each arm surrounded by a helix of wires, which, when the current passes, causes the armature to be attracted and give motion to machinery, by which a bell or a glass is rung.

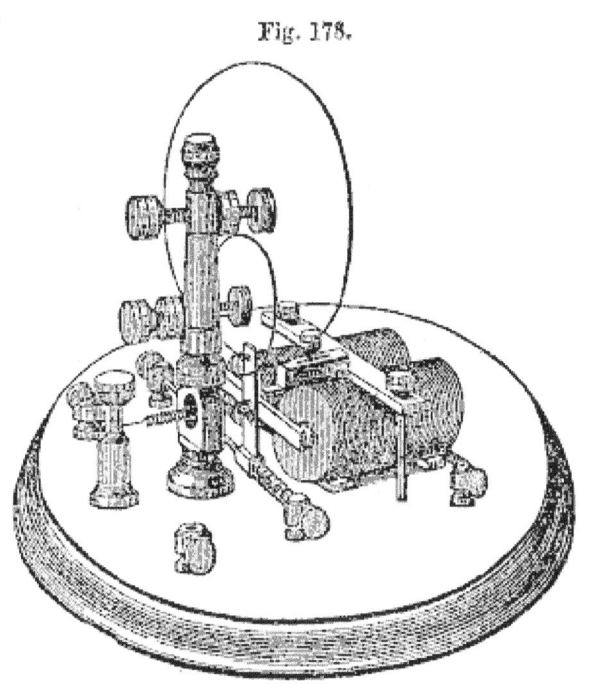

Fig. 178.

Explain Fig. 179. 1187. Fig. 179 represents the receiving magnet in its improved form. The armature is mounted on an upright bar, directly before the poles of the U magnet, which is surrounded by many coils of insulated wire. In this magnet the points of contact are preserved from oxidation by the use of platinum.

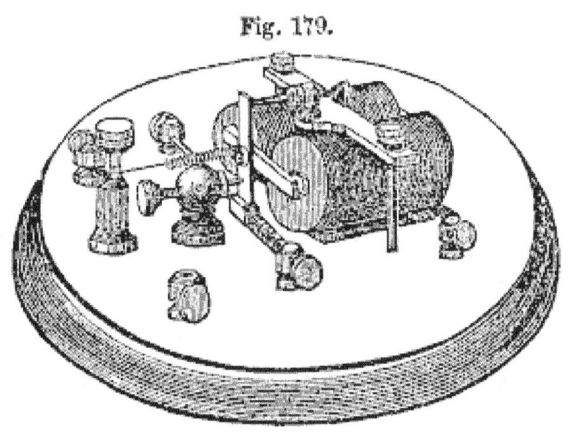

Fig. 179.

1188. House's Printing Telegraph. — This telegraph differs from the other principally in its *printing* with great rapidity the letters which form the message.

Explain Fig. 180. 1189. Fig. 180 represents the mechanical part of House's telegraph. The operator sits at a key-board, similar to that of a pianoforte or organ, and, by depressing a

Fig. 180.

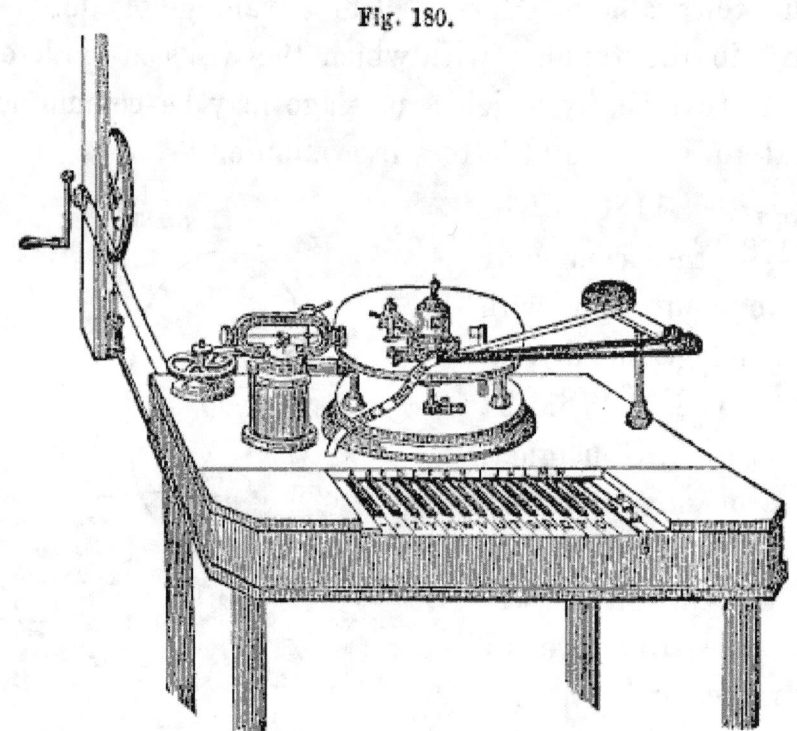

key, the letter corresponding with the key is made to appear at a little window at the top of the instrument, while it is at the same time printed on a strip of paper below. The principle by which this exceedingly ingenious operation is performed is simply this: A given number of electrical impulses are given for each letter. These impulses give motion to a wheel, so that on the depression of a key the circuit will be broken at precisely the point which corresponds with the letter. The machinery by which this is effected is necessarily complicated, and it falls not within the province of this work to go further into the explanation. The whole process is described in Davis' Book of the Telegraph, to which this volume is indebted for most of the particulars which have been given in relation to the subject.

ELECTRO-MAGNETIC TELEGRAPH.

The following history of the electric telegraph in this country is extracted from the *Portland Advertiser*, and deserves a place in this connexion:

"The electric telegraph, being used solely for the conveyance of news and communications, is so intimately connected with posts and post-offices, that a brief sketch of its rapid progress in the United States is here given.

"It is to American ingenuity that we owe the practical application of the magnetic telegraph for the purpose of communication between distant points, and it has been perfected and improved mainly by American science and skill. While the honor is due to Professor Morse for the practical application and successful prosecution of the telegraph, it is mainly owing to the researches and discoveries of Professor Henry, and other scientific Americans, that he was enabled to perfect so valuable an invention.

"The first attempt which was made to render electricity available for the transmission of signals, of which we have any account, was that of Lesage, a Frenchman, in 1774. From that time to the present there have been numerous inventions and experiments to effect this object; and, from 1820 to 1850, there were no less than sixty-three claimants for different varieties of telegraphs. We will direct attention only to those of Morse, Bain and House, they being the only kinds used in this country.

"During the summer of 1832, Professor S. F. B. Morse, an American, conceived the idea of an electric or electro-magnetic telegraph, and, after numerous experiments, announced his invention to the public in April, 1837.

"On the 10th of March, 1837, Hon. Levi Woodbury, then Secretary of the Treasury, issued a circular requesting information in regard to the propriety of establishing a system of telegraphs for the United States, to which Professor Morse replied, giving an account of his invention, its proposed advantages and probable expense. At that time 'he presumed five words could be transmitted in a minute.'

"In 1838, the American Institute reported that Morse could telegraph the words 'steamboat Caroline burnt' in six minutes. Now, a thousand such words are telegraphed in two minutes.

"In 1844, Congress built an experimental line from Baltimore to Washington, to test its practical operation. That line was soon continued on to Philadelphia and New York, and reached Boston the following year. Two branches diverge from this line, one from Philadelphia to St. Louis, 1000 miles, the other from New York, via Buffalo, to Milwaukie, 1300 miles long. One also, 1400 miles in length, goes from Buffalo to Lockport, and from thence through Canada to Halifax, N. S., whence there is a continuous line through Portland to Boston. The great Southern line, from Washington to New Orleans, is 1700 miles long. Another, 1200 miles, running to New Orleans from Cleveland, Ohio, via Cincinnati. The best paying line, it is said, is that between Washington and New York, which, during six months of last year, transmitted 154,514 messages, valued at $68,499; and the receipts for the year ending July, 1852, were $103,060. The average performance of the Morse instruments is from 8000 to 9000 letters per hour. The cost of construction, including wire, posts, labor, &c., is about $150 per mile. The Bain telegraph extends in the United States 2012 miles, and House's 2400 miles, making a total, with Morse's, of 89 lines, embracing 16,729 miles. At how many way stations the magnetic current is arrested and messages conveyed, we are not informed. Thus, in less than nine years, from a feeble beginning, under the fostering aid of government, have its wires and news communications spread all over the country.

"The astonishing results of the telegraph, victorious even in a run against time, are remarkable in the United States. The western cities, having a difference of longitude in their favor, actually receive news from New York sooner by the clock than it is sent. When the Atlantic made her first return voyage from Liverpool, a brief account of the news was telegraphed to New Orleans at a few minutes *after* noon (New York time), and reached its destination at a few minutes *before* noon (New Orleans time), and was published in the evening papers of both cities at the same hour. This is now a daily occurrence.

"Through its instrumentality (we mean no pun) Webster's death was simultaneously made known throughout the length and breadth of our land, and the next morning the pulpits from Maine to New Orleans were echoing in eulogies to his greatness, and mourning his departure.

"The great extent of the telegraph business, and its importance to the community, is shown by a statement of the amount paid for despatches by the associated press of New York, composed of the seven principal morning papers, — the *Courier and Enquirer*, *Tribune*, *Herald*, *Journal of Commerce*, *Sun*, *Times* and *Express*. During the year ending November 1, 1852, these papers paid nearly $50,000 for despatches, and about $14,000 for special and exclusive messages, not included in the expenses of the association.

"The difference between Morse's and House's telegraph is, principally, that the first traces at the distant end what is marked at the other; while House's does not trace at either end, but makes a signal of a letter at the distant end which has been made at the other, and thus, by new machinery, and a new power of air and axial magnetism, is enabled to print the signal letter at the last end, and this at the astonishing rate of sixty or seventy strokes or brakes in a second, and at once records the information, by its own machinery, in printed letters. Morse's is less complicated, and more easily understood; while House's is very difficult to be comprehended in its operations in detail, and works with the addition of two more powers, — one air, and the other called *axial magnetism*. One is a tracing or writing telegraph, the other a signal and printing telegraph.

"The telegraphs in England are next in importance and extent to those in this country. They were first established in 1845, and there are about 4000 miles of wire now in operation.

"The charge for transmission of despatches is much higher than in America, one penny per word being charged for the first fifty miles, and one farthing per mile for any distance beyond one hundred miles. A message of twenty words can be sent a distance of 500 miles in the United States for one dollar, while in England the same would cost seven dollars."

1190. THE ELECTRICAL FIRE ALARM. — The principle of the electric telegraph has recently been applied to a very ingenious piece of mechanism, by which an alarm of fire may be almost instantly communicated to every part of a large city. Wires, extending from the towers of the principal public buildings in which large bells are suspended, unite at a central point, where the operator is in constant attendance. On an alarm of fire in any locality, the watch or police of the district goes to a small box, kept in a conspicuous place, which he opens, and makes a telegraphic communication to the central operator, who, immediately recognizing the signal and the district from which it came, gives the alarm, by making each bell in connexion with the telegraph strike the number corresponding with the district in which the alarm commenced. By this means the alarm is communicated simultaneously to all parts of the city. This ingenious application of scientific principles has been in successful operation in the city of Boston long enough to prove its great value.

1191. The Atmospheric Telegraph.—An ingenious apparatus, called "*The Atmospheric Telegraph*," has recently been constructed by Mr. T. S. Richardson, of Boston, designed to send *packages* through continuous tubes by means of atmospheric pressure. An air-tight tube being laid between two places, either under or above ground, a piston, called by Mr. R. *a plunger*, is accurately fitted to its bore, behind which the package designed to be sent is attached. The air having been exhausted from the tube by engines at the opposite end, the pressure of the atmosphere will drive the piston, or plunger, with its load, forward to its proposed destination.

This ingenious application of atmospheric pressure operates with entire success in the model, and has been also successfully tested in tubes that have been laid to the extent of a mile. Patents have been secured for the invention in England, France, and other countries of Europe, as well as in this country; and a company is now forming for testing the principle between the cities of Boston and New York. The air is to be exhausted from the tubes by means of steam-engines, and there are to be intermediate stations between those two cities.

1192. The Electrotype Process.—This process, known by the various names electrotype, electro plating and gilding, galvanotype, galvano-plastic, electro-plastic and electro-metallurgy, is a process by which a coating of one metal is made to adhere to and take the form of another metal, by electrical agency.

1193. It is a process purely chemical and electrical, and the consideration of the subject pertains more properly to the science of Chemistry. As this volume has not professed to pursue a rigid classification, it may not be amiss to give this brief notice of the process.

1194. It consists in subjecting a chemical solution of one metal to electrical action with another metal. A solution of a salt or oxide, having a metallic base, forms part of the electric circuit, and, by the electrical action, the oxygen or acid will be drawn to the positive end of the circuit, while the pure metal will be forced to the negative pole, where it will either combine with the metal or adhere to it, taking its exact form. The thickness of the coating of the pure metal will depend on the length of time that the body to be coated is subjected to the combined action, chemical and electrical. Hence a mere film or a solid crust may be attached to any conducting substance.

When a substance not in itself a conductor is to be coated, it must first be made a conductor by covering its surface with some substance which will impart the conducting power. This is usually effected by means of finely-powdered black lead.

1195. When a part only of a body is to be coated by the electrotype process, the parts which are to remain uncoated must previously be protected by means of a thin covering of wax, tallow or some other non-conducting substance.

What is Magneto-electricity?

1196. MAGNETO-ELECTRICITY. — Magneto-electricity treats of the development of electricity by magnetism.

How is Magneto-electricity developed?

1197. Electric currents are excited in a conductor of electricity by magnetic changes taking place in its vicinity. Thus, the movement of a magnet near a metallic wire, or near an iron bar enclosed in a wire coil, occasions currents in the wire.

1198. When an armature, or any piece of soft iron, is brought into contact with one or both of the poles of a magnet, it becomes itself magnetic by induction, and by its reäction adds to the power of the magnet: on the contrary, when removed from the contact, it diminishes the power of the magnet, and these alternate changes in its magnetic state induce a current of electricity.

How are the most powerful effects of magneto-electricity obtained?

1199. The most powerful effects are obtained by causing a bar of soft iron, enclosed in a helix, to revolve *by mechanical means* near the poles of a steel magnet. As the iron approaches the poles in its revolution, it becomes magnetic; as it recedes from them, its magnetism disappears; and this alternation of magnetic states causes the flow of a current of electricity, which may be directed in its course to screw-cups, from which it may be received by means of wires connected with the cups.

Explain Fig. 181.

1200. THE MAGNETO-ELECTRIC MACHINE.— Fig. 181 represents the magneto-electric machine, in which an armature, bent twice at right angles, is made to revolve rapidly in front of the poles of a compound steel magnet of the U form. The U magnet, whose

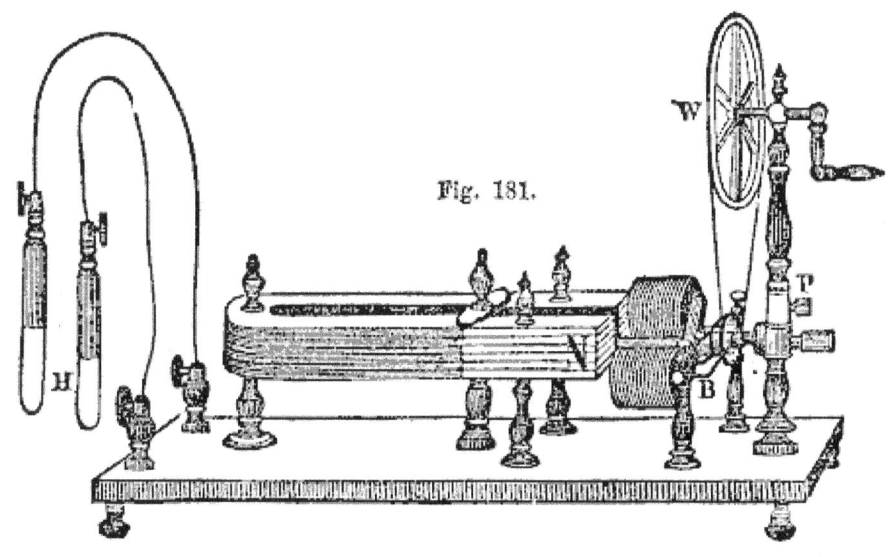

Fig. 181.

north pole is seen at N, is fixed in a horizontal position, with its poles as near the ends of the armature as will allow the latter to rotate without coming into contact with them. The armature is mounted on an axis, extending from the pillar P to a small pillar between the poles of the magnet. Each of its legs is enclosed in a helix of fine insulated wire. The upper part of the pillar P slides over the lower part, and can be fastened in any position by a binding screw. In this way the band connecting the two wheels may be tightened at pleasure, by increasing the distance between them. This arrangement also renders the machine more portable. By means of the multiplying-wheel W, which is connected by a band with a small wheel on the axis, the armature is made to revolve rapidly, so that the magnetism induced in it by the steel magnet is alternately destroyed and renewed in a reverse direction to the previous one. When the legs of the armature are approaching the magnet, the one opposite the north pole acquires south polarity, and the other north polarity. The magnetic power is greatest while the armature is passing in front of the poles. It gradually diminishes as the armature leaves this position, and nearly disappears when it stands at right angles with the magnet. As each leg of the armature approaches the other pole of the U magnet, by the continuance of the motion magnetism is again induced in it, but in the reverse direction to the previous one. These changes in the magnetic state of the armature excite electric currents in the surrounding helices, powerful in proportion to the rapidity with which the magnetic changes are produced.

1201. Shocks may thus be obtained from the machine, and, if the motion is very rapid, in a powerful machine the torrent of shocks becomes insupportable — the muscles of the hands which grasp the handles are involuntarily contracted, so that it is impossible to loosen the hold. The shocks, however, are instantly suspended by bringing the metallic handles into contact.

What is Thermo-electricity?

1202. THERMO-ELECTRICITY. — Thermo-electricity expresses a form of electricity developed by the agency of heat.

1203. In the year 1822, Professor Seebeck, of Berlin, discovered that currents of electricity might be produced by the partial application of heat to a circuit composed exclusively of *solid* conductors. The electrical current thus excited has been termed *Thermo-electric* (from the Greek *Thermos*, which signifies heat), to distinguish it from the common galvanic current; which, as it requires the intervention of a *fluid* element, was denominated a *Hydro-electric* current. The term *Stereo-electric* current has also been applied to the former, in order to mark its being produced in systems formed of solid bodies alone. It is evident that if, as is supposed in the theory of Ampère, magnets owe their peculiar properties to the continual circulation of electric currents in their minute parts, these currents will come under the description of the *stereo-electric* currents.

1204. From the views of electricity which have now been given, it appears that there are, strictly speaking, *three states* of electricity. That derived from the common electrical machine is in the highest degree of tension, and accumulates until it is able to force its way through the air, which is a perfect non-conductor. In the galvanic apparatus the currents have a smaller degree of tension; because, although they pass freely through the metallic elements, they meet with some impediments in traversing the *fluid* conductor. But in the thermo-electric currents the tension is reduced to nothing; because, throughout the whole course of the circuit, no impediment exists to its free and uniform circulation.

1205. If the junction of two dissimilar metals be heated, an electrical current will flow from the one to the other.

1206. Instead of two different metals, one metal in different conditions can be used to excite the current.

1207. Metals differ greatly in their power to excite a current when associated in thermo-electric pairs. A current may be excited with two wires of the same metal, by heating the end of one, and bringing it into contact with the other. This experiment is most successful when metals are used that have the lowest conducting power of heat.

1208. Thermo-electric batteries have been constructed with sufficient power to give shocks and sparks, and produce various magnetic phenomena, indicative of great magnetic power; but the limits of this volume will not allow a further consideration of the subject.

What is Astronomy?

1209. ASTRONOMY.— Astronomy treats of the heavenly bodies, the sun, moon, planets, stars and comets, and of the earth as a member of the solar system.

1210. The study of astronomy necessarily involves an acquaintance with mathematics, but there are many interesting facts, which have been fully established by distinguished astronomers, which ought to be familiar to those who have neither the opportunity nor the leisure to pursue the subject by the aid of mathematical light. To such the following brief notice of the subject will not be devoid of interest.

Who are some of the most distinguished Astronomers?

1211. Some of the most distinguished men who have contributed to the great mass of facts and laws which make up the science of Astronomy were Hipparchus, Ptolemy, Pythagoras, Copernicus, Tycho Brahe, Galileo, Kepler and Newton. The present century has added to this list many others whose fame will descend to posterity with great lustre.

1212. Hipparchus is usually considered the father of Astronomy. He was born at Nicæa, and died about a hundred and twenty-five years before the Christian era. He divided the heavens into constellations, twelve in the ecliptic, twenty-one in the northern, and sixteen in the southern hemisphere, and gave names to all the stars.

He discovered the difference of the intervals between the autumnal and vernal equinoxes, and, likewise, by viewing a tree on a plain, and noticing its apparent position from different places of observation, he was led to the discovery of the parallax of the heavenly bodies; that is, the difference between their real and apparent position, viewed from the centre and from the surface of the earth. He determined longitude and latitude, fixing the first degree of longitude at the Canaries.

1213. Ptolemy flourished in the second century of the Christian era. He was a native of Alexandria, or Pelusium. In his system he placed the earth in the centre of the universe,— a doctrine universally adopted and believed until the sixteenth century, when it was confuted and rejected by Copernicus. Ptolemy gave an account of the fixed stars, and computed the latitude and longitude of one thousand and twenty-two of them.

1214. Pythagoras was born at Samos, and his death is supposed to have taken place about five hundred years before the Christian era. He supposed the sun to be the centre of the universe, and that the planets revolved around him in elliptical orbits. This doctrine, however, was deemed absurd until it was established by Copernicus in the sixteenth century.

1215. Tycho Brahe, a Danish astronomer, flourished about the middle of the sixteenth century. His astronomical system was singular and absurd, but the science is indebted to him for a more correct catalogue of the fixed stars, and for discoveries respecting the motions of the moon and the comets, the refraction of the rays of light, and for many other important improvements. To him, also, was Kepler indebted for the principal facts which were the basis of his astronomical labors.

1216. Copernicus was born in Prussia, in the latter part of the fifteenth century. He revived the system of Pythagoras, which placed the sun in the centre of the system. He taught the true doctrine that the apparent motion of the heavenly bodies is caused by the real motion of the earth. But, for nearly a century after the publication of his system, he gained but few followers.

1217 Galileo, a native of Pisa, flourished in the latter part of the sixteenth century. By his observation of the planets Venus and Jupiter, he gained a decisive victory for the Copernican system. He was persecuted and imprisoned by the inquisition for holding what was thought, in that age of ignorance and superstition, to be heretical opinions, and compelled on his knees to abjure the truths which he had discovered, and which he had too much sense to disbelieve. Notice has already been taken of this distinguished philosopher in connexion with the laws of falling bodies (*see page* 52), for the discovery of which the world is indebted to him.

1218. Kepler, who, from his great discoveries, is called the *legislator* of the heavens, was a native of Wirtemberg, in 1571. Availing himself of the observations of Tycho Brahe, he discovered three great laws, known as Kepler's laws of the planetary motions, and on them were founded the discoveries of Newton, as well as the whole modern theory of the planets.

Kepler's laws could not have been discovered but for the observations of Tycho Brahe (as Kepler was not himself an observer), and no further discoveries could have been made than Kepler made but for the telescope of Galileo. It has elsewhere been stated that Galileo was indebted to Jansen, of Holland, for the idea of the telescope. But, since the days of Galileo, the telescope has been most wonderfully improved, and invested with almost inconceivable powers. Herschel computed that the power of his telescope was so great as to penetrate a space through which light (moving with the prodigious velocity of 200,000 miles in a second of time) would require 350,000 years to reach us. But the great telescope of Lord Rosse would probably reach an object ten times more remote.

1219. Sir Isaac Newton, who has been called the Creator of Natural Philosophy, was born in Lincolnshire, England, in 1642. His discovery of the universal law of gravitation, and many other valuable and important contributions which he made to science, place him among the foremost of those to whom the world is indebted for an insight into the magnificent displays of the material world.

Give an account of the solar system as now adopted.

1220. According to the system of Astronomy which is now universally adopted, the sun is the centre of a system of heavenly bodies, called planets, which revolve around him as a centre.

Secondly. The earth is one of these planets.

Thirdly. That some of these planets are attended by satellites or moons, which revolve around their respective planets, and with them around the sun.

Fourthly. That the size, distance and rapidity of motion of each of these planets is known to be different.

Fifthly. That the stars are all of them suns, with systems of their own, and probably many, if not all of them, having planets, with their moons revolving around them as centres.

Sixthly. That there is a central point of the universe, around which all systems revolve.

What is meant by the Solar System? 1221. OF THE SOLAR SYSTEM.—By the Solar System is meant the sun and all the heavenly bodies which revolve around it. These are the planets with their satellites or moons, our earth with its moon, together with an unknown number of comets.

What are Primary Planets? 1222. OF THE PRIMARY PLANETS.—Those bodies which revolve around the sun, without revolving, at the same time, around some other central body, are called Primary Planets.

Give the names of the eight primary planets. 1223. For many years the planets were considered to be six in number only, and they were all, except our earth, named after the gods of heathen mythology,—Mercury, Venus, Earth, Mars, Jupiter, and Saturn. In the year 1781, Sir William Herschel discovered another, to which the name of Uranus has been given; and in the year 1846 an eighth was discovered, to which the name of *Le Verrier* was at first given, from a distinguished French astronomer, by means of whom it was pointed out. It is now known by the name of Neptune.

How many minor primary planets have been discovered? 1224. Besides these primary planets, it was discovered, between the years 1800 and 1807, that between Mars and Jupiter there were four smaller planets, of such diminutive size, compared with the others, that they were called Asteroids. Since the year 1845 thirty-one more have been discovered, so that there are now known to be no fewer than thirty-five asteroids, or minor planets, between the orbits of Mars and Jupiter.

1225. THE MINOR PLANETS.—The following is a catalogue of the minor planets at present known, arranged in the order of their discovery, together with the other known planets of our solar system:

Name and Number by which the Minor Planets are known.	Date of Discovery.	Names of Discoverers.
SUN.		
MERCURY.		
VENUS.		
THE EARTH.		
MARS.		
1. Ceres	1801..Jan. 1	Piazzi, of Sicily.
2. Pallas	1802..March 28	Olbers, of Bremen.
3. Juno	1804..Sept. 1	Harding.
4. Vesta	1807..March 29	Olbers.
5. Astrea	1845..Dec. 8	Hencke, of Germany.
6. Hebe	1847..July 1	Hencke.
7. Iris	1847..August 13	Hind, of London.
8. Flora	1847..Oct. 18	Hind.
9. Metis	1848..April 26	Graham, of Ireland.*
10. Hygeia	1849..April 12	De Gasparis, of Naples.
11. Parthenope	1850..May 11	De Gasparis.
12. Clio	1850..Sept. 13	Hind.
13. Egeria	1850..Nov. 2	De Gasparis.
14. Irene	1851..May 19	Hind.
15. Eunomia	1851..July 29	De Gasparis.
16. Psyche	1852..March 17	De Gasparis.
17. Thetis	1852..April 17	Luther, of Germany.
18. Melpomene	1852..June 25	Hind.
19. Fortuna	1852..August 22	Hind.
20. Massilia	1852..Sept. 19	De Gasparis.
21. Lutetia	1852..Nov. 15	Goldschmidt, Paris.
22. Calliope	1852..Nov. 16	Hind.
23. Thalia	1852..Dec. 15	Hind.
24. Themis	1853..April 5	De Gasparis.
25. Phocæa	1853..April 6	Chacornac, of Marseilles.
26. Proserpina	1853..May 5	Luther.
27. Euterpe	1853..Nov. 8	Hind.
28. Bellona	1854..March 1	Luther.
29. Amphitrite	1854..March 1	Marth, of London.
30. Urania	1854..July 22	Hind.
31. Euphrosyne	1854..Sept. 1	Ferguson, of Washington.
32. Pomona	1854..Oct. 28	Goldschmidt.
33. Polhymnia	1854..Oct. 28	Chacornac.
34.*	1855..April 14	Chacornac.
35.*	1855..April 27	
JUPITER.		
SATURN.		
URANUS	1781	Sir William Herschel.
NEPTUNE	1846..Sept. 23	Dr. Galle, of Berlin, by direction of Le Verrier, of Paris.

* To the last two asteroids in the list no names have as yet been given. It is proper to be observed that the asteroids are frequently known better by their *numbers* than by their names. Thus ◯ represents Polhymnia, and ◯ Calliope, &c.

What is the difference between a planet and a star?

1226. The name *planet* properly means a *wandering star*, and was given to this class of the heavenly bodies because they are constantly moving, while those bodies which are called *fixed stars* preserve their relative positions. The planets may likewise be distinguished from the fixed stars by the eye by their steady light, while the fixed stars, on the contrary, appear to *twinkle*.

1227. The sun, the moon, the planets, and the fixed stars, which appear to us so small, are supposed to be large worlds, of various sizes, and at different but immense distances from us. The reason that they appear to us so small is, that on account of their immense distances they are seen under a small angle of vision.

What universal law keeps the planets and other heavenly bodies in their places?

1228. It has been stated, in the early pages of this book, that *every portion of matter* is attracted by every other portion, and that the force of the attraction depends upon the *quantity of matter* and the *distance*. As attraction is *mutual*, we find that all of the heavenly bodies attract the earth, and the earth likewise attracts all of the heavenly bodies. It has been proved that a body when actuated by several forces will be influenced by each one, and will move in a direction *between* them. It is so with the heavenly bodies; each one of them is attracted by every other one; and these attractions are so nicely balanced by creative wisdom, that, instead of rushing together in one mass, they are caused to move in regular paths (called *orbits*) around a central body, which, being attracted in *different* directions by the bodies which revolve around it, will itself revolve around the centre of gravity of the system. Thus, the sun is the centre of what is called the *solar* system, and the planets revolve around it in different times, at different distances, and with different velocities.

What is meant by an orbit?

1229. The paths or courses in which the planets move around the sun are called their orbits.

All of the heavenly bodies move in conic sections,* namely, the circle, the ellipse, the parabola and the hyperbola.

What is meant by a year? 1230. In obedience to the universal law of gravitation, the planets revolve around the sun as the centre of their system; and the time that each one takes to perform an entire revolution is called its year. Thus, the planet Mercury revolves around the sun in 87 of our days; hence a year on that planet is equal to 87 days. The planet Venus revolves around the sun in 224 days; that is, therefore, the length of the year of that planet. Our earth revolves around the sun in about 365 days and 6 hours. Our year, therefore, is of that length.

1231. The length of time that each planet takes in performing its revolution around the sun, or, in other words, the length of the year on each planet, is as follows. (*The fractional parts of the day are omitted.*) In the same connexion will also be found the mean distance of each planet from the sun, and the time of revolution around its axis; or, in other words, the length of the day on each.

	Length of the Year in Days.	Mean Distance from the Sun in millions of Miles.	Length of the Day in Hours and Minutes.
Mercury	87	36¾	24 5'
Venus	224	68¾	23 21'
Earth	365	95	24 00'
Mars	686	145	24 39'
1. Ceres	1,680		
2. Pallas	1,688		
3. Juno	1,592		
4. Vesta	1,325		
5. Astrea			
6. Hebe			
7. Iris			
8. Flora			
9. Metis	Between 1,400 and 2,100	About 266	
10. Hygeia			
11. Parthenope			
12. Clio			
13. Egeria			
14. Irene			
15. Eunomia			
16. Psyche	1,835		

* Conic sections are curvilinear figures, so called because they can all be formed by cutting a cone in certain directions. If a cone be cut perpendicular to its axis, the surface cut will be a *circle*. If cut oblique to the axis, the surface cut will be an *ellipse*. If cut parallel to the slope of the cone, the section will be a *parabola*. If cut parallel to the axis, the section will be an *hyperbola*.

	Length of the Year in Days.	Mean Distance from the Sun in millions of Miles.	Length of the Day in Hours and Minutes.
17. Thetis........................	1,430		
18. Melpomene................	1,269		
19. Fortuna.....................	1,396		
20. Massilia.....................	1,359	About 266	
21. Lutetia......................	1,387		
22. Calliope	1,815		
23. Thalia.......................	1,571		
24. Themis......................	2,037		
25. Phocæa.....................			
26. Proserpina.................			
27. Euterpe			
28. Bellona.....................			
29. Amphitrite			
30. Urania.......................			
31. Euphrosyne................			
32. Pomona.....................			
33. Polhymnia			
34. } Unnamed...............			
35. }			
JUPITER...........................	4,332	494	9 55'
SATURN...........................	10,759	906	10 16'
URANUS	30,686	1,824	
NEPTUNE	60,126	2,856	

The sun turns on its axis in about 25 days and 10 hours.

Give an account of Bode's law.

1232. There is a very remarkable law, discovered by Professor Bode, founded, it is true, on no known mathematical principle, but which has been found to accord so exactly with other calculations, that it is recognized as Bode's law for estimating the distances of the planets from the sun. Thus :

Write the arithmetical progression,

$$0, 3, 6, 12, 24, 48, 96, 192, 384.$$

To each of the series add 4, and we have the sums,

$$4, 7, 10, 16, 28, 52, 100, 196, 388,$$

which will represent very nearly the comparative distance of each planet. Now, the distance of the earth from the sun is 95 millions of miles, and as that distance is represented in the progression by 10, it follows that the distance of Mercury is $\frac{4}{10}$ of 95 millions, of Venus $\frac{7}{10}$, &c.

What led to the discovery of the minor planets?

1233. It is to be observed, however, that before the discovery of the minor planets, there was a very remarkable interval between the planets Mars and Jupiter, and that Bode's law, which seemed to accord with the distance of all the other planets, appeared here to fail in its application. Kepler had suspected that an undiscovered planet existed in the interval; but it was not certainly known until a number of distinguished observers assembled at Lilienthal, in Saxony, in 1800, who resolved to direct their observations especially to that part of the heavens where the unknown planet was supposed to be. The result of the labors of these observers, and others who have followed them, has been the discovery of the thirty-five minor planets, all situated between the planets Mars and Jupiter. But these *What opinion has been formed in relation to the minor planets?* minor planets are so small, and their paths or orbits vary so little, that it has been conjectured that they originally formed one large and resplendent orb, which, by the operation of some unknown cause, has exploded and formed the minor planets that revolve in orbits very near that of the original planet.

1234. Of these thirty-five small bodies, which are quite invisible without the aid of a good telescope, *ten* were discovered by Mr. Hind, of Mr. Bishop's private observatory, Regent's Park, London; *seven* by De Gasparis, of Naples; *three* by Chacornac, at Marseilles; *three* by Luther, at Bilk, Germany; *two* by Olbers, of Bremen; *two* by Hencke, of Driessen, Germany; *two* by Goldschmidt, at Paris; and *one* each by Piazzi, of Palermo; Harding, of Lilienthal, Germany; Graham, at Mr. Cooper's private observatory, Markree Castle, Ireland; Marth, of London; and Ferguson, of Washington.

What is the shape of the orbits of the planets? 1235. The paths or orbits of the planets are not exactly circular, but elliptical. They are, therefore, sometimes nearer to the sun than at others. The mean distance is the medium between their greatest and least distance. Those planets which are nearer to the sun than the earth are called inferior planets, because their orbits are *within* that of the earth; and those which are further from the sun are called superior planets, because their orbits are *outside* that of the earth.

Give the relative size of the sun, moon, and primary planets. 1236. The relative size of the sun, the moon and the larger planets, as expressed by the length of their diameters, is as follows:

344   NATURAL PHILOSOPHY.

Sun	882,000	Mars	4,100
Moon	2,153	Jupiter	88,640
Mercury	2,950	Saturn	75,000
Venus	7,800	Uranus	34,500
Earth	7,912	Neptune	37,500

How large are the minor planets? 1237. The size of the minor planets has been so variously estimated, that little reliance can be placed on the calculations. Some astronomers estimate them as a little over 1000 miles, while others place them much below that standard. Vesta has been described as presenting a pure white light; Juno, of a reddish tinge, and with a cloudy atmosphere; Pallas is also stated as having a dense, cloudy atmosphere; and Ceres, as of a ruddy color. These four undergo various changes in appearance, and but little is known of any of them, except their distance and time of revolution.

Explain Fig. 182. 1238. Fig. 182 is a representation of the comparative size of the larger planets.

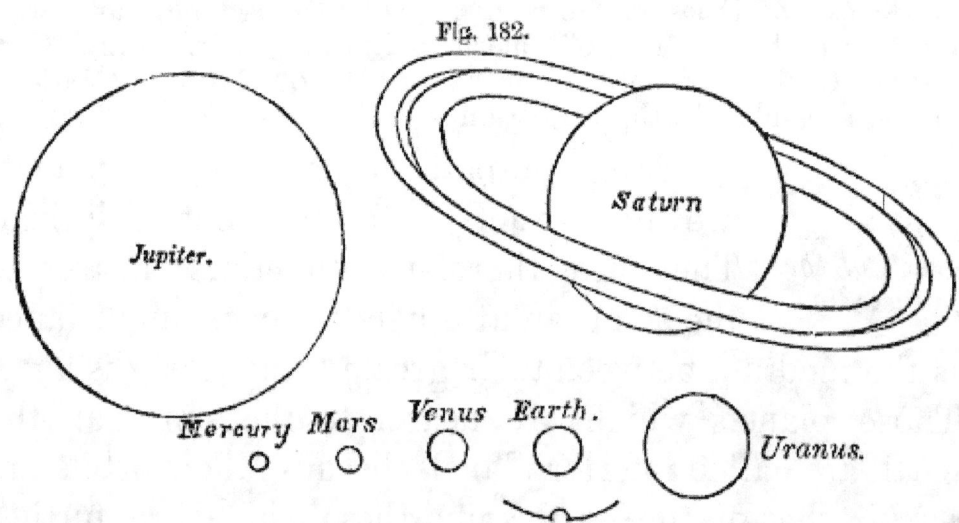

Fig. 182.

Sir J. F. W. Herschel gives the following illustration of the comparative size and distance of the bodies of the solar system. "On a well-levelled field place a globe two feet in diameter, to represent the Sun; Mercury will be represented by a grain of mustard-seed, on the circumference of a circle 164 feet in diameter for its orbit; Venus, a pea, on a circle 284 feet in diameter; the Earth, also a pea, on a circle of 430 feet; Mars, a rather large pin's head; on a circle of 654 feet; Juno, Ceres, Vesta, and Pallas, grains of sand, in orbits of from 1,000 to 1,200 feet; Jupiter, a moderate-sized orange, in a circle nearly half a mile in diameter; Saturn, a small orange, on a circle of four-fifths of a mile; Uranus, a full-sized cherry, or small plum, upon the circumference of a circle more than a mile and a half; and Neptune, a good-sized plum, on a circle about two miles and a half in diameter.

"To imitate the motions of the planets in the above-mentioned orbits, Mercury must describe its own diameter in 41 seconds; Venus, in 4 minutes and 14 seconds; the Earth, in 7 minutes; Mars, in 4 minutes and 48 seconds; Jupiter, in 2 hours, 56 minutes; Saturn, in 3 hours, 13 minutes; Uranus, in 12 hours, 16 minutes; and Neptune, in 3 hours, 30 minutes."

What is the Ecliptic, and why is it so called?

1239. The Ecliptic is the apparent path of the sun, or the real path of the earth.

It is called the ecliptic, because every *eclipse*, whether of the sun or the moon, must be in or near it.

What is the Zodiac?

1240. The Zodiac is a space or belt, sixteen degrees broad, eight degrees each side of the ecliptic.

It is called the *zodiac* from a Greek word, which signifies *an animal*, because all the stars in the twelve parts into which the ancients divided it were formed into constellations, and most of the twelve constellations were called after some animal.

1241. Sir J. F. W. Herschel, in his excellent treatise on Astronomy, says: "Uncouth figures and outlines of men and monsters are usually scribbled over celestial globes and maps, and serve, in a rude and barbarous way, to enable us to talk of groups of stars, or districts in the heavens, by names which, though absurd or puerile in their origin, have obtained a currency from which it would be difficult to dislodge them. In so far as they have really (as some have) any slight resemblance to the figures called up in imagination by a view of the more splendid 'constellations,' they have a certain convenience; but as they are otherwise entirely arbitrary, and correspond to no *natural* subdivisions or groupings of the stars, astronomers treat them lightly, or altogether disregard them, except for briefly naming remarkable stars, as '*Alpha Leonis*,' '*Beta Scorpii*,' &c., by letters of the Greek alphabet attached to them.

"This disregard is neither supercilious nor causeless. The constellations seem to have been almost purposely named and delineated to cause as much confusion and inconvenience as possible. Innumerable snakes twine through long and contorted areas of the heavens, where no memory can follow them; bears, lions, and fishes, large and small, northern and southern, confuse all nomenclature, &c. A better system of constellations might have been a material help as an artificial memory."

What are the signs of the zodiac, and how many degrees in each?   1242. The zodiac is divided into twelve signs, each sign containing thirty degrees of the great celestial circle. The names of these signs are sometimes given in Latin, and sometimes in English. They are as follows:

Latin.	English.	Latin.	English.
(1) Aries,	The Ram.	(7) Libra,	The Balance.
(2) Taurus,	The Bull.	(8) Scorpio,	The Scorpion.
(3) Gemini,	The Twins.	(9) Sagittarius,	The Archer.
(4) Cancer,	The Crab.	(10) Capricornus,	The Goat.
(5) Leo,	The Lion.	(11) Aquarius,	The Water-bearer.
(6) Virgo,	The Virgin.	(12) Pisces,	The Fishes.

1243. The signs of the zodiac and the various bodies of the solar system are often represented in almanacs and astronomical works, by signs or characters.

In the following list the characters of the planets, &c., are represented.

☉ The Sun.   ⊕ The Earth.   ⚳ Ceres.
☾ The Moon.   ♂ Mars.   ⚴ Pallas.
☿ Mercury.   ⚶ Vesta.   ♃ Jupiter.
♀ Venus.   ⚵ Juno.   ♄ Saturn.
              ⛢ Uranus.

The following characters represent the signs of the Zodiac.

♈ Aries.   ♌ Leo.   ♐ Sagittarius.
♉ Taurus.   ♍ Virgo.   ♑ Capricornus.
♊ Gemins.   ♎ Libra.   ♒ Aquarius.
♋ Cancer.   ♏ Scorpio.   ♓ Pisces.

From an inspection of Fig. 183 it appears that when the earth as seen from the sun, is in any particular constellation, the sun, as viewed from the earth, will appear in the opposite one.

Have the signs of the zodiac always remained the same, and why?

1244. The *constellations* of the zodiac do not now retain their original names. Each constellation is about 30 degrees eastward of the sign of the same name. For example, the constellation Aries is 30 degrees eastward of the sign Aries, and the constellation Taurus 30 degrees eastward of the sign Taurus, and so on. Thus the sign Aries lies in the constellation Pisces; the sign Taurus, in the constellation Aries; the sign Gemini, in the constellation Taurus, and so on. Hence the importance of distinguishing between the *signs* of the zodiac and the *constellations* of the zodiac. The cause of the difference is the precession of the equinoxes, a phenomenon which will be explained in its proper connexion.

How are the orbits of the planets situated with respect to that of the earth?

1245. The orbits of the other planets are inclined to that of the earth; or, in other words, they are not in the same plane.

Explain Fig. 183.

Fig. 183 represents an oblique view of the plane of the ecliptic, the orbits of all the primary planets, and of the comet of 1680. That part of each orbit which is above the plane is shown by a white line; that which is below it, by a dark line. That part of the orbit of each planet where it crosses the ecliptic, or, in other words, where the white and dark lines in the figure meet, is called the node of the planet, from the Latin *nodus*, a knot or tie.

Explain Fig. 184.

1246. Fig. 184 represents a section of the plane of the ecliptic, showing the inclination of the orbits of the planets. As the zodiac extends only eight degrees on each side of the ecliptic, it appears from the figure that the orbits of some of the planets are wholly in the zodiac, while those of others rise above and descend below it. Thus the orbits of Juno, Ceres and Pallas, rise above, while those of all the other planets are confined to the zodiac.

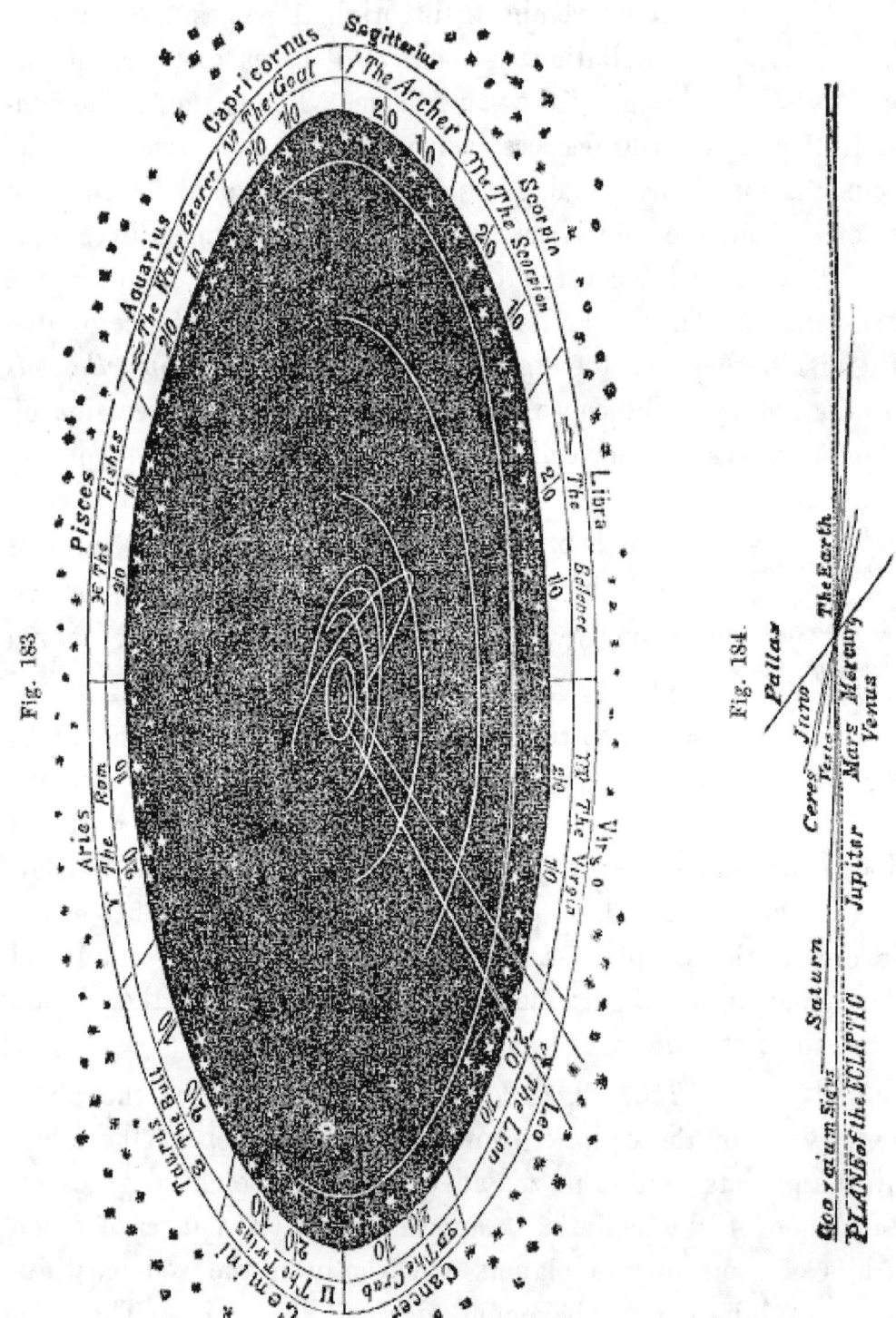

Fig. 183.

Fig. 184.

When is a heavenly body said to be in any constellation?

1247. When a planet or heavenly body is in that part of its orbit which appears to be near any particular constellation, it is said to be in that constellation.

Thus, in Fig. 147, the comet of 1680 appears to approach the sun from the constellation Leo.

What is meant by the perihelion and aphelion, the perigee and apogee, of a heavenly body?

1248. The perihelion* and aphelion* of a heavenly body express its situation with regard to the sun. When a body is nearest to the sun, it is said to be in its perihelion. When furthest from the sun, it is said to be in its aphelion.

1249. The earth is three millions of miles nearer to the sun in its perihelion than in its aphelion.

The apogee* and perigee* of a heavenly body express its situation with regard to the earth. When the body is nearest to the earth, it is said to be in perigee; when it is furthest from the earth, it is said to be in apogee.

Where is the perihelion and aphelion of the earth?

1250. The perihelia of the planets, *as seen from the sun*, are in the following signs of the zodiac, namely: Mercury in Gemini, Venus in Leo, the Earth in Cancer, Mars in Pisces, Vesta in Sagittarius, Juno in Taurus, Ceres in Leo, Pallas in Leo, Jupiter in Aries, Saturn in Cancer, Uranus in Virgo, and Neptune in Taurus.

What is meant by the inferior and superior conjunction and opposition of a planet?

1251. When a planet is so nearly on a line with the earth and the sun as to pass between them, it is said to be in its *inferior conjunction*; when behind the sun, it is said

---

* The plural of *Perihelion* is *Perihelia*, and of *Aphelion* is *Aphelia*. The words *perihelion, aphelion, apogee,* and *perigee*, are derived from the Greek language, and have the following meaning:

   Perihelion, *near* the sun.
   Aphelion, *from* the sun.
   Perigee, *near* the earth.
   Apogee, *from* the earth

to be in its *superior conjunction;* but when behind the earth, it is said to be in *opposition*.

What is the inclination of the axes of the planets to the plane of their orbits?

1252. The axes of the planets, in their revolution around the sun, are not perpendicular to their orbits, nor to the plane of the ecliptic, but are inclined in different degrees.

What causes the seasons?
What causes the differences in the length of the days and nights?

1253. This is one of the most remarkable circumstances in the science of Astronomy, because it is the cause of the different seasons, spring, summer, autumn and winter; and because it is also the cause of the difference in the length of the days and nights in the different parts of the world, and at the different seasons of the year.

1254. The motion of the heavenly bodies is not uniform. They move with the greatest velocity when they are *in perihelion*, or in that part of their orbit which is nearest to the sun; and slowest when *in aphelion*.

1255. It was discovered by Kepler, and proved by Newton, that if a line is drawn from the sun to either of the planets, this line passes over or describes equal areas in equal times. This line is called the *radius-vector*. This is one of Kepler's great laws.

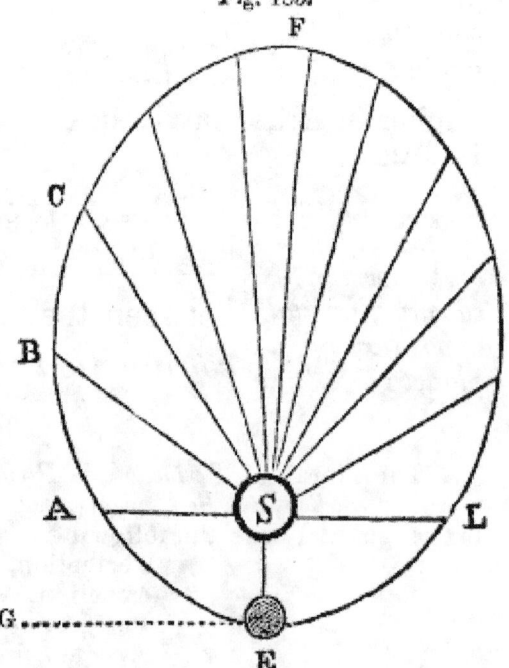

Fig. 185.

Explain Fig. 185. In Fig. 185, let S represent the sun, and E the earth, and the ellipse or oval, be the earth's orbit, or path around the sun. By lines drawn from the sun at S to the outer edge of the figure, the orbit is divided into twelve areas of different shapes, but each containing the same quantity of space. Thus, the spaces E S A, A S B, D S C, &c., are all supposed to be equal. Now, if the earth in the space of one month will move in its orbit from E to A, it will in another month move from A to B, and in the third month from B to C, &c., and thus its radius vector will describe equal areas in equal times.

The reason why the earth (or any other heavenly body) moves with a greater degree of velocity in its perihelion than in its aphelion may likewise be explained by the same figure. Thus:

The earth, in its progress from F to L, being *constantly* urged forward by the sun's attraction, must (as is the case with a falling body) move with an accelerated motion. At L, the sun's attraction becomes stronger, on account of the nearness of the earth; and consequently in its motion from L to E the earth will move with greater rapidity. At E, which is the perihelion of the earth, it acquires its greatest velocity. Let us now detain it at E, merely to consider the direction of the forces by which it is urged. If the sun's attraction could be destroyed, the force which has carried it from L to E would carry it off in the dotted line from E to G, which is a tangent to its orbit. But, while the earth has this tendency to move towards G, the sun's attraction is continually operating with a tendency to carry it to S. Now, when a body is urged by two forces, it will move between them; but, as the sun's attraction is constantly exerted, the direction of the earth's motion will not be in a straight line, the diagonal of *one* large parallelogram, but through the diagonal of a number of infinitely small parallelograms; which, being united, form the curve line E A.

It is thus seen that while the earth is moving from L to E the attraction of the sun is stronger than in any other part of its orbit, and will cause the earth to move rapidly. But in its motion from E to A, from A to B, from B to C, and from C to F, the attraction of the sun, operating in an opposite direction, will cause its motion from the sun to be retarded, until, at F, the direction of its motion is reversed, and it begins again to approach the sun. Thus it appears that in its passage from the perihelion to the aphelion the motion of the earth, as well as that of all the heavenly bodies, must be constantly retarded, while in moving from their aphelion to perihelion it is constantly accelerated, and at their perihelion the velocity will be the greatest. The earth, therefore, is about seven days longer in performing the aphelion part of its orbit than in traversing the perihelion part; and the revolution of all the other planets, being the result of the same cause, is affected in the same manner as that of the earth.

What are the three laws of Kepler? 1256. The other two great laws discovered by Kepler, on which the discoveries of Newton, as well as the whole modern theory of the planets, are based, are —

1257. (1.) That the planets do not move in circles, but in ellipses, of which the sun is in one of the foci.

1258. (2.) In the motion of the planets, the squares of the times of revolution are as the cubes of the mean distances from the sun.

It was by this law that, in the want of other means, the distance of the planet Uranus from the sun was estimated.

How much nearer is the earth to the sun in summer than in the winter? [Be careful not to be caught in this question.] 1259. The earth is about three millions of miles nearer to the sun in winter than in summer. The heat of summer, therefore, can be only partially affected by the distance of the earth from the sun.

The sun is nearest to the earth in the summer of the southern hemisphere, and the heat is more intense there than in corresponding latitudes of the north. This is due to the greater amount of land in the northern hemisphere, which by its radiating power diffuses the heat more equally.

When is the heat of the sun the greatest? 1260. On account of the inclination of the earth's axis, the rays of the sun fall more or less obliquely on different parts

ASTRONOMY. 353

of the earth's surface at different seasons of the year. The heat is always the greatest when the sun's rays fall *vertically;* and the more obliquely they fall, the fewer of them fall on any given space.

This is the reason why the days are hottest in summer, although the earth is further from the sun at that time.

Explain Fig. 186 1261. Fig. 186 represents the manner in which the rays of the sun fall upon the earth in summer and in winter. The north pole of the earth, at all seasons, constantly points to the north star N; and, when the earth is nearest to the sun, the rays from the sun fall as indicated by W in the figure; and, as their direction is very oblique, and they have a larger

Fig. 186.

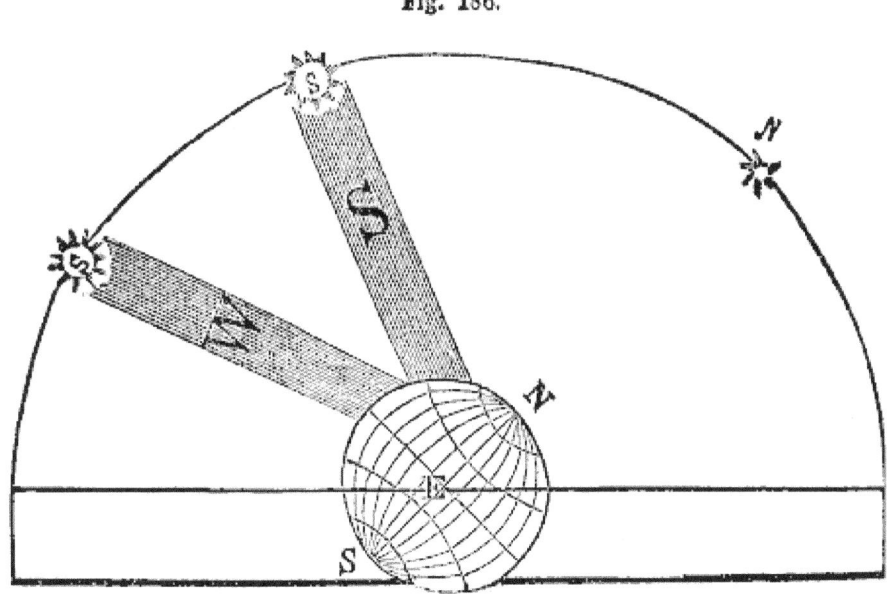

portion of the atmosphere to traverse, much of their power is lost. Hence we have *cold* weather when the earth is nearest to the sun. But when the earth is in aphelion the rays fall almost vertically or perpendicularly, as represented by S in the figure and, although the earth is then nearly three millions of miles further from the sun, the heat is greatest, because the rays fall more directly, and have a less portion of the atmosphere to traverse.

This may be more familiarly explained by comparing summer rays to a ball or stone thrown directly *at* an object, so as to

30*

strike it with all its force; and winter rays to the same ball or stone thrown obliquely, so as merely to *graze* the object.

Why is it cooler early in the morning than in the middle of the day? 1262. For a similar reason we find, even in summer, that early in the morning and late in the afternoon it is much cooler than at noon, because the sun then shines more obliquely. The heat is generally the greatest at about three o'clock in the afternoon; because the earth retains its heat for some length of time, and the additional heat it is constantly receiving from the sun causes an elevation of temperature, even after the rays begin to fall more obliquely.

What causes the different climates in different parts of the world? 1263. It is the same cause which occasions the variety of climate in different parts of the earth. The sun always shines in a direction nearly perpendicular, or vertical, on the equator, and with different degrees of obliquity on the other parts of the earth. For this reason, the greatest degree of heat prevails at the equator during the whole year. The further any place is situated from the equator, the more obliquely will the rays fall at different seasons of the year, and, consequently, the greater will be the difference in the temperature.

What places will have the coolest temperature? 1264. If the axis of the earth were perpendicular to its orbit, those parts of the earth which lie under the equator would be constantly opposite to the sun; and as, in that case, the sun would, at all times of the year, be vertical to those places equally distant from both poles, so the light and heat of the sun would be dispersed with perfect uniformity towards each pole; we should have no variety of seasons; day and night would be of the same length, and the heat of the sun would be of the same intensity every day throughout the year.

What effects are produced by the inclination of the earth's axis? 1265. *It is*, therefore, as has been stated, *owing to the inclination of the earth's axis that we have the agreeable variety of the seasons, days and nights of different lengths, and that wisely-ordered variety of climate which causes so great a variety of productions, and which has afforded so powerful a stimulus to human industry.*

1266. The wisdom of Providence is frequently displayed in *apparent* inconsistencies. Thus, the very circumstance which, to the *short-sighted* philosopher, appears to have thrown an insurmountable barrier between the scattered portions of the human race, has been wisely ordered to establish an interchange of blessings, and to bring the ends of the earth in communion. Were the same productions found in every region of the earth, the stimulus to exertion would be weakened, and the wide field of human labor would be greatly diminished. It is our mutual *wants* which bind us together.

1267. In order to understand the illustration of the causes of the seasons, &c., it is necessary to have some knowledge of the circles which are drawn on the artificial representations of the earth. It is to be remembered that all of these circles are wholly imaginary; that is, that there are on the earth itself no such circles or lines. They are drawn on maps merely for the purpose of illustration.

Explain Fig. 187. 1268. Fig. 187 represents the earth. N S is the axis, or imaginary line, around which it daily turns; N is the north pole, S is the south pole. These poles, it will be seen, are the extremities of the axis N S. C D represents the equator, which is a circle around the earth, at an equal distance from each pole. The curved lines proceeding from N to S are meridians. They are all circles surrounding the earth, and passing through the poles. These meridians may be multiplied at pleasure.

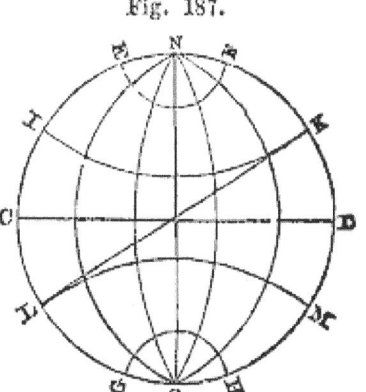

Fig. 187.

The lines E F, I K, L M, and G H, are designed to represent circles all of them parallel to the equator, and for this reason they are called parallels of latitude. These also may be multiplied at pleasure.

But in the figure these lines, which are parallel to the equator, and which are at a certain distance from it, have a different name, derived from the manner in which the sun's rays fall on the surface of the earth.

Thus the circle I K, 23½ degrees from the equator, is called the *tropic* of Cancer, and the circle L M is called the *tropic* of Capricorn. The circle E F is called the Arctic Circle. It represents the limit of perpetual day when it is summer in the northern hemisphere, and of perpetual night when it is winter.

On the 21st of March the rays of the sun fall vertically on the equator, and on each succeeding day on places a little to the north, until the 21st of June, when they fall vertically on places 23½ degrees north of the equator. Their vertical direction then *turns* back again towards the equator, where the rays again fall vertically on the 23d of September, and on the succeeding days a little to the south, until the 21st of December, when they fall vertically on the places 23½ south of the equator. Their vertical direction then again *turns* towards the equator. Hence the circles I K and L M are called the *tropics* of Cancer and Capricorn. The word *tropic* is derived from a word which signifies *to turn*. The tropics, therefore, are the boundaries of the sun's apparent path north and south of the equator, or the lines at which the sun *turns* back.

The circle G H is the Antarctic Circle, and represents the limit of perpetual day and night in the southern hemisphere. The line L K represents the circle of the ecliptic, which, as has already been stated, is the *apparent* path of the sun, or the *real* path of the earth. This circle, although it is generally drawn on the terrestrial globe, is, in reality, a circle in the heavens; and differs from the zodiac only in its width, — the zodiac extending eight degrees on each side of the ecliptic.

Explain Fig. 188. 1269. Fig. 188 represents the manner in which the sun shines on the earth in different parts of its orbit; or, in other words, the cause of the change in the seasons. S represents the sun, and the dotted oval, or ellipse, A B C D, the orbit of the earth. The outer circle represents the zodiac, with the position of the twelve signs or constellations. On the 21st of June, when the earth is at D, the whole northern polar region is continually in the light of the sun. As it turns on its axis, therefore, it will be day to all the parts which are exposed to the light of the sun. But, as the whole of the Antarctic circle is within the line of perpetual darkness, the sun can shine on no part of it. It will, therefore, be constant night to all places within that circle. As the whole of the Arctic circle is within

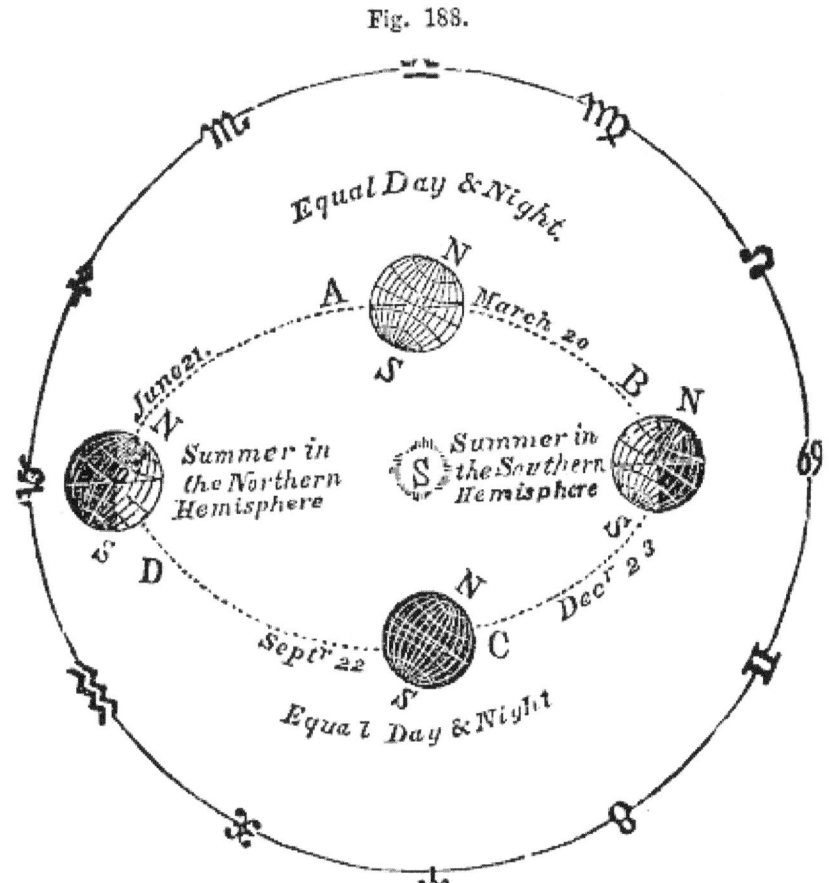

Fig. 188.

the line of perpetual light, no part of that circle will be turned from the sun while the earth turns on its axis. To all places, therefore, within the Arctic circle, it will be constant day.

On the 22d of September, when the earth is at C, its axis is neither inclined *to* nor *from* the sun, but is sidewise; and, of course, while one-half of the earth, from pole to pole, is enlightened, the other half is in darkness, as would be the case if its axis were perpendicular to the plane of its orbit; and it is this which causes the days and nights of this season of the year to be of equal length.

On the 23d of December the earth has progressed in its orbit to B, which causes the whole space within the northern polar circle to be continually in darkness, and more of that part of the earth north of the equator to be in the shade than in the light of the sun. Hence, on the 21st of December, at all places north of the equator the days are shorter than the nights, and at all places south of the equator the days are longer than the nights. Hence, also, within the Arctic circle it is uninterrupted night, the sun not shining at all; and within the Antarctic circle it is uninterrupted day, the sun shining all the time.

On the 20th of March, the earth has advanced still further, and is at A, which causes its axis, and the length of the days and nights, to be the same as on the 20th of September.

What is meant by the Equinoxes and the Solstices?

1270. From the explanation of figure 198, it appears that there are two parts of its orbit in which the days and nights are equal all over the earth. These points are in the sign of Aries and Libra, which are therefore called the equinoxes. Aries is the vernal (or spring) equinox, and Libra the autumnal equinox.

1271. There are also two other points, called solstices, because the sun appears to *stand* at the *same height* in the heavens in the middle of the day for several days. These points are in the signs Cancer and Capricorn. Cancer is called the summer solstice, and Capricorn the winter solstice.

How are day and night caused, and what is the reason of the difference in their length?

1272. Day and night are caused by the rotation of the earth on its axis every 24 hours. It is day to that side of the earth which is towards the sun, and night to the opposite side. The length of the days is in proportion to the inclination of the axis of the earth *towards* the sun. It may be seen, by the above figure, that in summer the axis is most inclined towards the sun, and then the days are the longest. As the north pole becomes less inclined, the days shorten, till on the 21st of December it is inclined 23½ degrees *from* the sun, when the days are the shortest. Thus, as the earth progresses in its orbit, after the days are the shortest, it changes its inclination towards the sun, till it is again inclined as in the longest days in the summer.

Which of the planets has the greatest difference in its seasons? 1273. As the difference in the length of the days and the nights, and the change of the seasons, &c., on the earth, is caused by the inclination of the earth's axis, it follows that all the planets whose axes are inclined must experience the same vicissitude, and that it must be in proportion to the degree of the inclination of their axes. As the axis of the planet Jupiter is nearly perpendicular to its orbit, it follows that there can be little variation in the length of the days and little change in the seasons of that planet.

1274. There can be little doubt that the sun, the planets, stars, &c., are all of them inhabited; and, although it may be thought that some of them, on account of their immense distance from the sun, experience a great want of light and heat, while others are so near, and the heat consequently so great, that water cannot remain on them in a fluid state, yet, as we see, even on our own earth, that creatures of different natures live in different elements,—as, for instance, fishes in water, animals in air, &c.,— creative wisdom could, undoubtedly, adapt the being to its situation, and with as little exertion of power form a race whose nature should be adapted to the nearest or the most remote of the heavenly bodies, as was required to adapt the fowls to the air, or the fishes to the sea.

What is the Sun, and what is its diameter? 1275. OF THE SUN. — The Sun is a spherical body, situated near the centre of gravity of the system of planets of which our earth is one.

How much larger is the earth than the sun? [*Answer carefully.*] 1276. Its diameter is 882,000 English miles, which is equal to 100 diameters of the earth; and, as spheres are to each other in the proportion of the cube of their respective diameters, therefore his cubic magnitude must exceed that of the earth one million of times. It revolves around its axis in 25 days and 8 hours. This has been ascertained by means of several dark spots which have been seen with telescopes on its surface.

1277. Sir Wm. Herschel supposed the spots on the sun to be the dark body of the sun, seen through openings in the luminous atmosphere which surrounds him.

1278. It is probable that the sun,* like all the other heavenly bodies (excepting, perhaps, comets), is inhabited by beings whose nature is adapted to their peculiar circumstances.

1279. Many theories have been advanced with regard to the nature of the sun. By some it has been regarded as an immense ball of fire; but the theory which seems most in accordance with facts is, that the light and heat are communicated from a luminous atmosphere, or atmosphere of flame, which surrounds the sun, at a considerable distance above the surface.

What is the zodiacal light, and its cause?

1280. The zodiacal light is a singular phenomenon, accompanying the sun. It is a faint light which often appears to stream up from the sun a little after sunset and before sunrise. It appears nearly in the form of a cone, its sides being somewhat curved, and generally but ill defined. It extends often from 50° to 100° in the heavens, and always nearly in the direction of the plane of the ecliptic. It is most distinct about the beginning of March, but is constantly visible in the torrid zone. The cause of this phenomenon is not known.

1281. The sun, as viewed from the different planets, appears of different sizes according to their respective distances. Fig. 189 affords a comparative view of his apparent magnitude, as seen from all except the last twenty of the minor planets.

* In almanacs the sun is usually represented by a small circle, with the face of a man in it, thus : ☉

ASTRONOMY.

Fig. 189.

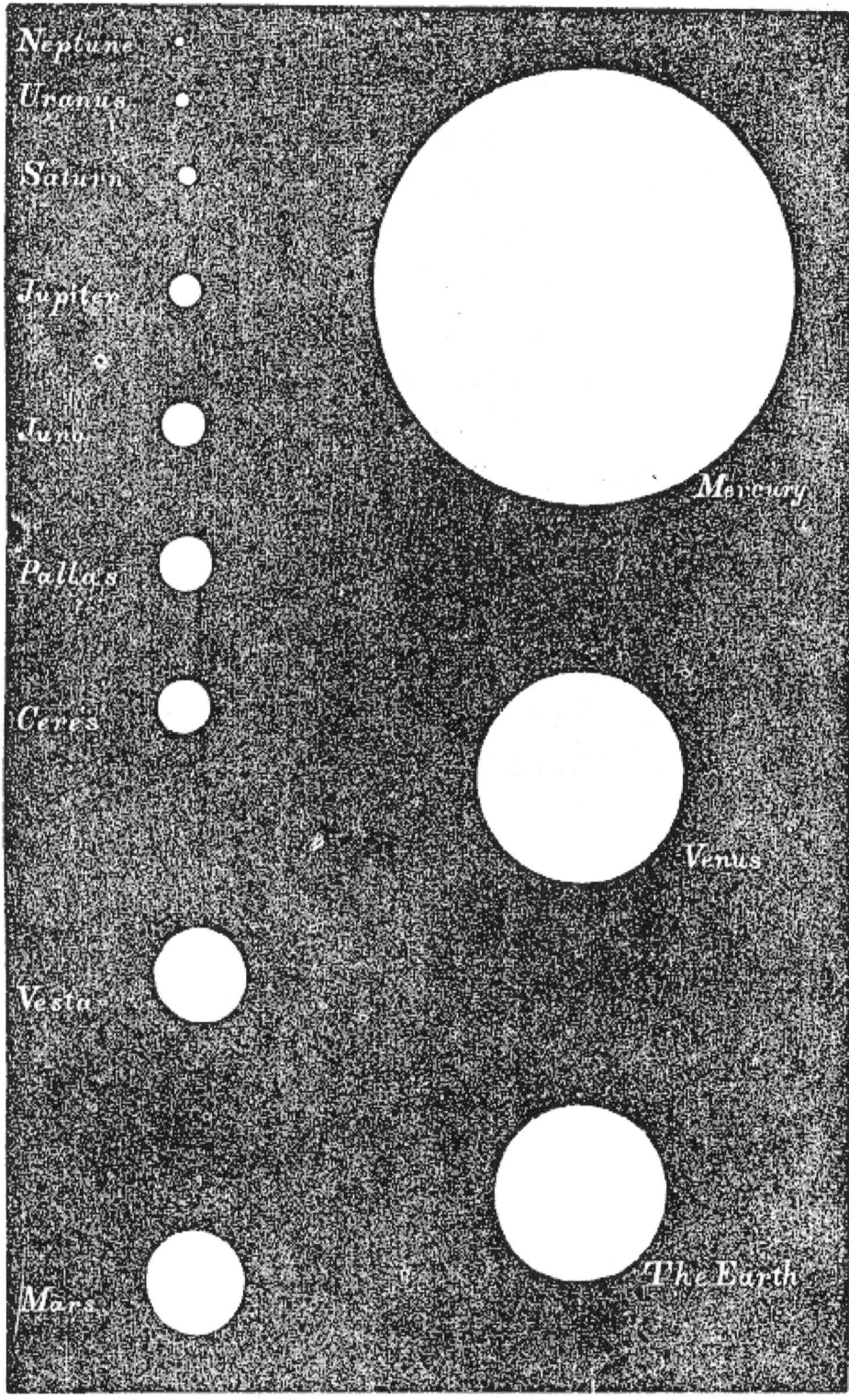

Apparent Magnitude of the Sun as seen from the Planets

NATURAL PHILOSOPHY.

Describe the planet Mercury.

1282. OF MERCURY.—Mercury is the nearest planet to the sun, and is seldom seen; because his vicinity to the sun occasions his being lost in the brilliancy of the sun's rays.

How many seas are there on the planet Mercury?

1283. The heat of this planet is so great that water cannot exist there, except in a state of vapor, and metals would be melted. The intensity of the sun's heat, which is in the same proportion as its light, is seven times greater in Mercury than on the earth, so that water there would be carried off in the shape of steam; for, by experiments made with a thermometer, it appears that a heat seven times greater than that of the sun's beams in summer will make water boil.

How late at night may Mercury be seen?

1284. Mercury, although in appearance only a small star, emits a bright white light, by which it may be recognized when seen. It appears a little before the sun rises, and again a little after sunset; but, as its angular distance from the sun never exceeds twenty-three degrees, it is never to be seen longer than one hour and fifty minutes after sunset, nor longer than that time before the sun rises.

How does Mercury appear when seen through a telescope?

1285. When viewed through a good telescope, Mercury appears with all the various phases, or increase and decrease of light, with which we view the moon, except that it never appears quite full, because its enlightened side is turned directly towards the earth only when the planet is so near the sun as to be lost to our sight in its beams. Like that of the moon, the crescent or enlightened side of Mercury is always towards the sun. The time of its rotation on its axis has been estimated at about 24 hours

Describe the planet Venus.

1286. OF VENUS. — Venus, the second planet in order from the sun, is the nearest to the earth, and on that account appears to be the largest and most beautiful of all the planets. During a part of the year it rises before the sun, and it is then called the morning star; during another part of the year it rises after the sun, and it is then called the evening star. The heat and light at Venus are nearly double what they are at the earth.

1287. By the ancient poets Venus was called *Phosphor*, or *Lucifer*, when it appeared to the west of the sun, at which time it is morning star, and ushers in the light of day; and *Hesperus*, or *Vesper*, when eastward of the sun, or evening star.

Why is Venus never seen late at night?

1288. Venus, like Mercury, presents to us all the appearances of increase and decrease of light common to the moon. Spots are also sometimes seen on its surface, like those on the sun. By reason of the great brilliancy of this planet, it may sometimes be seen even in the day-time by the naked eye. But it is never seen late at night, because its angular distance from the sun never exceeds forty-five degrees. In the absence of the moon it will cast a shadow behind an opaque body.

What is meant by the transit of a planet?

1289. Both Mercury and Venus sometimes pass directly between the sun and the earth. As their illuminated surface is towards the sun, their dark side is presented to the earth, and they appear like dark spots on the sun's disk. This is called the transit of these planets.

1290. The reason why we cannot see the stars and planets in the day-time is, that their light is so faint compared with the light of the sun reflected by our atmosphere.

Describe the Earth as a planet.

1291. OF THE EARTH. — The Earth on which we live is the next planet in the solar system, in the order of distance, to Venus. It is a large globe or ball, nearly eight thousand miles in diameter, and about twenty-five thousand miles in circumference. It is known to be round,—*first*, because it casts a circular shadow, which is seen on the moon during an eclipse; *secondly*, because the upper parts of distant objects on its surface can be seen at the greatest distance; *thirdly*, it has been circumnavigated. It is situated in the midst of the heavenly bodies which we see around us at night, and forms one of the number of those bodies; and it belongs to that system which, having the sun for its centre, and being influenced by its attraction, is called the *solar* system.

How much longer is the polar than the equatorial diameter of the earth? [Think before you speak.]

It is not a perfect sphere, but its figure is that of an *oblate spheroid*, the equatorial diameter being about twenty-six miles longer than its polar diameter.

It is attended by one moon, the diameter of which is about two thousand miles. Its mean distance from the earth is about 240,000 miles, and it turns on its axis in precisely the same time that it performs its revolution round the earth; namely, in twenty-seven days and seven hours.

Describe the earth as a moon.

1292. The earth, when viewed from the moon, exhibits precisely the same phases that the moon does to us, but in opposite order. When the moon is full to us, the earth will be dark to the inhabitants * of the moon; and when the moon is dark to us, the earth will be full to them. The earth appears to them about thirteen times larger than the moon does to us.

---

* This observation should be qualified by the condition that the moon is inhabited. Although there is abundant reason for the belief that the planets are "*the green abodes of life*," there are many reasons to believe that the moon, in its present state, is neither inhabited nor habitable.

As the moon, however, always presents nearly the same side to the earth, there is one-half of the moon which we never see, and from which the earth cannot be seen.

1293. As this book may possibly incite the inquiry how it is that the astronomer is able to measure the size and distances of those immense bodies the consideration of which forms the subject of Astronomy, the process will here be described by which the diameter of the earth may be ascertained.

1294. All circles, as has already been stated, are divided into 360 degrees, and, by means of instruments prepared for the purpose, the number of degrees in any arc or part of a circle can be correctly ascertained. Let us now suppose that an observer, standing upon any fixed point, should notice the position of a particular star, — the north or polar star, for instance. Let him then advance from his station, and travel towards the north, until he has brought the star exactly one degree higher over his head. Let him then measure the distance over which he has travelled between the two points of observation, and that distance will be exactly the length of one degree of the earth's circumference. Let him multiply that distance by 360, and it will give him the circumference of the earth. Having thus found the circumference, the diameter may readily be found by the common rules of arithmetic.

This calculation is based on the supposition that the earth is a perfect sphere, which is not the case, the equatorial diameter being about twenty-six miles longer than the polar. But it is sufficiently near the truth for the present purpose. The design of this work not admitting rigid mathematical demonstrations, this instance of the *commencement* of a calculation is given merely to show that what the astronomer and the mathematician tell us, wonderful as it may appear, is neither bare assertion nor unfounded conjecture.

What motions have the inhabitants of the earth on the earth as a planet? See, also, No. 1296.

1295. It has been stated that the earth revolves upon its axis every day. Now, as the earth is about 25,000 miles in circumference, it follows that the inhabitants of the equator are carried around this whole distance in about twenty-four hours, and every hour they are thus carried through space in the direction of the diurnal motion of the earth at the rate of $\frac{1}{24}$th of 25,000 miles, which is more than 1000 miles in an hour.

1296. But this is not all. Every inhabitant travels with the earth through its immense orbit, the diameter of which is about 190 millions of miles, or through a space of more than 570 millions of miles every year. This will give him, at the same time, a motion of more than 68 000 miles in an hour in a different direction. If the question be asked, why each individual is not sensible of these tremendously rapid motions, the answer is, that no one ever knew what it is to be without them. We cannot be sensible that we have moved without feeling our motion, as when in a boat a current takes us in one direction, while a gentle wind carries us, at the same time, in another direction It is only when our progress is arrested by obstacles of some kind that we can perceive the difference between a state of motion and a state of rest.

What would be the consequence if the earth should revolve on its axis once in an hour?

1297. The rapid motion of a thousand miles in an hour is not sufficient to overcome the centripetal force caused by gravity; but, if the earth should revolve around its axis seventeen times in a day, instead of once, all bodies at the equator would be lifted up, and the attraction of gravitation would be counterbalanced, if not wholly overcome.

1298. Certain irregularities in the orbit of the earth have been noticed by astronomers, which show that it is deviating from its elliptical form, and approaching that of a circle. In this fact, it has been thought, might be seen the seeds of decay. But Laplace has demonstrated that these irregularities proceed from causes which, in the lapse of immensely long periods, counterbalance each other, and give the assurance that there is no other limit to the present order of the universe than the will of its great Creator.

Describe the planet Mars.

1299. OF MARS.—Next to the earth is the planet Mars. It is conspicuous for its fiery-red appearance, which is supposed by Sir John Herschel* to be caused by the color of its soil.

---

* Sir John Herschel is the son of Sir William Herschel, the discoverer of the planet Uranus.

The degree of heat and light at Mars is less than half of that received by the earth.

1300. OF THE MINOR PLANETS.—It has already been mentioned that between the orbits of Mars and Jupiter thirty-five small bodies have been discovered, which are called the minor planets. It is a remarkable fact, that before the discovery of Bode's law (see No. 1232) certain irregularities observed in the motions of the old planets induced some astronomers to suppose that a planet existed between the orbits of Mars and Jupiter. The opinion has been advanced that these small bodies originally composed one larger one, which, by some unknown force or convulsion, burst asunder. This opinion is maintained with much ingenuity and plausibility by Sir David Brewster. (See *Edin. Encyc.*, art. ASTRONOMY.) Dr. Brewster further supposes that the bursting of this planet may have occasioned the phenomena of meteoric stones; that is, stones which have fallen on the earth from the atmosphere.

Describe the planet Jupiter. 1301. OF JUPITER.—Jupiter is the largest planet of the solar system, and the most brilliant, except Venus. The heat and light at Jupiter are about twenty-five times less than that at the earth. This planet is attended by four moons, or satellites, the shadows of some of which are occasionally visible upon his surface.

1302. The distance of those satellites from the planet are two, four, six and twelve hundred thousand miles, *nearly*.

The nearest revolves around the planet in less than two days; the next, in less than four days; the third, in less than eight days; and the fourth, in *about* sixteen days.

These four moons must afford considerable light to the inhabitants of the planet; for the nearest appears to them four times the size of our moon, the second about the same size, the third somewhat less, and the fourth about one-third the diameter of our moon.

1303. As the axis of Jupiter is nearly perpendicular to its orbit, it has no sensible change of seasons.

What use has been made of the eclipses of Jupiter's satellites? 1304. The satellites of Jupiter often pass behind the body of the planet, and also into its shadow, and are eclipsed. These eclipses are of use in ascertaining the longitude of places on the earth. By these eclipses, also, it has been ascertained that light is about eight minutes in coming from the sun to the earth; for an eclipse of one of these satellites appears to us to take place sixteen minutes sooner when the earth is in that part of its orbit nearest Jupiter than when in the part furthest from that planet. Hence, light is sixteen minutes in crossing the earth's orbit, and of course half of that time, or eight minutes, in coming from the sun to the earth.

What is the appearance of Jupiter as seen through a telescope? 1305. When viewed through a telescope, several belts or bands are distinctly seen, sometimes extending across his disk, and sometimes interrupted and broken. They differ in distance, position, and number. They are generally dark; but white ones have been seen.

On account of the immense distance of Jupiter from the sun and also from Mercury, Venus, the Earth and Mars, observers on Jupiter, with eyes like ours, can never see either of the above named planets, because they would always be immersed in the sun's rays.

Describe the planet Saturn. 1306. OF SATURN.—Saturn is the second in size, and the last but two in distance from the sun. The degree of heat and light at this planet is eighty times less than that at the earth.

How is Saturn particularly distinguished from the other planets? 1307. Saturn is distinguished from the other planets by being encompassed by three large luminous rings. They reflect the sun's light in the same manner as his moons. They are entirely detached from each other, and from the body of the planet. They turn on nearly the same axis with the planet, and in nearly the same time.

1308. These rings move together around the planet, but are about *three minutes* longer in performing their revolution about him than Saturn is in revolving about his axis. The edge of these rings is constantly at right angles with the axis of the planet. Stars are said to have been seen between the rings, and also between the inner ring and the body of the planet. The breadth of the two outer rings is about 27,000 miles, and the distance of the second ring from the planet is about 19,000 miles. As they cast shadows on the planet, Sir Wm Herschel thought them solid.

1309. The surface of Saturn is sometimes diversified, like that of Jupiter, with spots and belts. Saturn has eight satellites, or moons, revolving around him at different distances, and in various times, from less than one to eighty days.

1310. Saturn may be known by his pale and steady light. The eight moons of Saturn revolve at different distances around the outer edge of his rings. Sir William Herschel saw them moving along it, like bright beads on a white string. They do not often suffer eclipse by passing into the shadow of the planet, because the ring is in an oblique direction.

Describe the planet Uranus. 1311. OF URANUS.—Uranus, the fourth in size, is the most remote of all the old planets. It is scarcely visible to the naked eye. The light and heat at Uranus are about 360 times less than that of the earth.

1312. This planet was long known by the name of Herschel, the discoverer, who, in announcing his discovery, named it the "Georgium Sidus," in honor of King George III. The name of Uranus was given to it by the continental astronomers.

It was formerly considered a small star, but Sir Wm. Herschel, in 1781, discovered, from its motion, that it is a planet.

By how many moons is Uranus attended? 1313. Uranus is attended by six moons, or satellites, all of which were discovered by Sir Wm. Herschel, and all of them revolve in orbits nearly perpendicular to that of the planet. Their motion is retrograde.

What is the general law of the rotation of satellites? 1314. It appears to be a general law of satellites, or moons, that *they turn on their axis in the same time in which they revolve around their primaries.* On this account, the inhabitants of secondary planets observe some singular appearances, which the inhabitants of primary planets do not. Those who dwell on the side of a secondary planet next to the primary will always see that primary; while those who live on the opposite side will never see it. Those who always see the primary will see it constantly in very nearly the same place. For example, those who dwell near the edge of the moon's disk will always see the earth near the horizon, and those in or near the centre will always see it directly or nearly overhead. Those who dwell in the moon's south limb will see the earth to the northward; those in the north limb will see it to the southward; those in the east limb will see it to the westward; while those in the west limb will see it to the eastward; and all will see it nearer the horizon in proportion to their own distance from the centre of the moon's disk. Similar appearances are exhibited to the inhabitants of all secondary planets. These observations are predicated on the supposition that the moon is *inhabited*. But it is not generally believed that our moon is inhabited, or in its present condition fitted for the residence of any class of beings.

1315. It is a singular circumstance, that before the discovery of

Uranus some disturbances and deviations were observed by astronomers in the motions of Jupiter and Saturn, which they could account for only on the supposition that these two planets were influenced by the attraction of some more remote and undiscovered planet. The discovery of Uranus completely verified their opinions, and shows the extreme nicety with which astronomers observe the motions of planets.

What led to the discovery of the planet Neptune?

1316. OF NEPTUNE.—The discovery of the planet Neptune (named originally Le Verrier, from its discoverer, in 1846) is one of the greatest triumphs which the history of science records. As certain perturbations of the movements of Saturn led astronomers to suspect the existence of a remoter planet, which suspicions were fully confirmed in the discovery of Uranus, so also, after the discovery of Uranus, certain irregularities were perceived in his motions, that led the distinguished astronomers of the day to the belief that even beyond the planet Uranus still another undiscovered planet existed, to reward the labors of the discoverer. Accordingly Le Verrier, a young French astronomer, urged by his friend Arago, determined to devote himself to the attempt at discovery. With indefatigable industry he prepared new tables of planetary motions, from which he determined the perturbations of the planets Jupiter, Saturn, and Uranus, and as early as June, 1846, in a paper presented to the Academy of Sciences in Paris, he pointed out *where* the suspected planet would be on the 1st of January, 1847. He subsequently determined the mass and the elements of the orbits of the planet, and that, too, before it had been seen by a human eye. On the 18th of September of 1846, he wrote to his friend, M. Galle, of Berlin, requesting him to direct his telescope to a certain point in the heavens, where he suspected the stranger to be. His friend complied with his request, and on the first evening of examination discovered a strange star of the eighth magnitude, which had not been laid down in any of the maps of that portion of the heavens. The following evening it was found to have moved in a direction and with a velocity very nearly like that which Le Verrier had pointed out. The planet was found within less than one degree of the place where Le Verrier had located it. It was subsequently ascertained that a young English mathematician, Mr. Adams, of Cambridge, had been engaged in the same computations, and had arrived at nearly the same results with Le Verrier.

1317. What shall we say of science, then, that enables its devoted followers to reach out into space, and feel successfully in the dark for an object more than twenty-eight hundred millions of miles distant?

1318. In conclusion of this brief notice of the planets, a plate is here presented showing the relative appearance of the planets as viewed through a telescope. It will be observed that the planets Mercury and Venus have similar phases to those of our moon.

Fig 190

Relative Telescopic appearance of the Planets.

What is a Comet? 1319. OF COMETS.—The word Comet is derived from a Greek word, which means *hair;* and this name is given to a numerous class of bodies, which occasionally visit and appear to belong to the solar system. These bodies seem to consist of a nucleus, attended with a lucid haze, sometimes resembling flowing hair; from whence the name is derived. Some comets appear to consist wholly of this hazy or hairy appearance, which is frequently called the *tail* of the comet.

Fig. 191

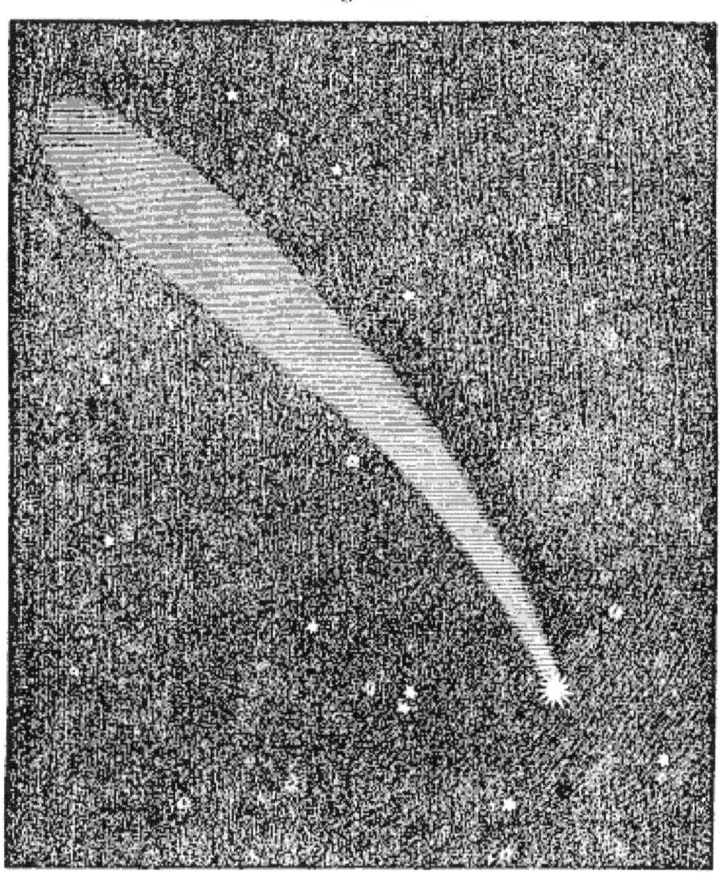

Comet of 1811, one of the most brilliant of modern times. Period, 2888 years.

1320. In ancient times the appearance of comets was regarded with superstitious fear, in the belief that they were the forerunners of some direful calamity. These fears have now been banished, and the comet is viewed as a constituent member of the system, governed by the same harmonious and unchanging laws which regulate and control all the other heavenly bodies.

1321. The number of comets that have occasionally appeared within the limits of the solar system is variously stated from 350 to 500. The paths or orbits of about 98 of these have been calculated from observation of the times at which they most nearly approached the sun; their distance from it and from the earth at those times; the direction of their movements, whether from east to west, or from west to east; and the places in the starry sphere at which their orbits crossed that of the earth and their inclination to it. The result is, that, of these 98, 24 passed *between* the sun and Mercury, 33 passed between Mercury and Venus, 21 between Venus and the Earth, 16 between the Earth and Mars, 3 between Mars and Ceres, and 1 between Ceres and Jupiter: that 50 of these comets moved from east to west; that their orbits were inclined at every possible angle to that of the earth. The greater part of them ascended above the orbit of the earth when very near the sun; and some were observed to dash down from the upper regions of space, and, after turning round the sun, to mount again.

What is the shape of the orbits of comets? 1322. Comets, in their revolution, describe long narrow ovals. They approach very near the sun in one of the ends of these ovals, and when they are in the opposite end of the orbit their distance from the sun is immensely great.

1323. The extreme nearness of approach to the sun gives to a comet, when in perihelion, a swiftness of motion prodigiously great. Newton calculated the velocity of the comet of 1680 to be 880,000 miles an hour. This comet was remarkable for its near approach to the sun, being no further than 580,000 miles from it, which is but little more than half the sun's diameter. Brydone calculated that the velocity of a comet which he observed at Palermo, in 1770, was at the rate of two millions and a half of miles in an hour.

1324. The luminous stream, or tail, of a comet, follows it as it approaches the sun, and goes before it when the comet recedes from the sun. Newton, and some other astronomers, considered the tails of comets to be vapors, produced by the excessive heat of the sun. Others have supposed them to be caused by a repulsive influence of the sun. Of whatever substance they may be, it is certain that it is very *rare*, because the stars may be distinctly seen through it.

1325. The tails of comets differ very greatly in length,

and some are attended apparently by only a small cloudy light, while the length of the tail of others has been estimated at from 50 to 80 millions of miles.

Fig. 192.

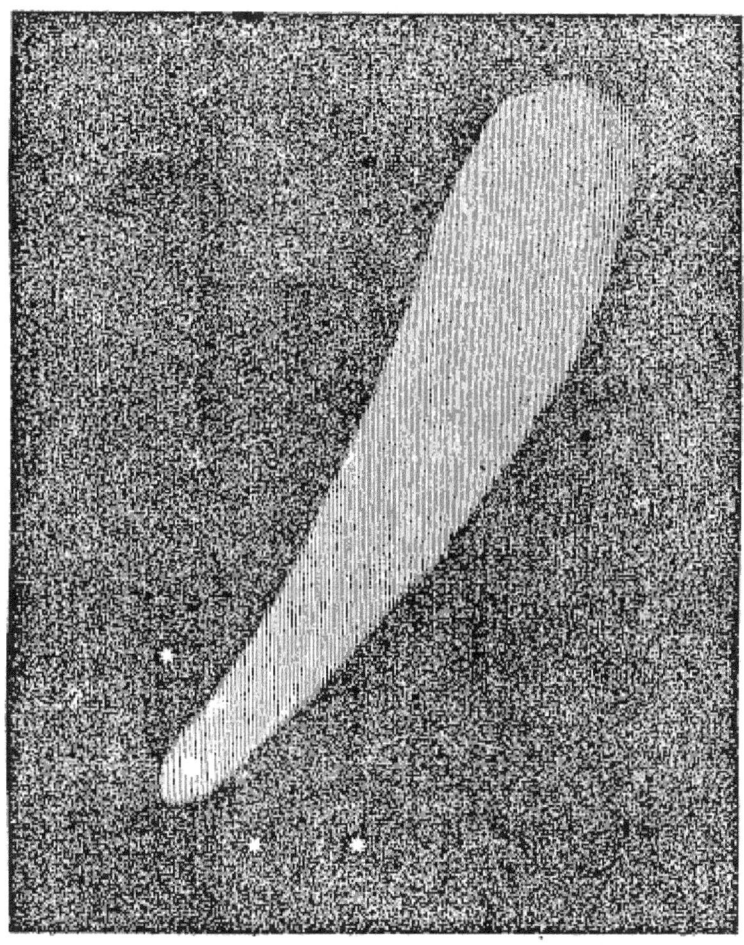

The comet of 1680, observed by Newton. Rapidity of its motion around the sun, a million of miles in an hour.
Length of tail, 100 millions of miles. Period, 600 years. It has never reappeared.

1326. It has been argued that comets consist of very little solid substance, because, although they sometimes approach very near to the other heavenly bodies, they appear to exert no sensible attractive force upon those bodies. It is said that in 1454 the moon was eclipsed by a comet. The comet, must, therefore, have been very near the earth (less than 240,000 miles); yet it produced no sensible effect on the earth or the moon; for it did not cause them to make any perceptible deviation from their accustomed paths round the sun. It has been ascertained that comets are disturbed by the gravitating power of the planets; but it does not appear that the planets are in like manner affected by comets.

Some comets have exhibited the appearance of two or more tails, and the great comet of 1744 had six

The great comet of 1744

1327. Many comets escape observation because they traverse that part of the heavens only which is above the horizon in the day-time. They are, therefore, lost in the brilliancy of the sun, and can be seen only when a *total* eclipse of the sun takes place. Seneca, 60 years before the Christian Era, states that a large comet was actually observed very near the sun, during an eclipse.

1328. Dr. Halley, Professor Encke and Gambart, are the first astronomers that ever successfully predicted the return of a comet. The periodical time of Halley's comet is *about* 76 years. It appeared last in the fall of 1835, and presented different ap-

Fig. 194.

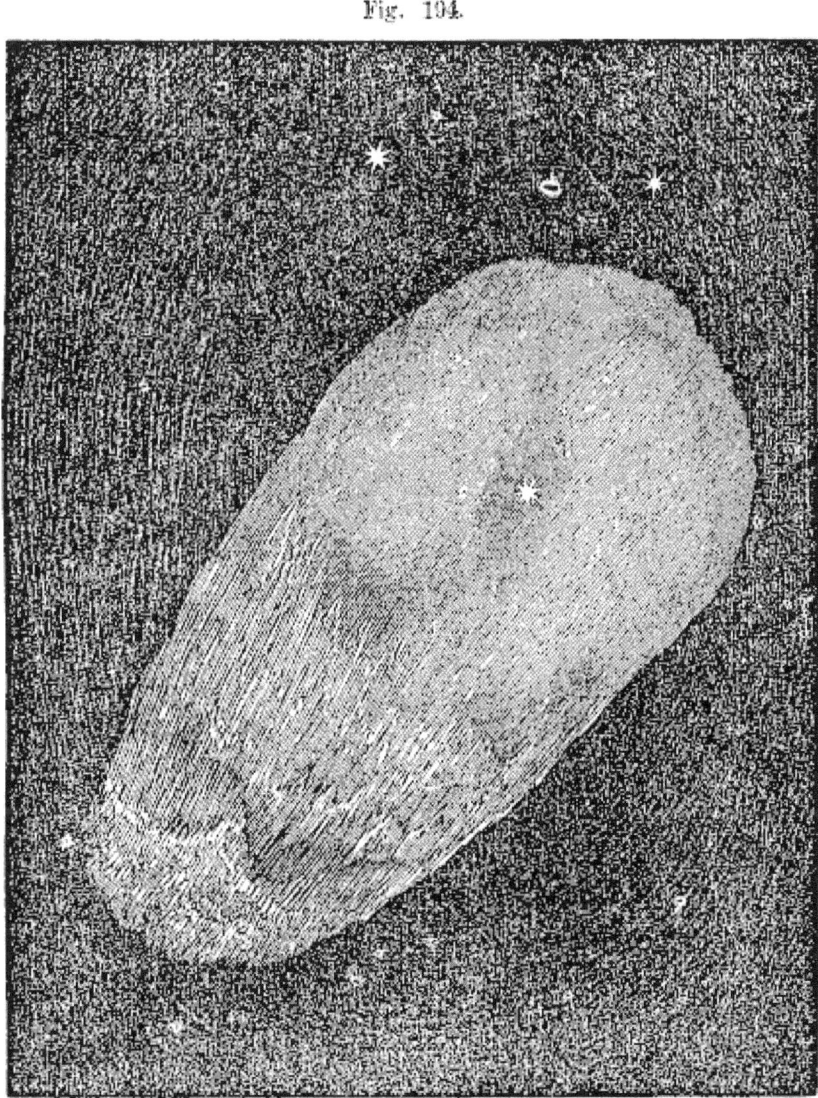

Halley's comet, as seen by Sir John Herschel, Oct. 29th, 1835. Very changeable in its appearance First recognized by Halley in 1682. Period, 76 years.

pearances from different points of observation. That of Encke is about 1200 days; that of Biela, about 6¾ years. This last comet appeared in 1832 and in 1838.

Fig. 195.

Halley's comet, as seen by Struve, Oct. 12th, 1835. First seen by Halley in 1682. Period, 75 years.

1329. The comet of 1758, the return of which was predicted by Dr. Halley, was regarded with great interest by astronomers, *because its return was predicted*. But four revolutions before, in 1456, it was looked upon with the utmost horror. Its long tail spread consternation over all Europe, already terrified by the rapid success of the Turkish arms. Pope Callixtus, on this occasion, ordered a prayer, in which both the comet and the Turks were included in one anathema. Scarcely a year or a month now elapses without the appearance of a comet in our system. But it is now known that they are bodies of such extreme rarity that our clouds are massive in comparison with them. They have no more density than the air under an exhausted receiver. Herschel saw stars of the 6th magnitude through a thickness of 30,000 miles of *cometic matter*. The number of comets in existence within the compass of the solar system is stated by some astronomers *as over seven millions*.

1330. Fig. 194 represents Halley's comet, as seen by Sir John Herschel, while Fig. 195 represents the same comet as seen only a few days before by Struve.

1331. THE COMET OF 1856.—The following interesting details in relation to a comet expected in 1856 are given by Babinet, an eminent French astronomer. It is translated from the *Courier des Etats Unis*.

"This comet is one of the grandest of which historians make mention. Its period of revolution is about three hundred years. It was seen in the years 104, 392, 683, 975, 1264, and the last time in 1556. Astronomers agreed in predicting its return in 1848; but it failed to appear, and continues to shine still unseen by us. Already the observatories began to be alarmed for the fate of their beautiful wandering star, when a learned calculator of Middlebourg, M. Bomme, reässured the astronomical world of the continued existence of the venerable and magnificent comet.

"Disquieted, as all other astronomers were, by the non-arrival of the comet at the expected time, M. Bomme, aided by the preparatory labors of Mr. Hind, has revised *all* the calculations and estimated *all* the actions of *all* the planets upon the comet for three hundred years of revolution. The result of this patient labor gives the arrival of the comet in August, 1858, with an uncertainty of two years, more or less; so that from 1856 to 1860 we may expect the great comet which was the cause of the abdication of the Emperor Charles V., in 1556.

"It is known that, partaking of the general superstition, which interpreted the appearance of a comet as the forerunner of some fatal event, Charles V. believed that this comet addressed its menaces particularly to him, as holding the first rank among sovereigns. The great and once wise but now wearied and shattered monarch, had been for some time the victim of cruel reverses. There were threatening indications in the political, if not in the physical horizon, of a still greater tempest to come. He was left to cry in despair, 'Fortune abandons old men.' The appearance of the blazing star seemed to him an admonition from Heaven that he must cease to be a sovereign if he would avoid a fatality from which one without authority might be spared. It is known that the emperor survived his abdication but a little more than two years.

"Another comet, which passed near us in 1835, and which has appeared 25 times since the year 13 before the Christian Era, has been associated by the superstitious with many important events which have occurred near the periods of its visitation.

"In 1066, William the Conqueror landed in England at the head of a numerous army about the time that the comet appeared which now bears the name of Halley's comet. The circumstance was regarded by the English as a prognostic of the victory of the Normans. It infused universal terror into the minds of the people, and contributed not a little towards the submission of the country after the battle of Hastings, as it had served to discourage the soldiers of Harold before the combat. The comet is represented upon the famous tapestry of Bayeux, executed by Queen Matilda, the wife of the conqueror.

"This celebrated tapestry is preserved in the ancient episcopal palace at Bayeux. It represents the principal incidents, including the appearance of the comet, in the history of the conquest of England by William, Duke of Normandy. It is supposed to have been executed by Matilda, the conqueror's wife, or by the Empress Matilda, daughter of Henry I. It consists of a linen web, 214 feet in length and 20 inches broad; and is divided into 72 compartments, each having an inscription indicating its subject. The figures are all executed by the needle.

"The same comet, in 1451, threw terror among the Turks under the command of Mahomet II., and into the ranks of the Christians during the terrible battle of Belgrade, in which forty thousand Mussulmans perished. The comet is described by historians of the time as 'immense, terrible, of enormous length, carrying in its train a tail which covered two celestial signs (60 degrees), and producing universal terror.' Judging from this portrait, comets have singularly degenerated in our day. It will be remembered, however, that in 1811 there appeared a comet of great brilliancy, which inspired some superstitious fears. Since that epoch science has noted nearly 80 comets, which, with few exceptions, were visible only by the aid of the telescope. Kepler, when asked how many comets he thought there were in the heavens, answered, 'As many as there are fish in the sea.'

"Thanks to the progress of astronomical science, these singular stars are no longer objects of terror. The theories of Newton, Halley, and their successors, have completely destroyed the imaginary empire of comets. As respects their physical nature, it was for a long time believed that they were composed of a compact centre, surrounded by a luminous atmosphere. On this subject the opinion of M. Babinet, who must be regarded as good authority on such questions, is as follows: 'Comets cannot exercise any material influence upon our globe; and the earth, should it traverse a comet in its entire breadth, would perceive it no more than if it should cross a cloud a hundred thousand millions of times lighter than our atmosphere, and which could no more make its way through our air than the slightest puff of an ordinary bellows could make its way through an anvil.' It would be difficult to find a comparison more reässuring."*

What are the Fixed Stars supposed to be?

1332. OF THE FIXED STARS.—The Fixed Stars are all supposed to be immensely large bodies, like our own sun, shining by their own light, which they dispense to systems of their own.

How are the fixed stars classified?

1333. They are classed by their apparent magnitudes, those of the sixth magnitude being the smallest that can be seen by the naked eye. Stars which can be seen only by means of the telescope

---

* THE COMET OF 1853.— Mr. Hind, in a letter to the *London Times*, gives the following particulars with regard to the comet which appeared during the year now closing (1853):

"The comet which has been so conspicuous during the last week was very favorably seen here on Saturday, and again on Sunday evening. On the latter occasion, allowing for the proximity of the comet to the horizon, and the strong glow of twilight, its nucleus was fully as bright as an average star of the first magnitude; the tail extended about three degrees from the head. When viewed in the comet-seeker, the nucleus appeared of a bright gold color, and about half the diameter of the planet Jupiter, which was shining at the same time in the southern heavens, and could be readily compared with the comet. The tail proceeds directly from the head in a single stream, and not, as sometimes remarked, in two branches. The distance of this body from the earth at 8 o'clock last evening, was 80,000,000 miles; and hence it results, that the actual diameter of the bright nucleus was 8000 miles, or about equal to that of the earth, while the tail had a real length of 4,500,000 miles, and a breadth of 250,000, which is rather over the distance separating the moon from the earth. It is usual to assume that the intensity of a comet's light varies as the reciprocal of the products of the squares of the distance from the earth and sun; but the present one has undergone a far more rapid increase of brilliancy than would result from this hypothesis. The augmentation of light will go on till the 3rd of September, and it will be worth while to look for the comet in the day-time about that date; for this purpose an equatorially mounted telescope will be required, and I would suggest the addition of a light green or red glass, to take off the great glare of sunlight, the instrument being adjusted to a focus on the planet Venus. This comet was discovered on the 10th of June, by Mr. Klinkenfues, of the Observatory at Gottingen, but was not bright enough to be seen without a telescope until about August 13. In a letter, copied into the *Times* a few days since, Sir William Hamilton hints at the possibility of this being the comet I had been expecting; but I avail myself of the present opportunity of stating that such is not the case, the elements of the orbits having no resemblance. The comet referred to will probably reäppear between the years 1858 and 1861; and, if the perihelion passage takes place during the summer months, we may expect to see a body of far more imposing aspect than the one at present visible."

are called telescopic stars. They, also, are classified; the classes reaching even to the seventeenth or twentieth magnitude.

How many stars are there of the first and second magnitude? 1334. The number of the stars of the first magnitude is about twenty-four; of the second magnitude, fifty; of the third, two hundred. The number of the smallest, visible without a telescope, is from twelve to fifteen thousand.

How many of the fixed stars have had their distances very nearly ascertained? 1335. Within a few years the distances of nine of the fixed stars have been calculated. This distance is so immense, that light, travelling with the inconceivable velocity of nearly two hundred thousand miles in a second of time, from *Sirius*, is more than fourteen years in reaching the earth; from Arcturus, more than twenty-five years; and from the Pole Star, more than forty-eight years.

1336. Tens of thousands of years must roll away before the most swiftly-flying of all the fixed stars can complete even a small fragment of its mighty orbit; but such has been the advance of science, that if a star move so slowly as to require five millions of years to complete its revolution, its motion could be perceived in one year; and in ten years its velocity can be computed, and its period will become known in the lifetime of a single observer.

Who first divided the stars into constellations? 1337. The stars are the fixed points to which we must refer in observations of the motions of all the heavenly bodies. Hence the stars were grouped in the earliest ages, (but by whom we know not), numbered and divided into constellations. the names of which have survived the fall of empires.

What probably causes the difference in the apparent size of the stars?

1338. It is generally supposed that part, if not all, of the difference in the apparent magnitudes of the stars is owing to the difference in their distance.

1339. The distance of the stars, according to Sir J. Herschel, cannot be less than 19,200,000,000,000 miles. How much greater it really is, we know not, except in a few cases.

1340. Although the stars generally appear fixed, they all have motion; but their distance being so immensely great, a rapid motion would not perceptibly change their relative situation in two or three thousand years. Some have been noticed alternately to appear and disappear. Several that were mentioned by ancient astronomers are not now to be seen; and some are now observed which were unknown to the ancients.

1341. Many stars which appear single to the naked eye, when viewed through powerful telescopes, appear double, treble, and even quadruple. Some are subject to variation in their apparent magnitude, at one time being of the second or third, and at another of the fifth or sixth magnitude.

What is the Galaxy?

1342. The Galaxy, or Milky Way, is a remarkably light, broad zone, visible in the heavens, passing from north-east to south-west. It is known to consist of an immense number of stars, which, from their apparent nearness, cannot be distinguished from each other by the naked eye.

1343. Sir Wm. Herschel saw, in the course of a quarter of an hour, the astonishing number of 116,000 stars pass through the field of his telescope, while it was directed to the milky way.

1344. The ancients, in reducing astronomy to a science, formed the stars into *clusters*, or *constellations*, to which they gave particular names.

1345. The number of constellations among the ancients was about 50. The moderns have added about 50 more.

1346. Our observations of the stars and nebulæ are confined principally to those of the northern hemisphere. Of the constellations near the south pole we know but little.

What effect has the atmosphere on the length of the day?

1347. In determining the true place of *any* of the celestial bodies, the refractive power of the atmosphere must always be taken into consideration. This property of the atmosphere adds to the length of the days, by causing the sun to appear *before* it has actually risen, and by detaining its *appearance* after it has actually set.

1348. On a celestial globe, the largest star in each constellation is usually designated by the first letter of the Greek alphabet, and the next largest by the second, &c. When the Greek alphabet is exhausted, the English alphabet, and then numbers, are used.

Why are the stars never seen in their true position?

1349. The stars, and other heavenly bodies, are never seen in their true situation, because the motion of light is *progressive;* and, during the time that light is *coming* to the earth, the earth is constantly in motion. In order, therefore, to see a star, the telescope must be turned somewhat *before* the star, and in the direction in which the earth moves.

What is meant by the aberration of light?

1350. Hence, a ray of light passing through the centre of the telescope to the observer's eye does not coincide with a direct line from his eye to the star, but makes an angle with it; and this is termed the *aberration of light.*

What is the Polar Star?

1351. The daily rotation of the earth on its axis causes the whole sphere of the fixed stars, &c., to appear to move round the earth every twenty-four hours from east to west. To the inhabitants of the northern hemisphere, the immovable point on which the whole seems to turn is the *Pole Star.* To the inhabitants of the southern hemisphere there is another and a corresponding point in the heavens.

What is the circle of perpetual apparition and of

1352. Certain of the stars surrounding the north pole never set to us. These are included in a circle parallel with the equator, *perpetual occultation?* and in every part equally distant from the north pole star. This circle is called the circle of *perpetual apparition*. Others never rise to us. These are included in a circle equally distant from the south pole; and this is called the circle of *perpetual occultation*. Some of the constellations of the southern hemisphere are represented as inimitably beautiful, particularly *the cross*.

What is parallax? 1353. The parallax of a heavenly body is the angular distance between the *true* and the *apparent* situation of the body.

Describe Fig. 196. 1354. In Fig. 196, A G B represents the earth, and C the moon. To a spectator at A, the moon

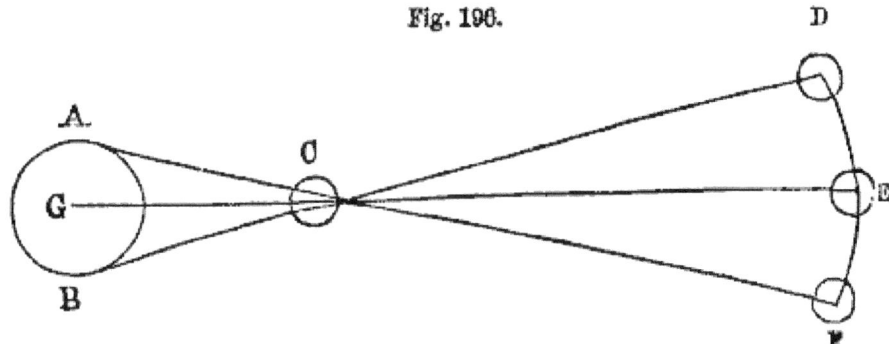

Fig. 196.

would appear at F; while to another, at B, the moon would appear at D; but, to a third spectator, at G, the centre of the earth, the moon would appear at E, which is the true situation. The distance from F to E is the parallax of the moon when viewed from A, and the distance from E to D is the parallax when viewed from B.

1355. From this it appears that the situation of the heavenly bodies must always be calculated from the centre of the earth; and the observer must always know the distance between the place of his observation and the centre of the earth, in order to make the necessary calculations to determine the true situation of the body. Allowance, also, must be made for the refraction of the atmosphere.

Describe the Moon.

1356. OF THE MOON.—The Moon is a secondary planet, revolving about the earth in about twenty-seven days, seven hours. Its distance from the earth is about 240,000 miles. It turns on its axis in precisely the same time that it performs its revolution about the earth. Consequently it always presents the same side to the earth; and as its apparent diameter in different parts of its orbit is different, it follows that it must be sometimes nearer to the earth than at others.

1357. The surface of the moon appears to be volcanic and very mountainous. Occasional volcanoes have been seen in action on the dark side. No heat has been detected in the moon's rays, even when most powerfully concentrated, that will affect the most delicate thermometer; and hence it has been inferred that the heat is absorbed in traversing the upper regions of our atmosphere.

What is one of the most common errors with regard to the moon, and how has it been proved an error?

1358. One of the most common errors with regard to the moon is that which ascribes to it an influence over the weather. Tables of the weather have been compared with the lunar phases for a period of a hundred years, and over a thousand lunations, during which time about 491 new or full moons have been attended by a change of the weather, and 509 have not.

1359. The moon is equally innocent of putrefaction, notwithstanding the popular belief that it hastens that process, especially in fish. The same cause which produces dew causes moisture on substances exposed to it, and this moisture is the real cause of putrefaction.

1360. Dr. Olbers, of Bremen, by a comparison of a great number of cases, arrived at the conclusion that the moon has no effect on insanity; although the popular belief is that the fits are aggravated or affected by the lunar phases.

What is the density of the moon compared with that of the earth?

1361. The force of gravity at the surface of the moon is about one-fifth that of the earth; hence ten pounds on the earth will be equal to two on the moon. The days and nights on the moon are each equal to fourteen of our days. The axis of the moon is perpendicular to its orbit, and therefore the moon can have no variety of seasons. The moon likewise has no atmosphere, and therefore it cannot be inhabited; for there can be no vegetation, no clouds, no ocean, no liquids, no light in dwellings, no twilight; in short, nothing that could fit it for the habitation of any order of beings with which we are acquainted.

1362. In connexion with what has now been stated with regard to the moon and its volcanic appearances, it will be proper to notice the subject of *aërolites*, or *meteoric stones;* because, according to the opinion of some, they are of lunar origin. Three theories have been broached with regard to them: 1st, that they are formed in the air, from materials existing there in a sublimated state; 2d, that they are parts of an exploded planet; 3d, that they are thrown from the volcanoes in the moon.

To the first of these theories there is a material objection in the fact that gases, when in contact, must mix, and gases necessary to form these substances cannot, therefore, remain in the air unmixed.

To the second hypothesis it may be objected, that if they were parts of a broken planet they would probably be composed of more heterogeneous materials. But it is well known that all of them are composed of the same constituent parts, namely, sulphur, magnesia, manganese, iron, nickel, chromium, and, in one recorded instance only, charcoal.

In favor of the third supposition, which refers them to a lunar origin, it may be remarked that a body thrown seventy miles from the moon would escape from the moon's attraction; and that a velocity six times greater than that of a cannon-ball would be sufficient to throw a body beyond the moon's attraction. As terrestrial volcanoes have thrown bodies with this velocity, it is not improbable that lunar volcanoes may do the same.

What is the most obvious fact in relation to the moon?

1363. The most obvious fact in relation to the moon is that its disc is constantly changing its appearance: sometimes only a semi-circular edge being illuminated, while the rest is dark;

388 NATURAL PHILOSOPHY.

at another time, the whole surface appearing resplendent. This is caused by the relative position of the moon with regard to the sun and the earth. The moon is an opaque body, and shines only by the light of the sun. When, therefore, the moon is between the earth and the sun, it presents its dark side to the earth; while the side presented to the sun, and on which the sun shines, is invisible to the earth. But when the earth is between the sun and the moon, the illuminated side of the moon is visible at the earth.

Describe Fig. 197. 1364. In Fig. 197, let S be the sun, E the earth, and A B C D the moon in different parts of her

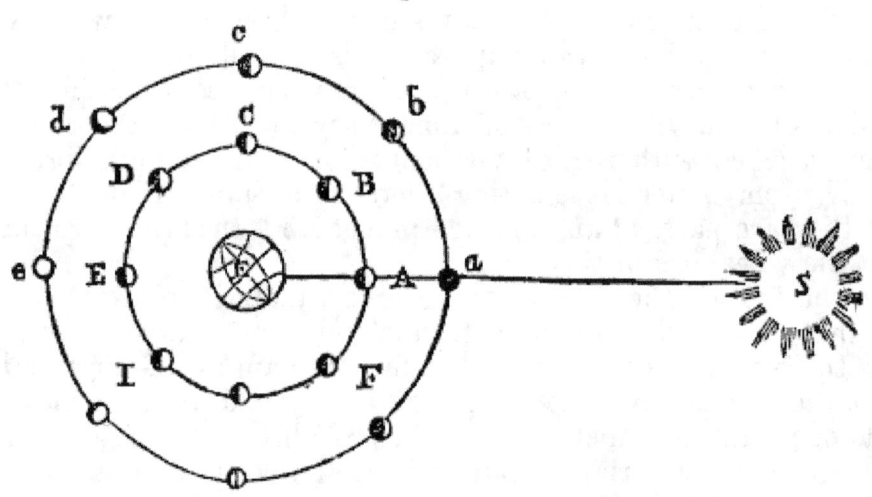

Fig. 197.

orbit. When the moon is at A, its dark side will be towards the earth, its illuminated part being always towards the sun. Hence the moon will appear to us as represented at *a*. But when it has advanced in its orbit to B, a small part of its illuminated side coming in sight, it appears as represented at *b*, and is said to be *horned*. When it arrives at C, one-half its illuminated side is visible, and it appears as at *c*. At C, and in the opposite point of its orbit, the moon is said to be in *quadrature*. At D its appearance is as represented at *d*, and it is said to be *gibbous*. At E all the illuminated side is towards us, and we have a full moon. During the other half of its revolution, less and less of its illuminated side is seen, till it again becomes invisible at A.

What is the mean difference in the rising of the moon from day to day?

1365. The mean difference in the rising of the moon, caused by its daily motion, is a little less than an hour. But, on account of the different angles formed with the horizon by different parts of the ecliptic, it happens that for six or eight nights near the full moons of September and October the moon rises nearly as soon as the sun is set. As this is a great convenience to the husbandman and the hunter, inasmuch as it affords them light to continue their occupation, and, as it were, lengthens out their day, the first is called the *harvest* moon, and the second the *hunter's* moon. These moons are always most beneficial when the moon's ascending node is in or near *Aries*.

What is meant by the Harvest and the Hunter's Moon, and when do they occur?

1366. The following signs are used in our common almanacs to denote the different positions and phases of the moon. ☽ or 𝔇 denote the moon in the *first* quadrature, that is, the quadrature between change and full; ☾ or ℭ denotes the moon in the *last* quadrature, that is, the quadrature between full and change. ● denotes new moon; ○ denotes full moon.

1367. When viewed through a telescope, the surface of the moon appears wonderfully diversified. Large dark spots, supposed to be excavations, or valleys, are visible to the eye; some parts also appear more lucid than the general surface. These are ascertained to be mountains, by the shadows which they cast. Maps of the moon's surface have been drawn, on which most of these valleys and mountains are delineated, and names are given to them. Some of these excavations are thought to be four miles deep, and forty wide. A high ridge generally surrounds them, and often a mountain rises in the centre. These immense depressions probably very much resemble what would be the appearance of the earth at the moon were all the seas and lakes dried up. Some of the mountains are supposed to be volcanic.

What are the Tides?   1368. OF THE TIDES.—The tides are the regular rising and falling of the water of the ocean twice in about twenty-five hours. They are occasioned by the attraction of the moon upon the matter of the earth; and they are also affected by that of the sun.

Explain Fig. 198.   1369. Let M, Fig. 198, be the moon revolving in her orbit; E, the earth covered with water; and S,

Fig. 198.

the sun. Now, the point of the earth's surface, which is nearest to the moon, will gravitate towards it more, and the remoter point less, than the centre, inversely as the squares of their respective distances. The point A, therefore, tends away from the centre, and the centre tends away from the point B; and in each case the fluid surface must rise, and in nearly the same degree in both cases. The effect must be diminished in proportion to the distance from these points in any direction; and at the points C and D, ninety degrees distant, it ceases. But there the level of the waters must be lowered, because of the exhaustion at those places, caused by the overflow elsewhere. Thus the action of the moon causes the ocean to assume the form of a spheroid, elongating it in the direction of the moon.

Thus any particular place, as A, while passing from under the moon till it comes under the moon again, has two tides. But the moon is constantly advancing in its orbit, so that the earth must a little more than complete its rotation before the place A comes under the moon. This causes high water at any place about fifty minutes later each successive day.

As the moon's orbit varies but little from the ecliptic, the moon is never more than 29° from the equator, and is generally much less. Hence the waters about the equator, being nearer the moon, are more strongly attracted, and the tides are higher than towards the poles.

1370. The sun attracts the waters as well as the moon. When the moon is at full or change, being in the same line of direction, (see Fig. 198), the sun acts with it; that is, the sun and moon tend to raise the tides at the same place, as seen in the figure. The tides are then very high, and are called *spring* tides.

Explain Fig. 199. But when the moon is in its quarters, as in Fig. 199, the sun and moon being in lines at

Fig. 199.

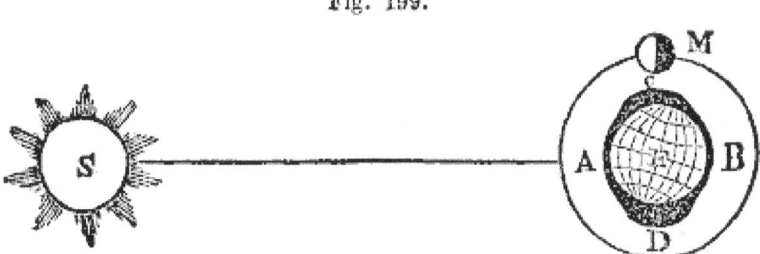

right angles tend to raise tides at different places; namely the moon at C and D, and the sun at A and B. Tides that are produced when the moon is in its quarters, are low, and are called *neap* tides.

1371. There are so many natural difficulties to the free progress of the tides, that the theory by which they are accounted for is, in fact, and necessarily, the most imperfect of all the theories connected with astronomy. It is, however, indisputable that the moon has an effect upon the tides, although it be not equally felt in all places, owing to the indentations of the coast, the obtructions of islands, continents, &c., which prevent the free motion of the waters. In narrow rivers the tides are frequently very high and sudden, from the resistance afforded by their banks to the free ingress of the water, whence what would otherwise be a tide, becomes an accumulation. It has been constantly observed, that the spring tides happen at the new and full moon, and the neap tides at the quarters. This circumstance is sufficient in itself to prove the connexion between the influence of the moon and the tides.

What is an Eclipse? 1372. An Eclipse is a total or partial obscuration of one heavenly body by the intervention of another.

When does an eclipse of the sun or of the moon take place? The situation of the earth with regard to the moon, or rather of the moon with regard to the earth, occasions eclipses both of the sun and moon. Those of the sun take place when the moon, passing between the sun and earth, intercepts his rays. Those of the moon take place when the earth, coming between the sun and moon, deprives the moon of his light. Hence, an eclipse of the sun can take place only when the moon changes, and an eclipse of the moon only when the moon fulls; for, *at the time of an eclipse*, either of the sun or the moon, *the sun earth, and moon, must be in the same straight line.*

Why is there not an eclipse at every new and full moon? If the moon revolved around the earth in the same plane in which the earth revolves around the sun, that is, in the ecliptic, it is plain that the sun would be eclipsed at every new moon, and the moon would be eclipsed at every full. For, at each of these times, these three bodies would be in the same straight line. But the moon's orbit does not coincide with the ecliptic, but is inclined to it at an angle of about 5° 20′. Hence, since the apparent diameter of the sun is but about $\frac{1}{2}$ a degree, and that of the moon about the same, no eclipse will take place at new or full moon, unless the moon be within ½ a degree of the ecliptic, that is, in or near one of its nodes. It is found that if the moon be within 16½° of a node at time of change, it will be so near the ecliptic, that the sun will be more or less eclipsed; if within 12° at time of full, the moon will be more or less eclipsed.

Why are there more eclipses of the sun than of the moon in a given course of years?

1373. It is obvious that the moon will be oftener within 16½° at the time of new moon, than within 12° at the time of full; consequently, there will be more eclipses of the sun than of the moon in a course of years. As the nodes commonly come between the sun and earth but twice in a year, and the moon's orbit contains 360°, of which 16½°, the *limit* of solar eclipses, and 12°, the *limit* of lunar eclipses, are but small portions, it is plain there must be many new and full moons without any eclipses.

Explain Fig. 200.

Although there are more eclipses of the sun than of the moon, yet more eclipses of the moon will be visible at a particular place, as Boston, in a course of years, than of the sun. Since the sun is very much larger than either the earth or moon, the shadow of

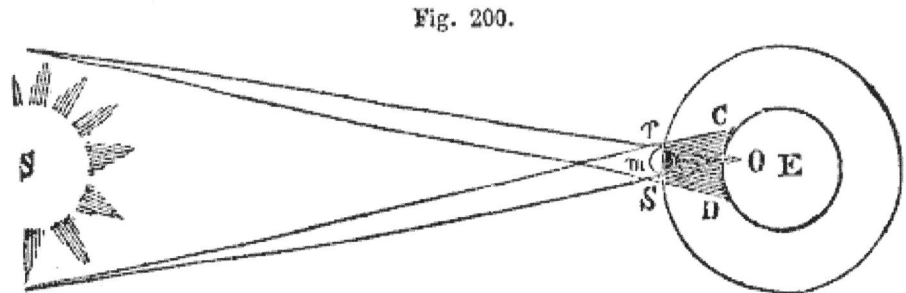

Fig. 200.

these bodies must always terminate in a point; that is, it must always be a cone. In Fig. 200, let S be the sun, *m* the moon, and E the earth. The sun constantly illuminates half the earth's surface, that is, a hemisphere; and consequently it is visible to all in this hemisphere. But the moon's shadow falls upon a part only of this hemisphere; and hence the sun appears eclipsed to a part only of those to whom it is visible. Sometimes, when the moon is at its greatest distance, its shadow, O

394   NATURAL PHILOSOPHY.

m, terminates before it reaches the earth. In eclipses of this kind, to an inhabitant directly under the point O, the outermost edge of the sun's disc is seen, forming a bright ring around the moon; from which circumstance these eclipses are called *annular*, from *annulus*, a Latin word for ring.

Besides the dark shadow of the moon, *m* O, in which all the light of the sun is intercepted (in which case the eclipse is called *total*), there is another shadow, *r* C D S, distinct from the former, which is called the *penumbra*. Within this, only a part of the sun's rays are intercepted, and the eclipse is called *partial*. If a person could pass, during an eclipse of the sun, from O to D, immediately on emerging from the dark shadow, O *m*, he would see a small part of the sun; and would continually see more and more till he arrived at D, where all shadow would cease, and the whole sun's disc be visible. Appearances would be similar if he went from O to C. Hence the penumbra is less and less dark (because a less portion of the sun is eclipsed), in proportion as the spectator is more remote from O, and nearer C or D. Though the penumbra be continually increasing in diameter, according to its length, or the distance of the moon from the earth, still, under the most favorable circumstances, it falls on but about half of the illuminated hemisphere of the earth. Hence, by half the inhabitants on this hemisphere, no eclipse will be seen.

Explain Fig. 201.

1374. Fig. 201 represents an eclipse of the moon. The instant the moon enters the earth's shadow at *x*, it is deprived of the sun's light,

Fig. 201.

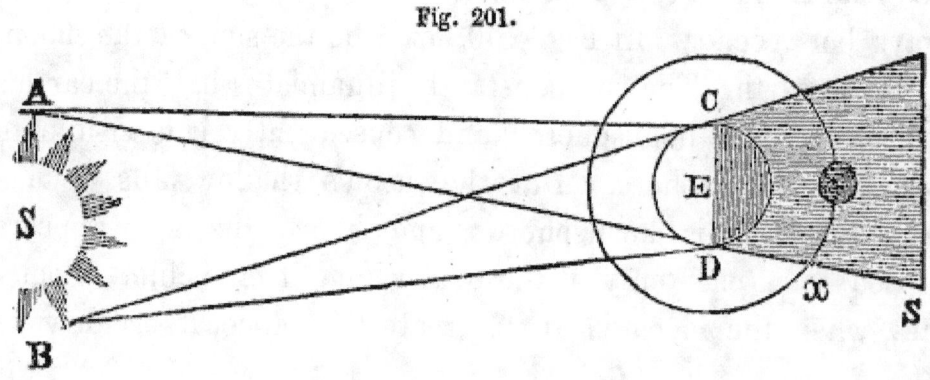

and is eclipsed to all in the unilluminated hemisphere of the earth. Hence, eclipses of the moon are visible to at least twice as many inhabitants as those of the sun can be; generally the proportion is much greater. Thus, the inhabitants at a particular place, as Boston, see more eclipses of the moon than of the sun.

The reason why a *lunar* eclipse is visible to all to whom the moon at the time is visible, and a *solar* one is not so to all to whom the sun at the time is visible, may be seen from the nature of these eclipses. We speak of the sun's being eclipsed; but, properly, it is the earth which is eclipsed. No change takes place in the sun; if there were, it would be seen by all to whom the sun is visible. The sun continues to diffuse its beams as freely and uniformly at such times as at others. But these beams are intercepted, and the earth is eclipsed only where the moon's shadow falls, that is, on only a part of a hemisphere. In eclipses of the moon, that body ceases to receive light from the sun, and, consequently, ceases to reflect it to the earth. The moon undergoes a change in its appearance; and, consequently, this change is visible at the same time to all to whom the moon is visible; that is, to a whole hemisphere of the earth.

1375. The earth's shadow (like that of the moon) is encompassed by a penumbra, C R S D, which is faint at the edges towards R and S, but becomes darker towards F and G. The shadow of the earth is but little darker than the region of the penumbra next to it. Hence it is very difficult to determine the exact time when the moon passes from the penumbra into the shadow, and from the shadow into the penumbra; that is, when the eclipse begins and ends. But the beginning and ending of a solar eclipse may be determined almost instantaneously.

What is meant by digits, as applied to eclipses of the sun and of the moon?

1376. The diameters of the sun and moon are supposed to be divided into twelve equal parts, called *digits*. These bodies are said to have as many digits eclipsed as there are of those parts involved in darkness

1377. There must be an eclipse of the sun as often, at least, as the moon, being near one of its nodes, comes between the sun and the earth.

The greatest number of both solar and lunar eclipses that can take place during the year is seven. The usual number is four, two solar and two lunar.

1378. A total eclipse of the sun is a very remarkable phenomenon.

June 16, 1806, a very remarkable total eclipse took place at Boston. The day was clear, and nothing occurred to prevent accurate observation of this interesting phenomenon. Several stars were visible; the birds were greatly agitated; a gloom spread over the landscape, and an indescribable sensation of fear or dread pervaded the breasts of those who gave themselves up to the simple effects of the phenomenon, without having their attention diverted by efforts of observation. The first gleam of light, contrasted with the previous darkness, seemed like the usual meridian day, and gave indescribable life and joy to the whole creation. A total eclipse of the sun can last but little more than three minutes. An annular eclipse of the sun is still more rare than a total one.

What is the difference between the solar and the sidereal year?

1379. OF TIME.—When time is calculated by the sun, it is called solar time, and the year a solar year; but when it is calculated by the stars, it is called sidereal time, and the year a sidereal year. The sidereal year is 20 minutes and 24 seconds longer than the solar year.

Which is the longer, a solar or a sidereal year, and by how much?

1380. The solar year consists of 365 days, 5 hours, 48 minutes, and 48 seconds; but our common reckoning gives 365 days only to the year. As the difference amounts to nearly a quarter of a day every year, it is usual every fourth year to add a day. Every fourth year the Romans reckoned the *6th of the calends of March, and the following day as one day;* which, on that account, they called bissextile, or twice the 6th day; whence we derive the name of bissextile for the leap year, in which we give to February, for the same reason, 29 days every fourth year.

1381. A solar year is measured from the time the earth sets out from a particular point in the ecliptic, as an equinox, or solstice, until it returns to the same point again. A sidereal year is measured by the time that the earth takes in making an entire revolution in its orbit; or, in other words, from the time that the sun takes to return into conjuction with any fixed star.

What is the precession of the equinoxes?

1382. Every equinox occurs at a point, 50" of a deg. of the great circle, preceding the place of the equinox, 12 months before; and this is called the *precession of the equinoxes*. It is this circumstance which has caused the change in the situation of the signs of the zodiac, of which mention has already been made.

1383. The earth's diurnal motion on an inclined axis, together with its annual revolution in an elliptic orbit, occasions so much complication in its motion as to produce many irregularities; therefore, true equal time cannot be measured by the sun. A clock which is always perfectly correct will, in some parts of the year, be before the sun, and in other parts after it. There are

When do the sun and clock agree?

but four periods in which the sun and a perfect clock will agree. These are the 15th of April, the 15th of June, the 1st of September, and the 24th of December.

What is the greatest difference between true and apparent time?

1384. The greatest difference between true and apparent time amounts to between sixteen and seventeen minutes. Tables of equation are constructed for the purpose of pointing out and correcting these differences between solar time and equal or mean time, the denomination given by astronomers to true time.

1385. As it may be interesting to those who have access to a celestial globe to know how to find any particular star or constellation, the following directions are subjoined.

There is always to be seen, on a clear night, a beautiful cluster of seven brilliant stars, which belong to the constellation "*Ursa Major*," or the Great Bear. Some have supposed that they will aptly represent a plough; others say that they are more like a wagon and horses, the four stars representing the body of the wagon, and the other three the horses. Hence they are called by some the *plough*, and by others they are called *Charles' wain*, or *wagon*.

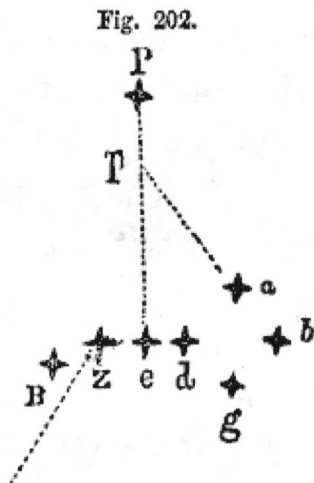

Fig. 202.

Fig. 202 represents these seven stars; *a b d g* represent the four, and *e z* B the other three stars. Perhaps they may more properly be called a large dipper, of which *e z* B represent the handle. If a line be drawn through the stars *b* and *a*, and carried upwards, it will pass a little to the left, and nearly touch a star represented in the figure by P. This is the polar star, or the north pole star; and the stars *b* and *a*, which appear to point to it, are called the *pointers*, because they appear to point to the polar star.

The polar star shines with a steady and rather dead kind of light. It always appears in the same position, and the north pole of the earth always points to it *at all seasons of the year*. The other stars seem to move round it as a centre. As this star is always in the north, the cardinal points may at any time be found by starlight.

By these stars we can also find any other star or constellation.

Thus, if we conceive a line drawn from the star *z*, leaving B a little to the left, it will pass through the very brilliant star A. By looking on a celestial globe for the star z, and supposing the line drawn on the globe, as we conceive it done on the heavens, we shall find the star and its name, which is Arcturus.

Conceiving another line drawn through g and b, and extended some distance to the right, it will pass just above another very brilliant star. On referring to the globe we find it to be Capella, or the goat.

In this manner the student may become acquainted with the appearance of the whole heavens.

Table I.

	Diameter in Miles.	Volume.	Gravity.	Light and Heat.		Time of revolution on its Axis.	Revolution around the Sun.	Distance from the Sun in Millions of Miles.
Mercury	3,140	.062	.47	6.680	Mercury	24 hrs.	87.98 ds.	37
Venus	7,800	.952	.93	1.911	Venus	23.36 "	224.7 "	68
Earth	7,926	1.000	1.00	1.000	Earth	24 "	365.25 "	95
Mars	4,100	.138	.50	.431	Mars	24.64 "	687. "	144
Jupiter	87,000	1233.412	2.85	.037	Jupiter	9.94 "	4332.60 "	494
Saturn	79,160	900.000	1.03	.011	Saturn	10.27 "	10759. "	906
Uranus	34,500	82.759	.76	.003	Uranus	9.5 "	30688. "	1,822
Neptune	41,500	144.008	.69	.001	Neptune			2,850
Sun	887,870	1410366.376	28.65		Moon			
Moon	2,163	.020	.15		Sun	25.3 "		

TABLE II.—Of Secondary Planets.

Of the Moon.

Distance from the Earth.	Inclination of Orbit to the Ecliptic.		Revolution around the Earth.		
Miles.	Deg.	Min.	Days.	Hours.	Min.
240,000	5	5	27	7	43

Moon's diameter, 2159. Bulk (that of the earth being 1), 1—49. Period from change to change, 29 days, 12 hours, 44 minutes.

Of the Satellites of Jupiter.

	Distance from Jupiter.	Inclination of Orbits to the Orbit of Jupiter.			Revolution around Jupiter.		
	Miles.	Deg.	Min.	Sec.	Days.	Hours.	Min.
I	269,800	3	18	38	1	18	27
II	426,500	3	18	00	3	14	58
III	680,000	3	13	58	7	3	42
IV	1,152,000	2	36	00	16	16	32

Of the Satellites of Saturn.

	Distance from Saturn.	Inclination of Orbits to the Orbit of Saturn.		Revolution around Saturn.		
	Miles.	Deg.	Min.	Days.	Hours.	Min.
VII	126,020	30		0	22	37
VI	161,720	30		1	8	53
I	200,235	30		1	21	18
II	256,449	30		2	17	44
III	358,225	30		4	12	25
IV	830,440	30		15	22	41
V	2,414,660	42	45	79	7	54

Of the Satellites of Uranus.

	Distance from Uranus.	Inclination of Orbits to the Orbit of Uranus.			Revolution around Uranus.		
	Miles.	Deg.	Min.	Sec.	Dys.	Hours.	Min.
I	226,320	99 or 80	43 16	53 7	5	21	25
II	293,600	do.			8	16	57
III	342,416	do.			10	23	4
IV	392,450	do.			13	10	56
V	785,060	do.			38	1	48
VI	1,570,000	do.			107	16	42

TABLE III.

	Latitude.			Longitude.				Latitude.			Longitude.		
London	51 30 49 N.			0 5 48 W.			Washington (U. S.)	38 53 34 N.			77 1 30 W.		
Greenwich	51 28 39 N.			0 0 0			New York	40 42 40 N.			74 1 8 W.		
Paris	48 50 13 N.			2 20 24 E.			Philadelphia	39 56 59 N.			75 9 54 W.		
Berlin	52 31 13 N.			13 23 52 E.			Boston	42 21 23 N.			71 4 9 W.		
Dublin	52 23 13 N.			6 20 30 W.			Baltimore	39 17 23 N.			76 37 30 W.		
Edinburgh	55 57 23 N.			3 10 54 W.			Albany	42 39 3 N.			73 44 49 W.		
Bremen	53 4 36 N.			8 48 58 E.			Brunswick (Maine)	43 53 0 N.			69 55 1 W.		
Copenhagen	55 40 53 N.			12 34 57 E.			Charleston (S. C.)	32 46 33 N.			79 57 27 W.		
Dorpat (Russia)	58 22 47 N.			26 43 45 E.			Cincinnati	39 5 54 N.			84 27 0 W.		
Milan	45 28 1 N.			9 11 48 E.			New Haven	41 18 30 N.			72 56 45 W.		
Marseilles	43 17 50 N.			5 22 15 E.			New Orleans	29 57 45 N.			90 6 49 W.		
Naples	40 51 47 N.			14 15 4 E.			Princeton (N. J.)	40 20 41 N.			74 39 33 W.		
Petersburgh	59 56 31 N.			30 18 57 E.			Providence	41 49 22 N.			71 24 48 W.		
Rome	41 53 52 N.			12 28 40 E.			Richmond (Va.)	37 32 17 N.			77 27 28 W.		
Turin	45 4 6 N.			7 42 6 E.			Charlottesville (Va.)	38 2 3 N.			78 31 29 W.		
Vienna	48 12 35 N.			16 23 0 E.			Savannah (Ga.)	32 4 56 N.			81 8 16 W.		
Wardhas (Lapland)	70 22 36 N.			31 7 54 E.			Schenectady (N. Y.)	42 48 0 N.			73 55 0 W.		
Canton	23 8 9 N.			113 16 54 E.									
Cape Horn	55 58 41 S.			67 10 53 W.									
Cape of Good Hope	33 56 3 S.			18 28 45 E.									

ILLUSTRATIONS.

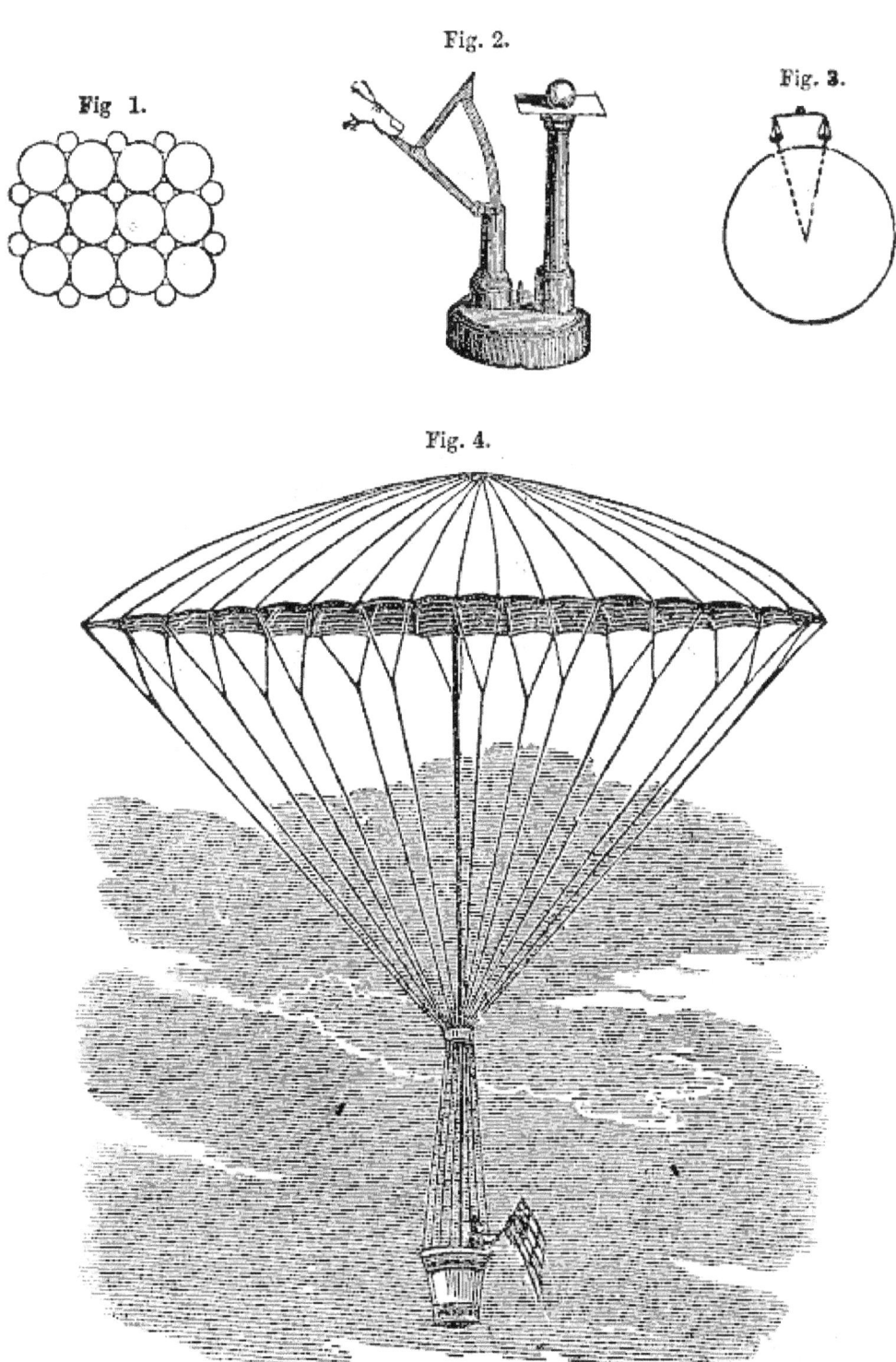

404 NATURAL PHILOSOPHY.

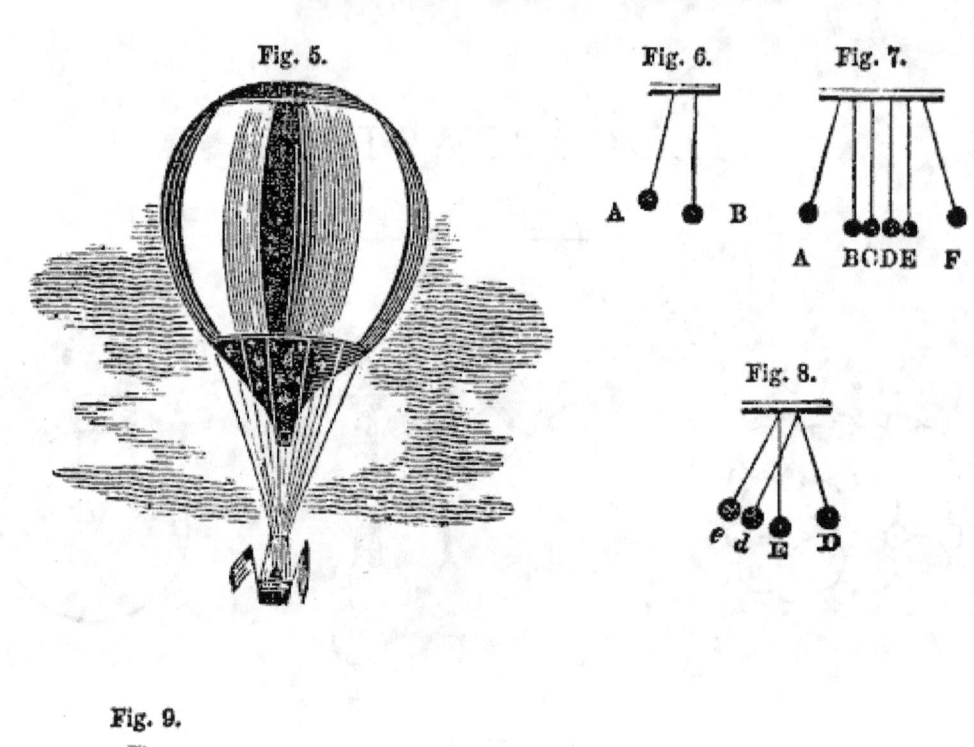

Fig. 5. Fig. 6. Fig. 7.

Fig. 8.

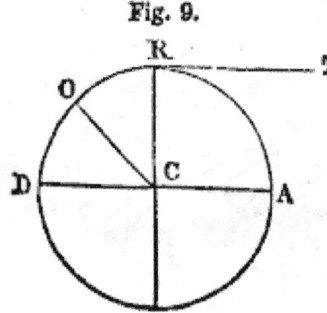
Fig. 9.

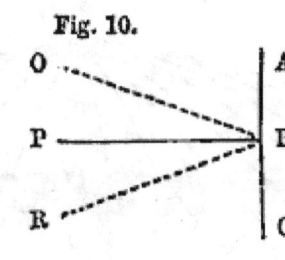
Fig. 10.

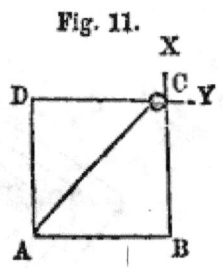
Fig. 11.

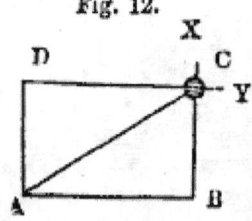
Fig. 12.

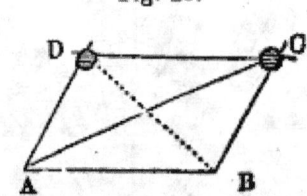
Fig. 13.

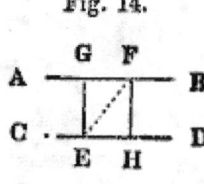
Fig. 14.

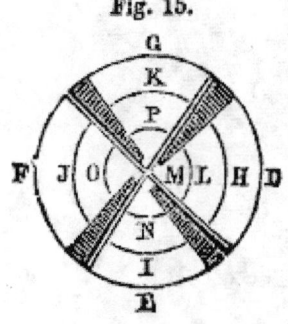
Fig. 15.

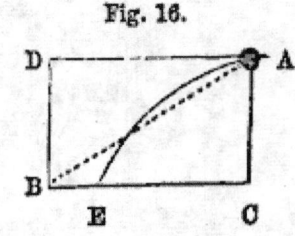
Fig. 16.

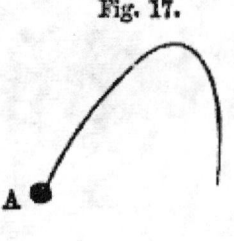
Fig. 17.

Fig. 18.

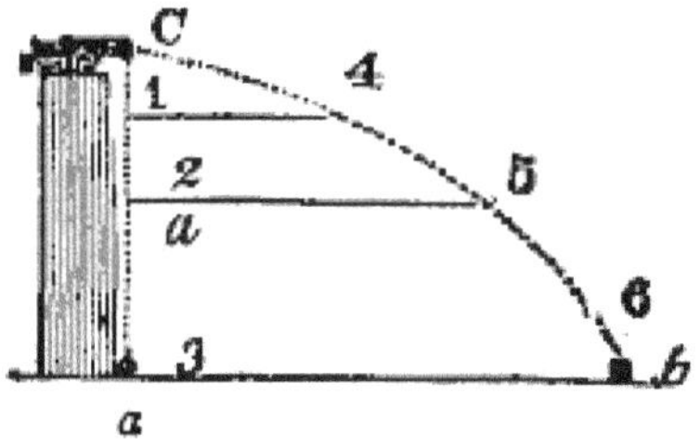

Fig. 19.

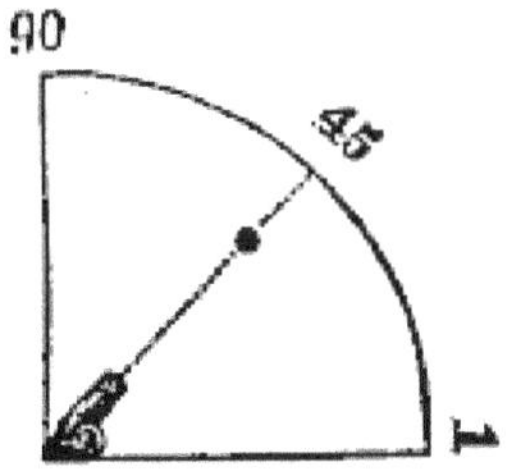

Fig. 20.

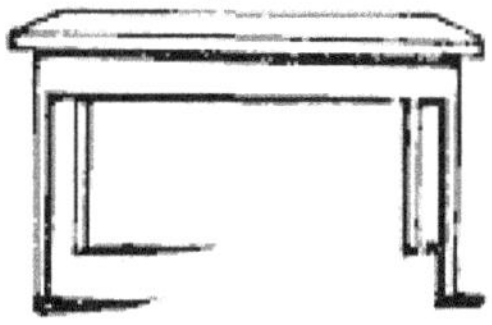

Fig. 21.

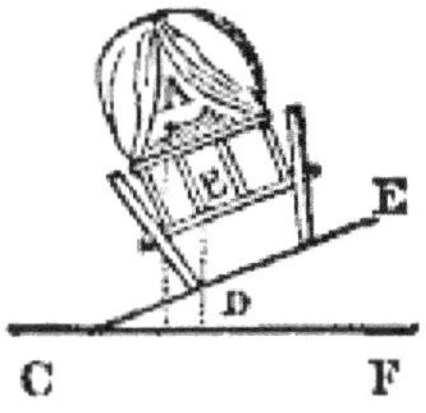

Fig. 22.

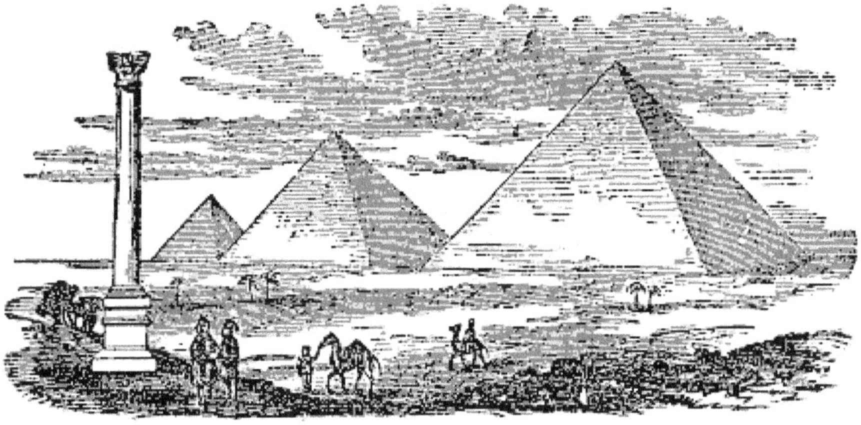

Fig. 23

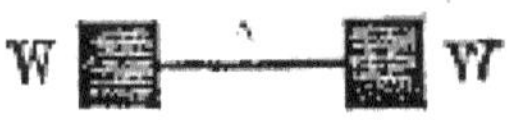

Fig. 24.

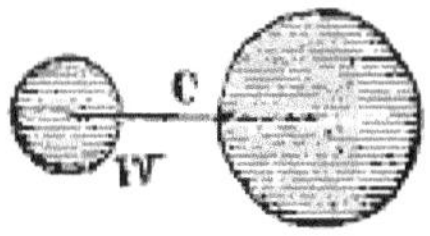

Fig. 25.

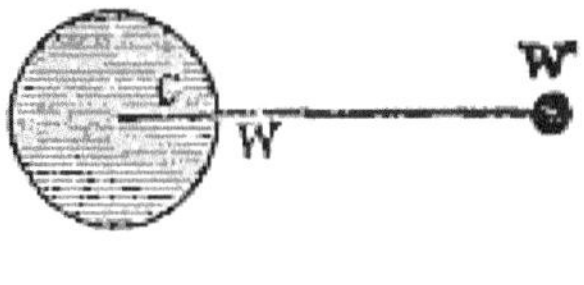

Fig. 26

Fig. 27.

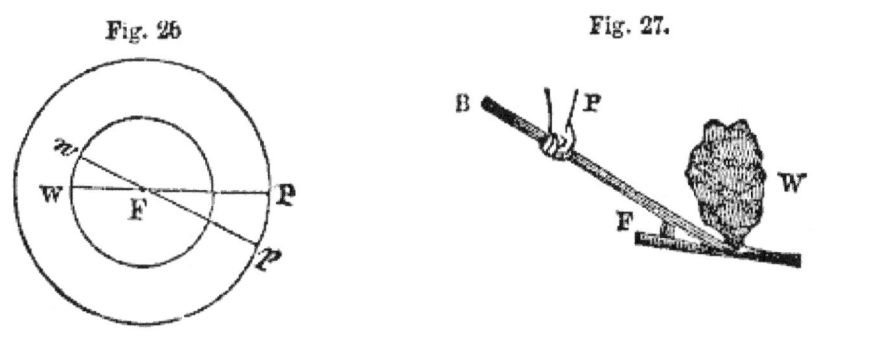

Fig. 28.

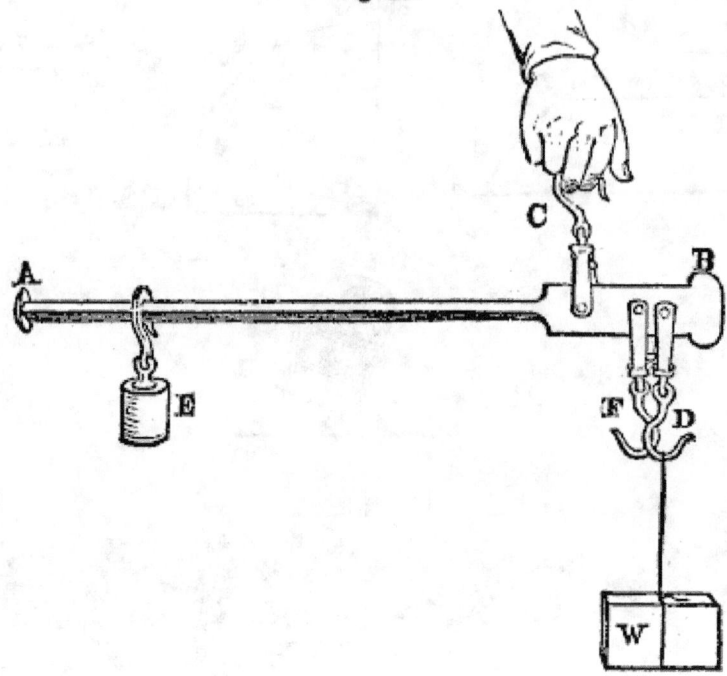

Fig. 29.

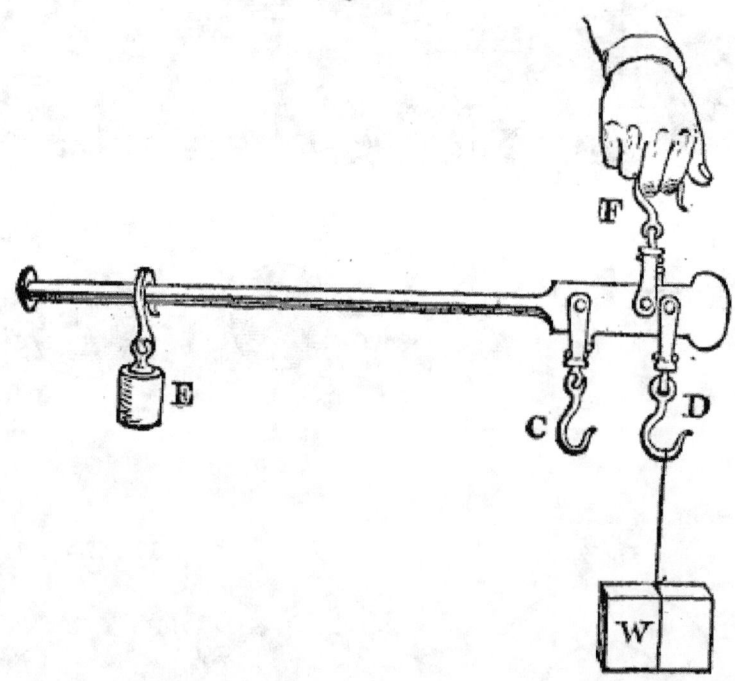

Fig. 30.

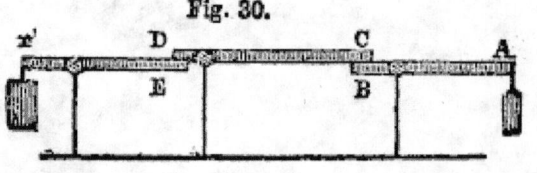

Fig. 31.

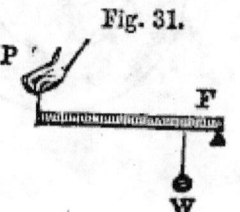

ILLUSTRATIONS. 407

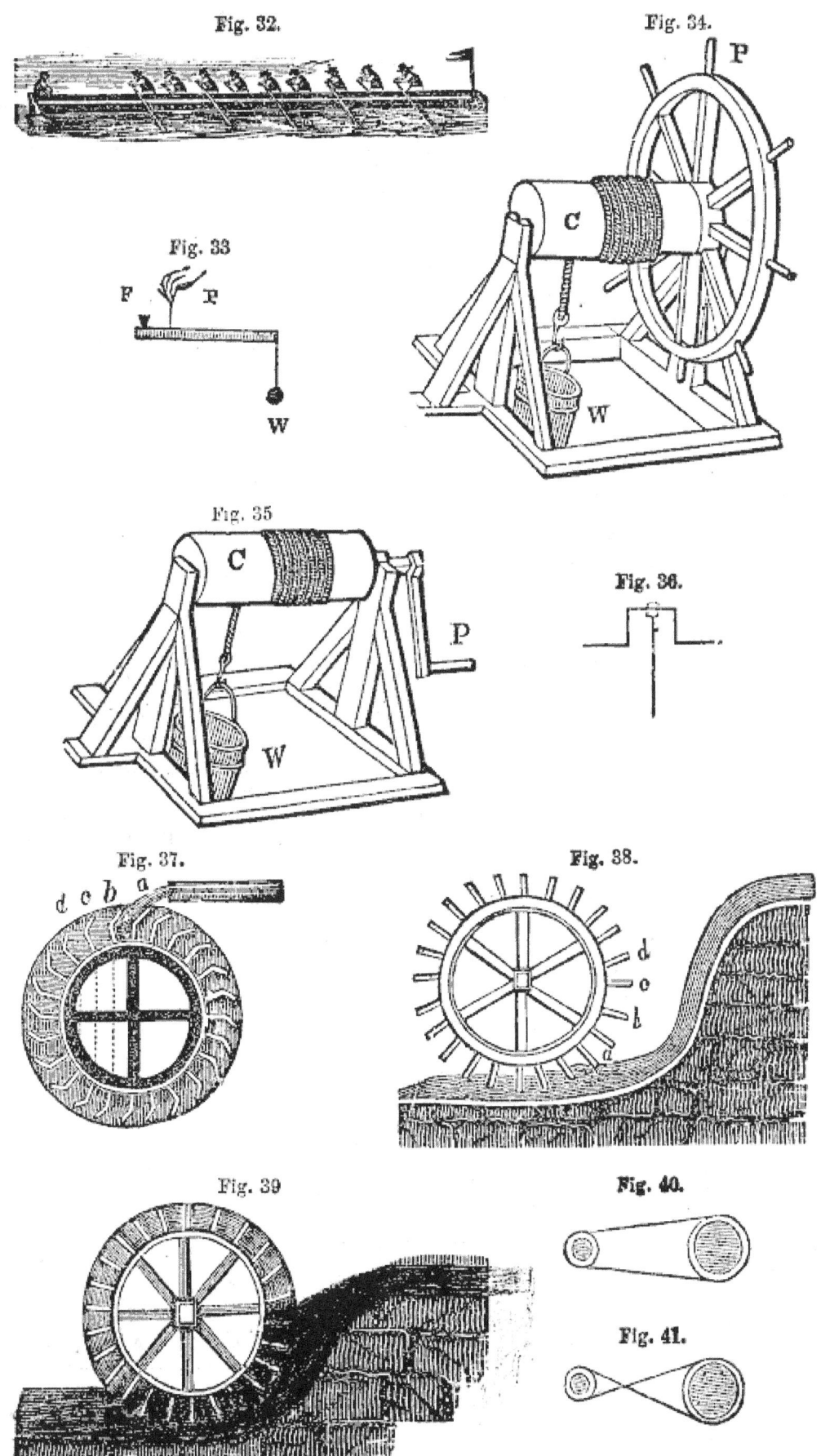

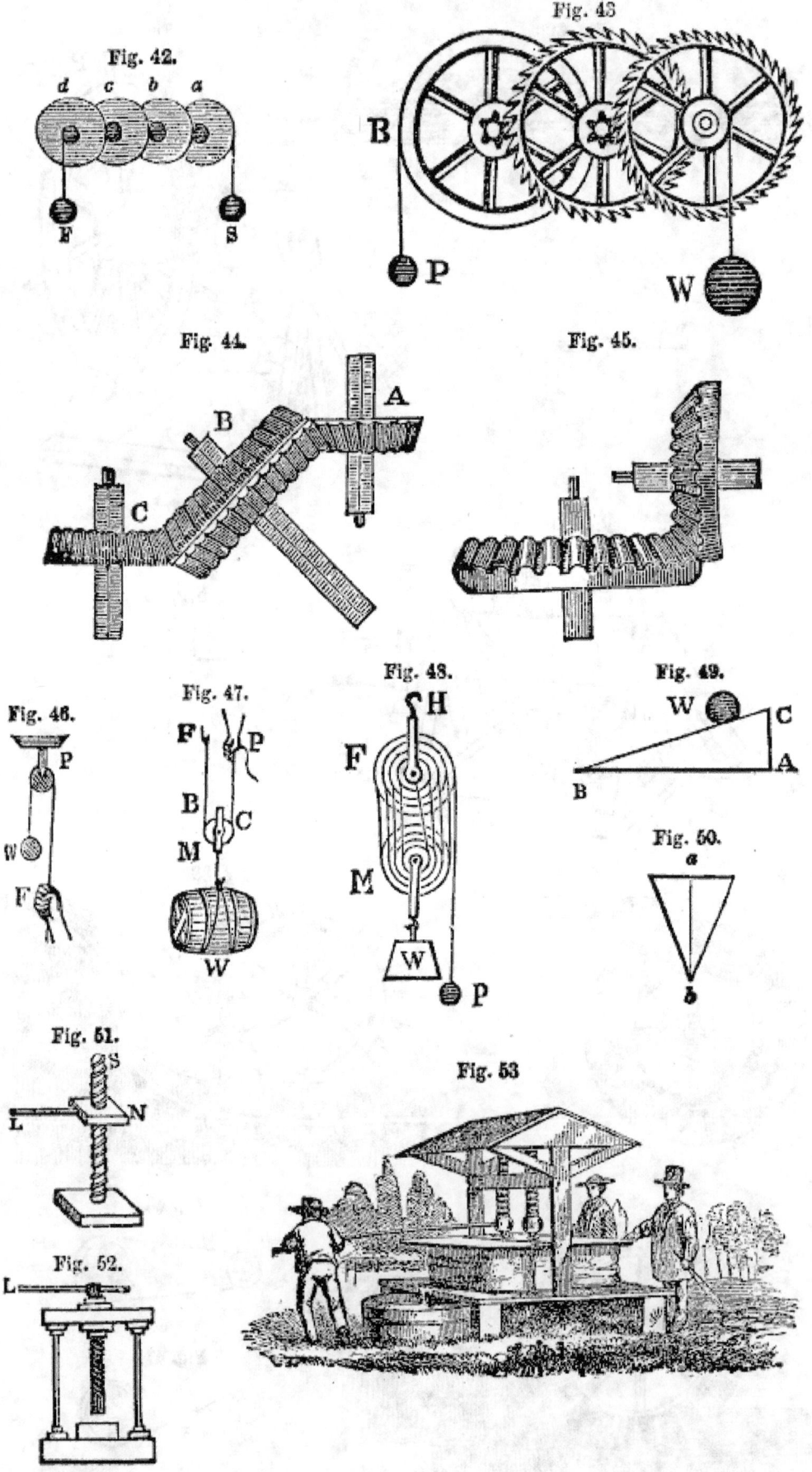

ILLUSTRATIONS.

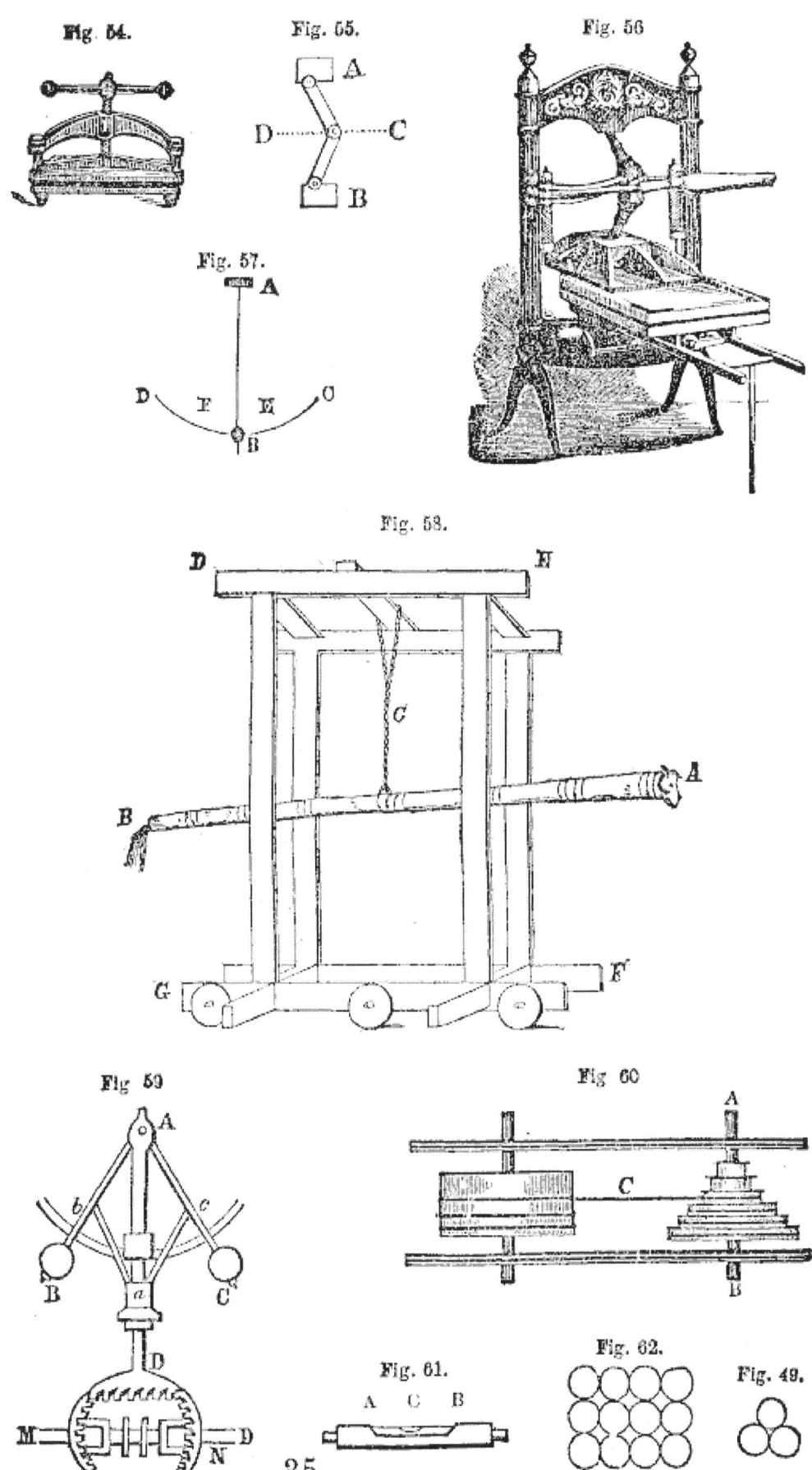

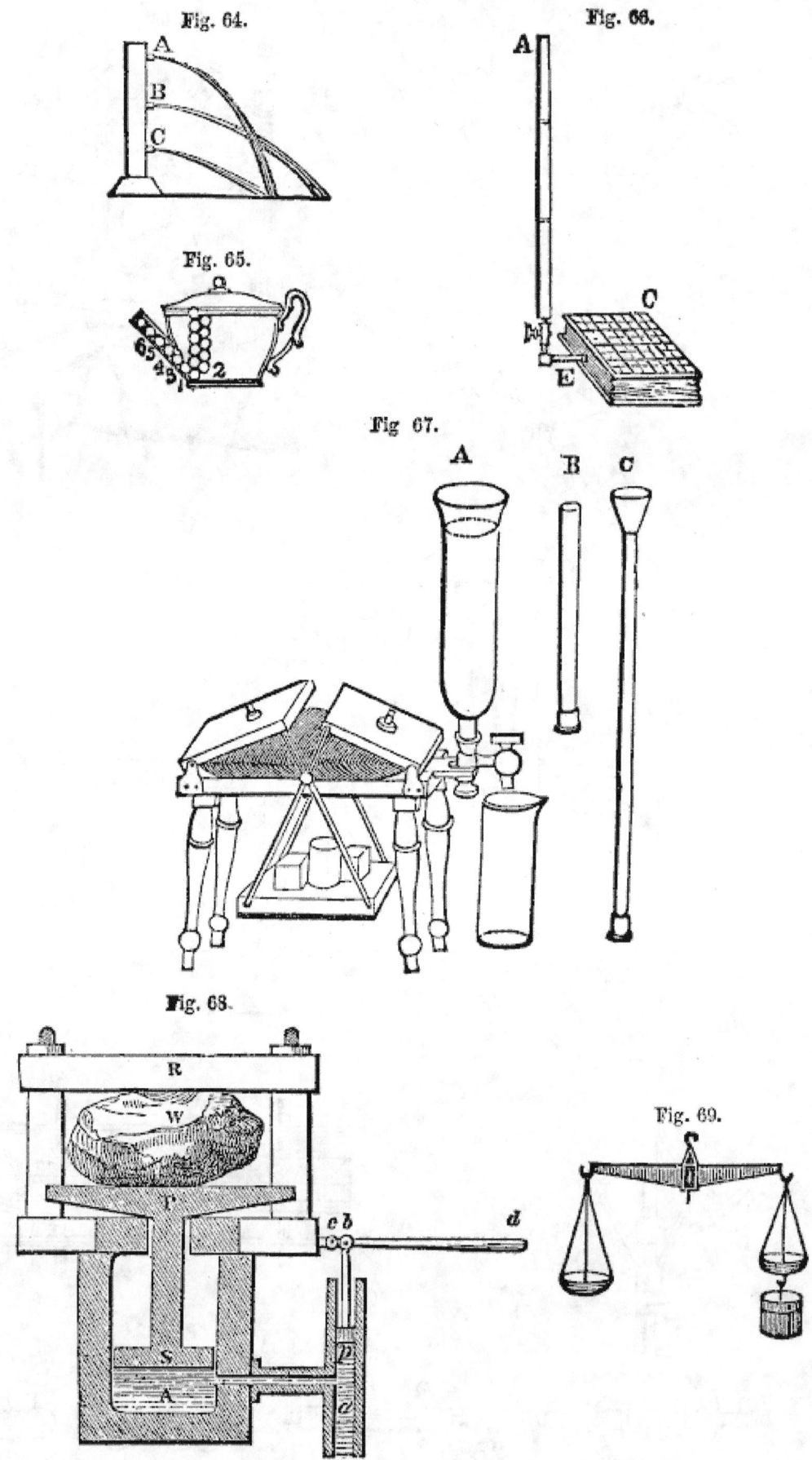

ILLUSTRATIONS. 411

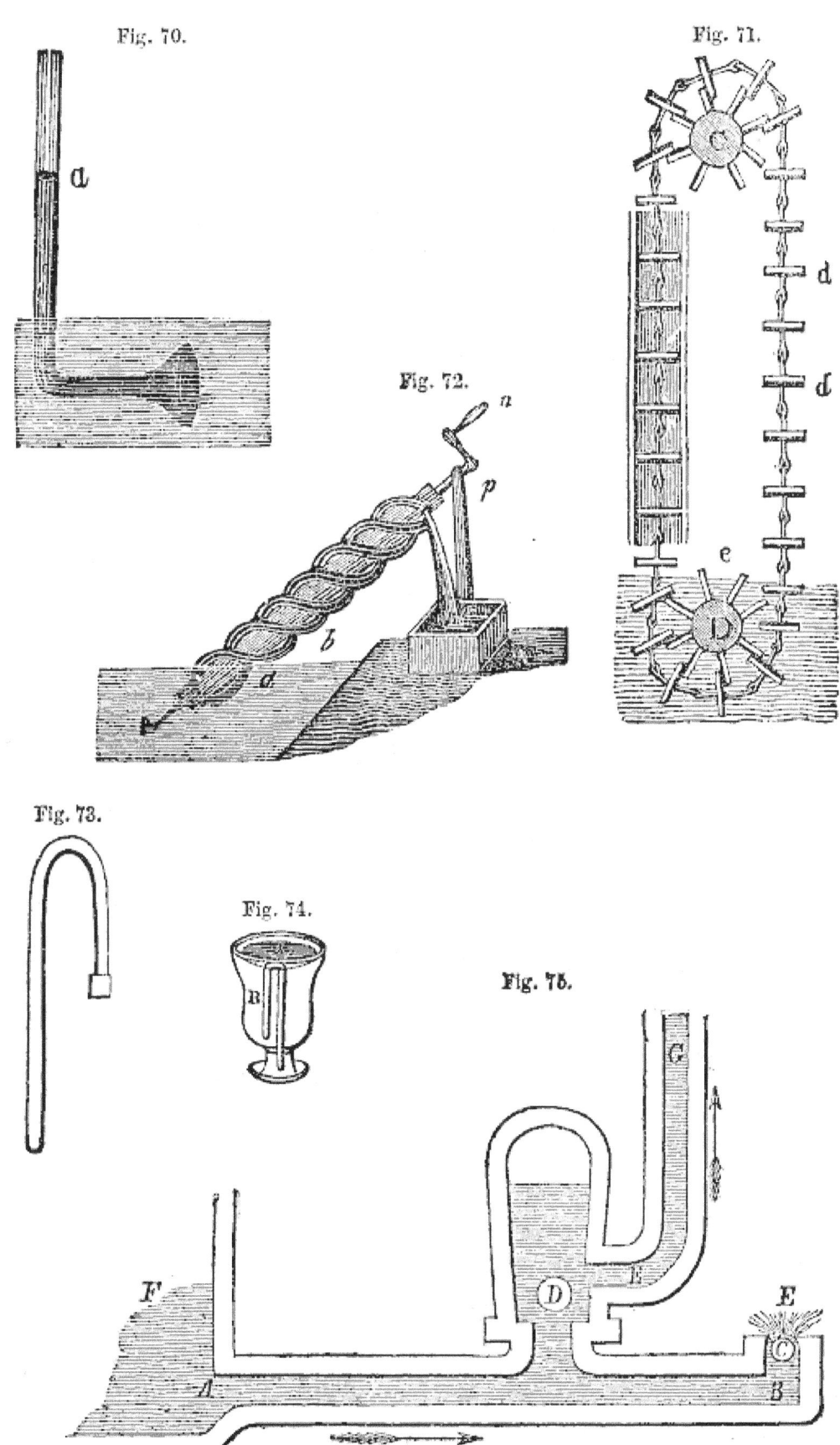

Fig. 70. Fig. 71. Fig. 72. Fig. 73. Fig. 74. Fig. 75.

412 NATURAL PHILOSOPHY.

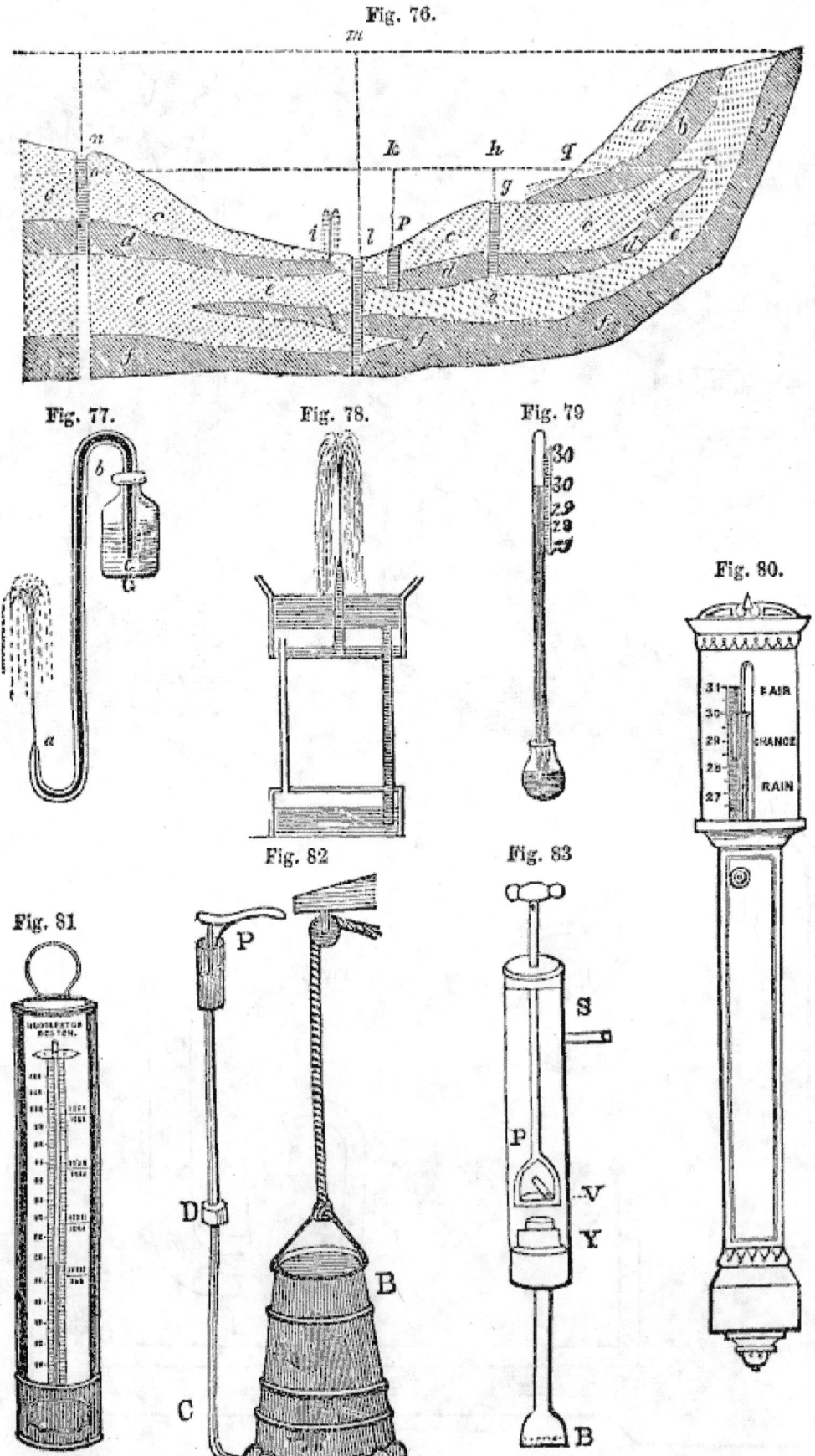

Fig. 76.

Fig. 77. Fig. 78. Fig. 79. Fig. 80.

Fig. 81. Fig. 82. Fig. 83.

ILLUSTRATIONS. 413

Fig. 84.

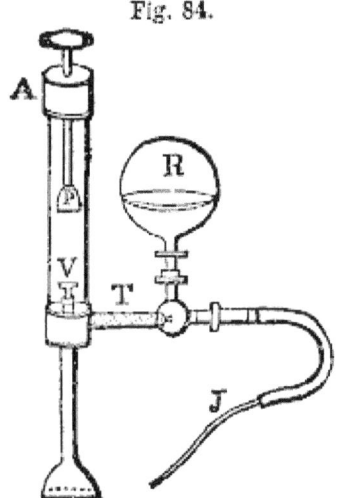

Fig. 85.

Fig. 86.

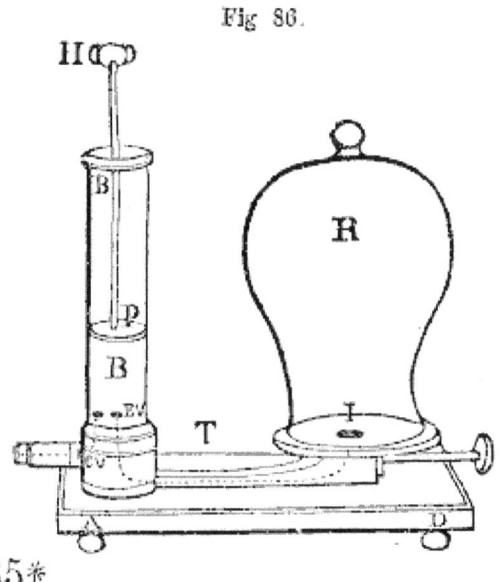

35*

414   NATURAL PHILOSOPHY.

Fig. 87.

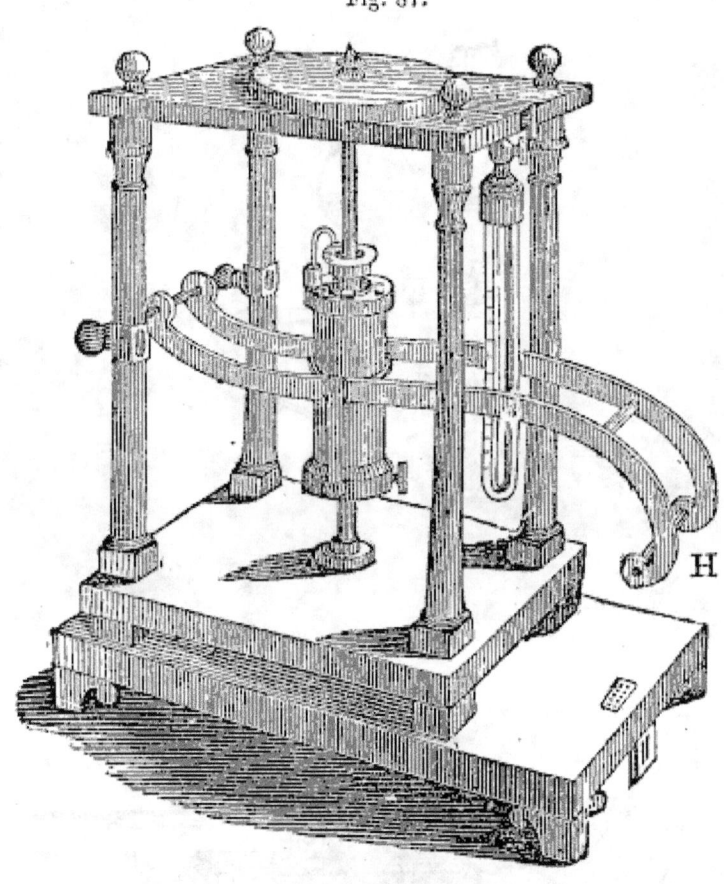

Fig. 88.

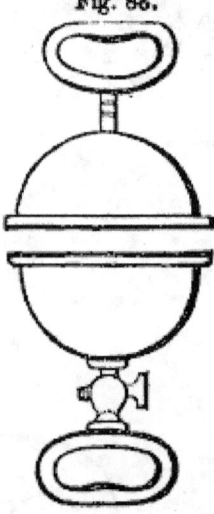

Fig. 89.

ILLUSTRATIONS. 415

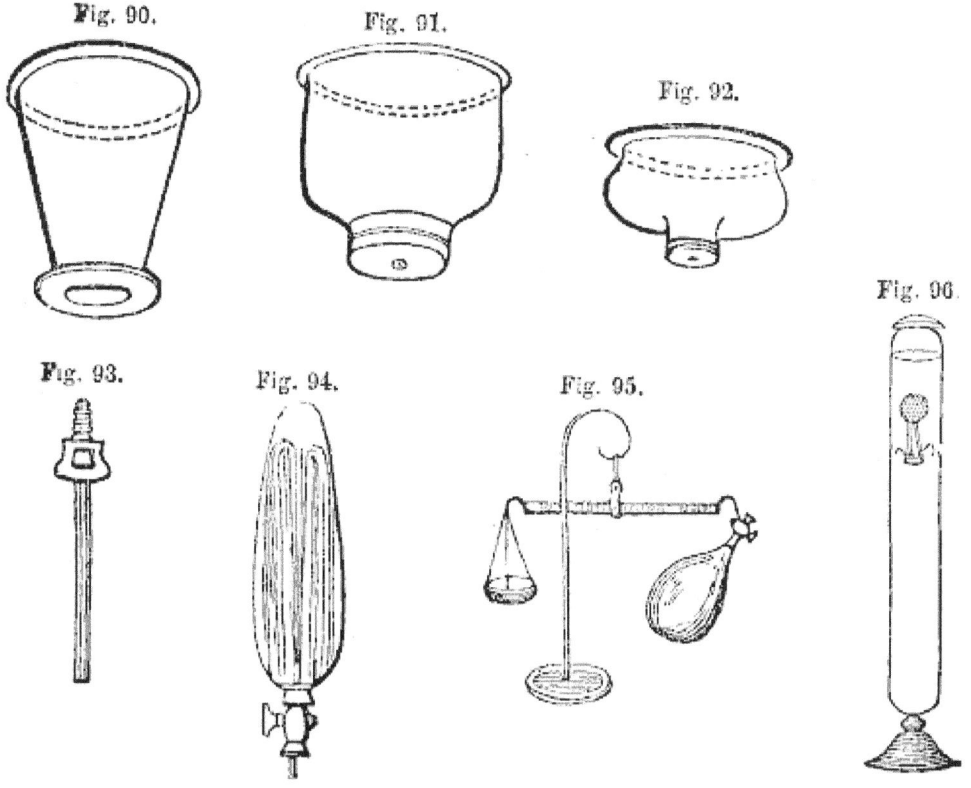
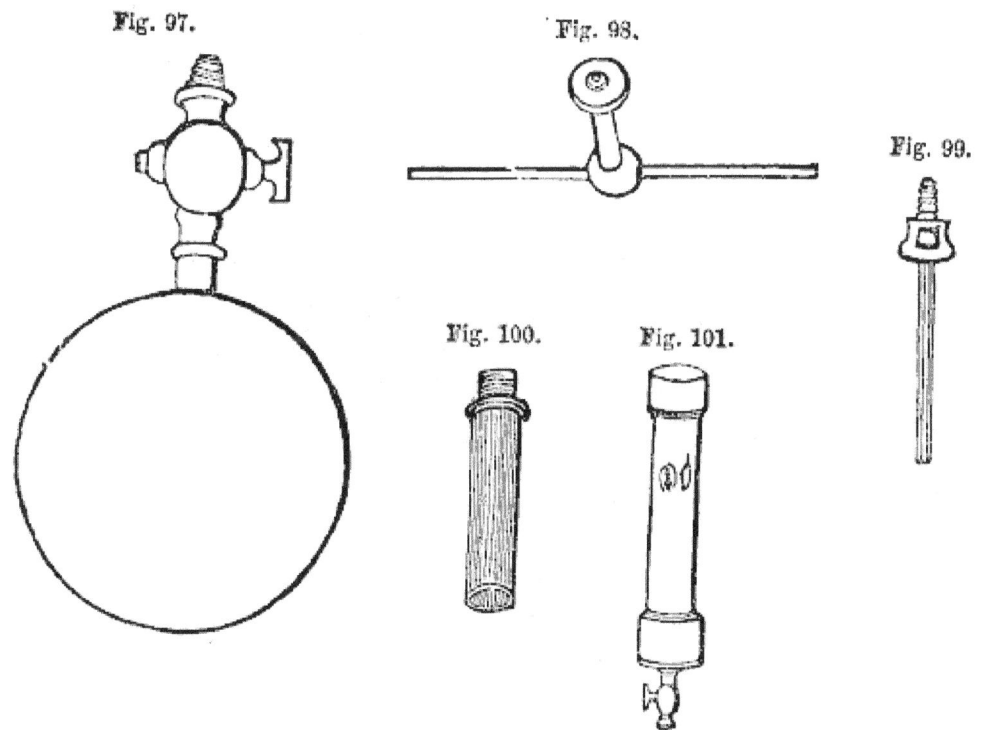

416 NATURAL PHILOSOPHY.

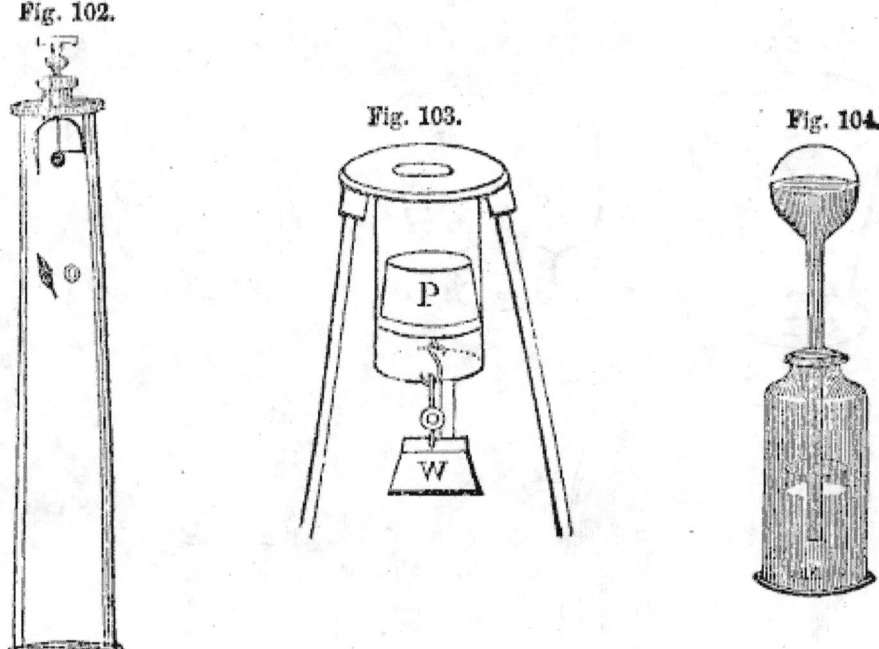

Fig. 102.   Fig. 103.   Fig. 104.

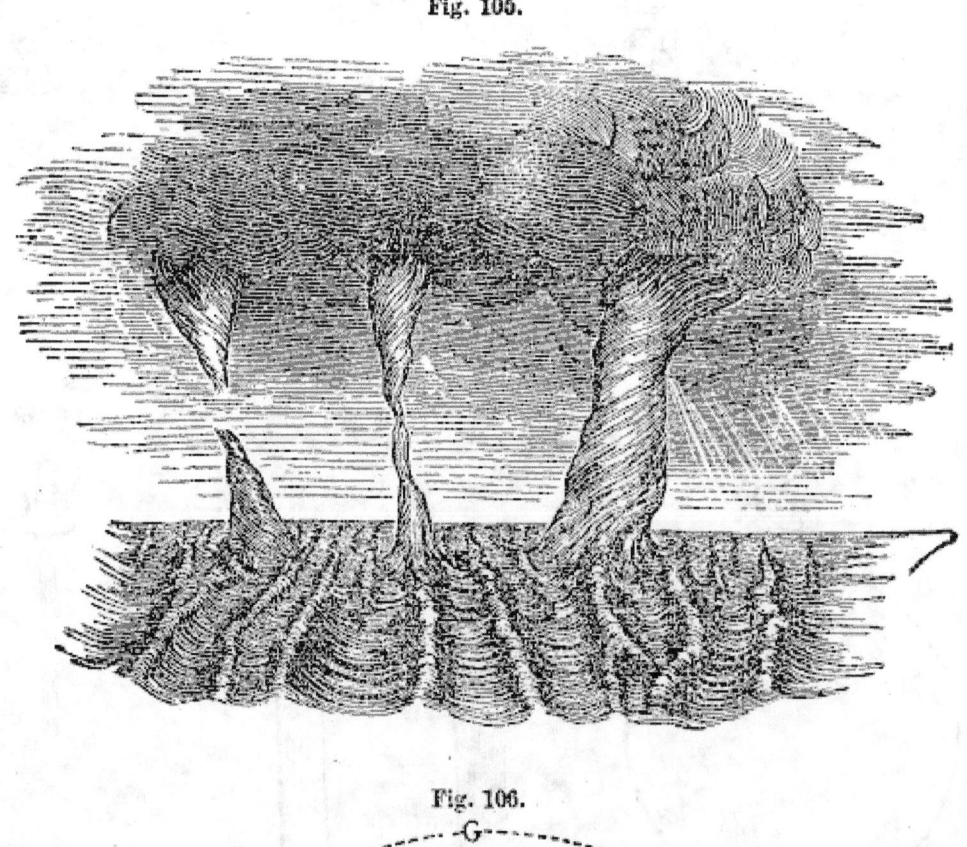

Fig. 105.

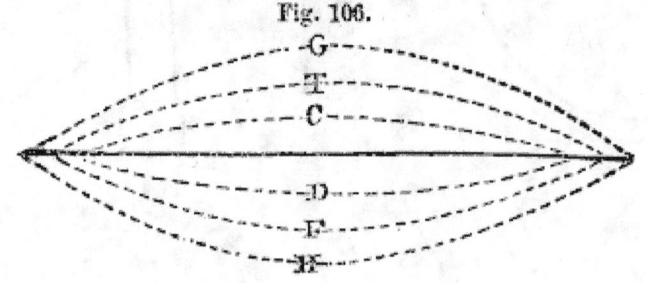

Fig. 106.

ILLUSTRATIONS. 417

Fig. 107.

Fig. 108.

Fig. 109.

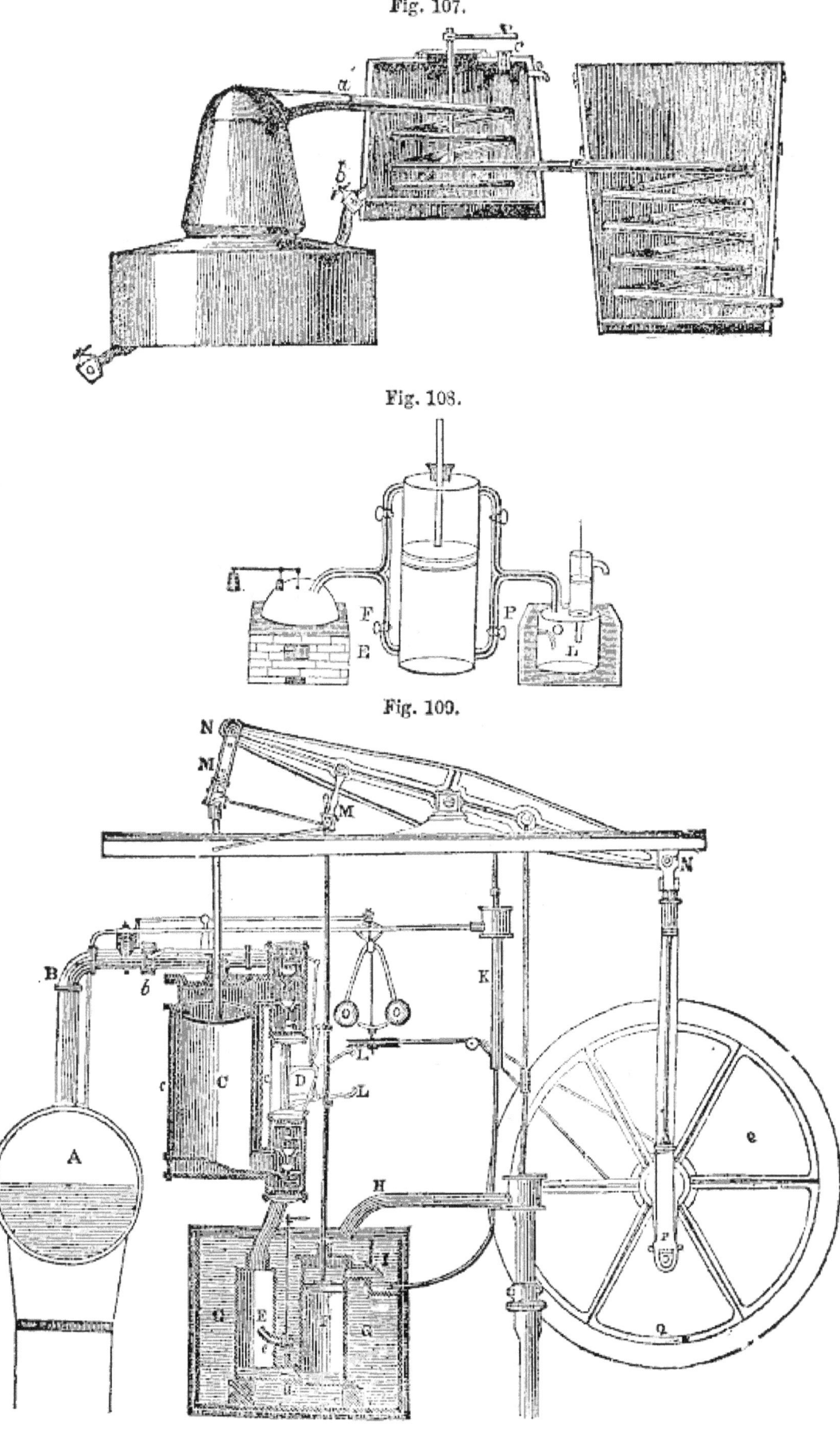

418 NATURAL PHILOSOPHY.

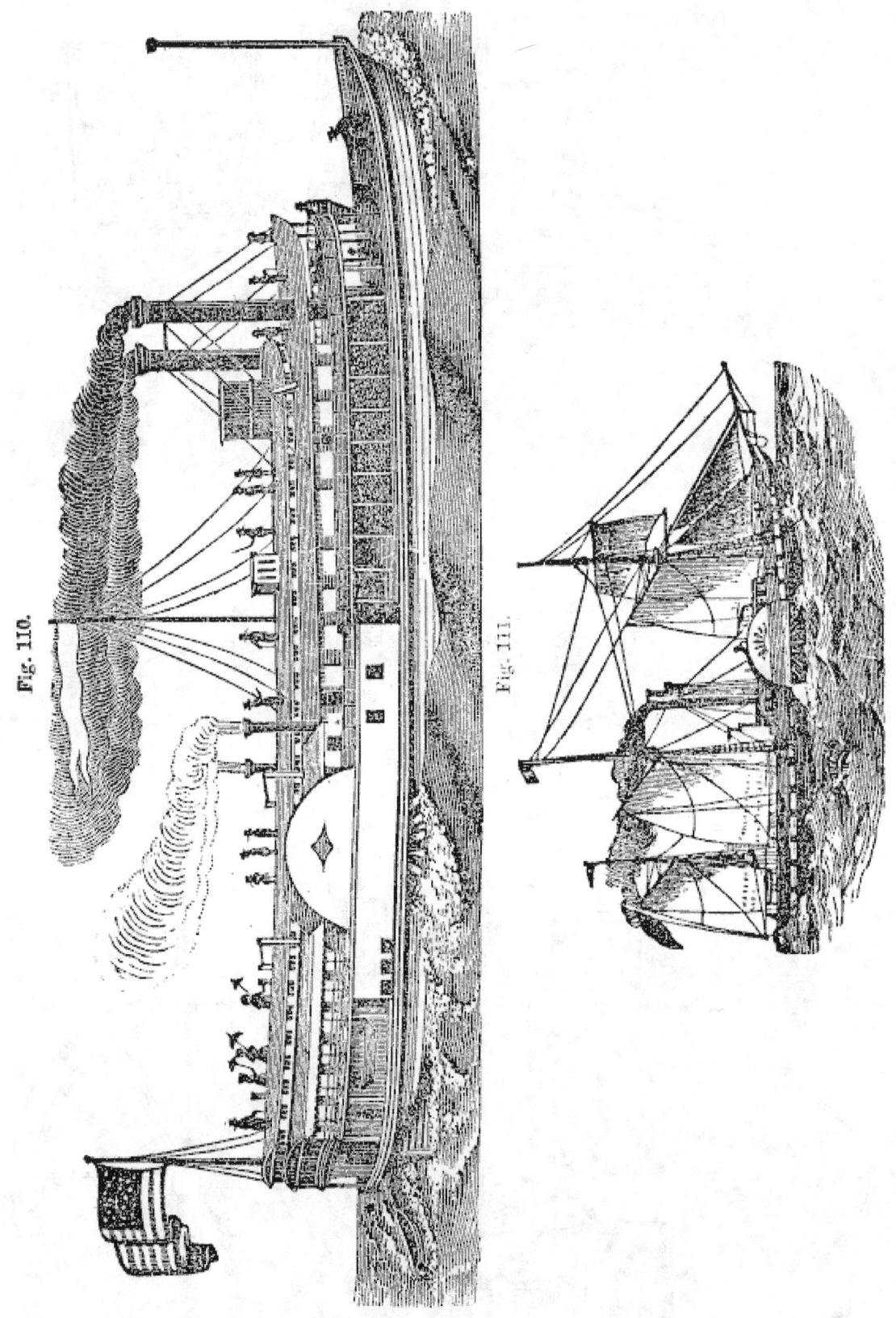

Fig. 110.

Fig. 111.

ILLUSTRATIONS. 419

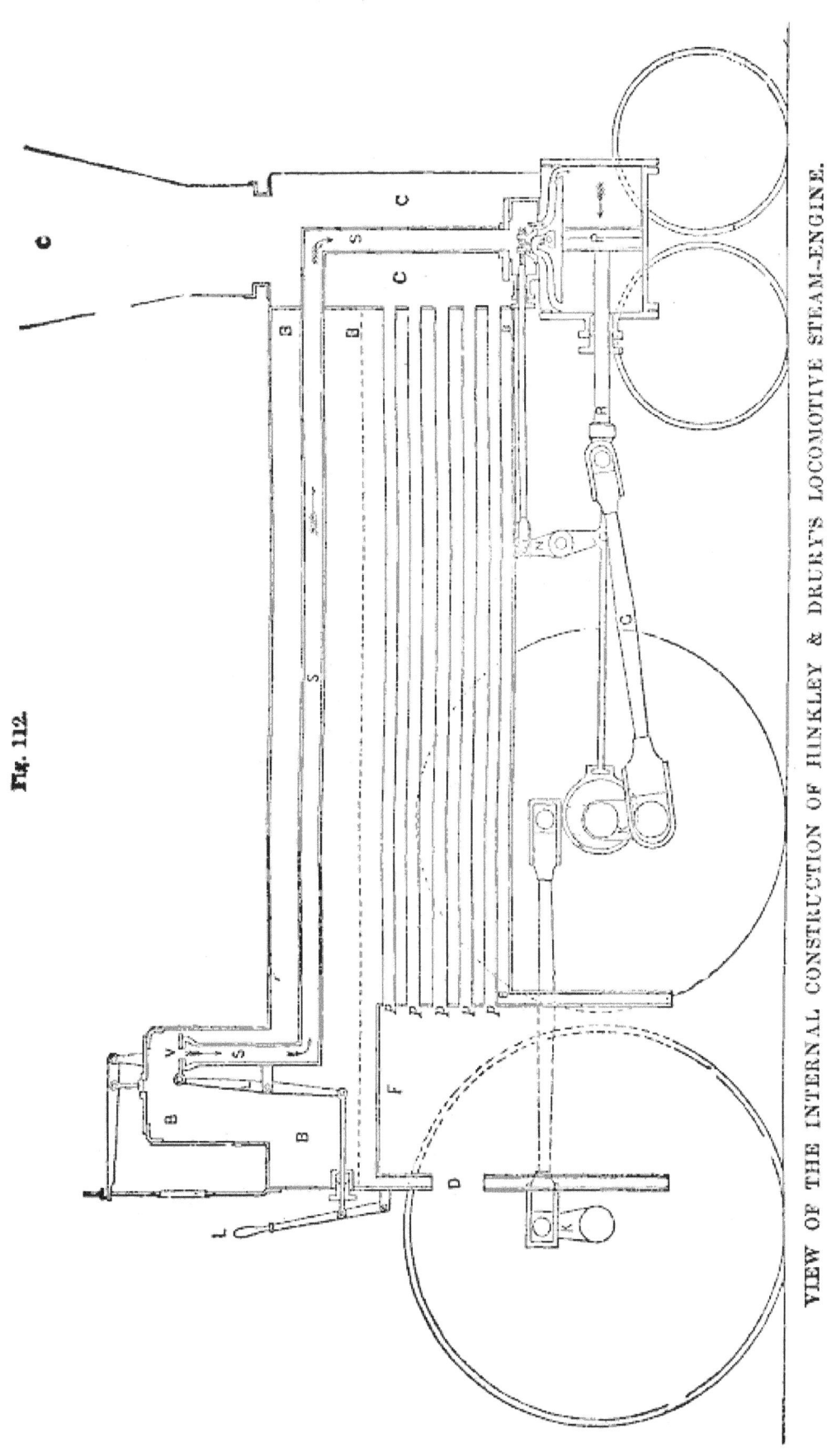

Fig. 112.

VIEW OF THE INTERNAL CONSTRUCTION OF HINKLEY & DRURY'S LOCOMOTIVE STEAM-ENGINE.

420 NATURAL PHILOSOPHY.

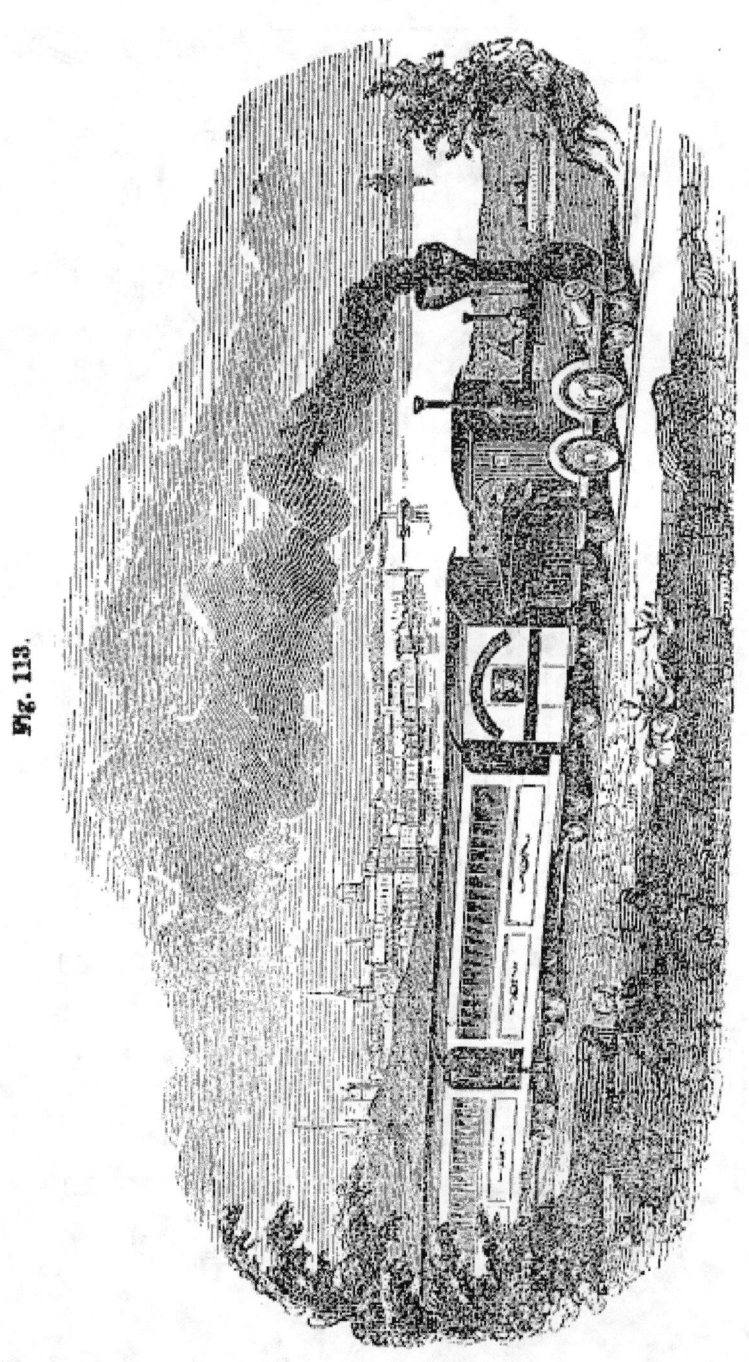

Fig. 113.

ILLUSTRATIONS. 421

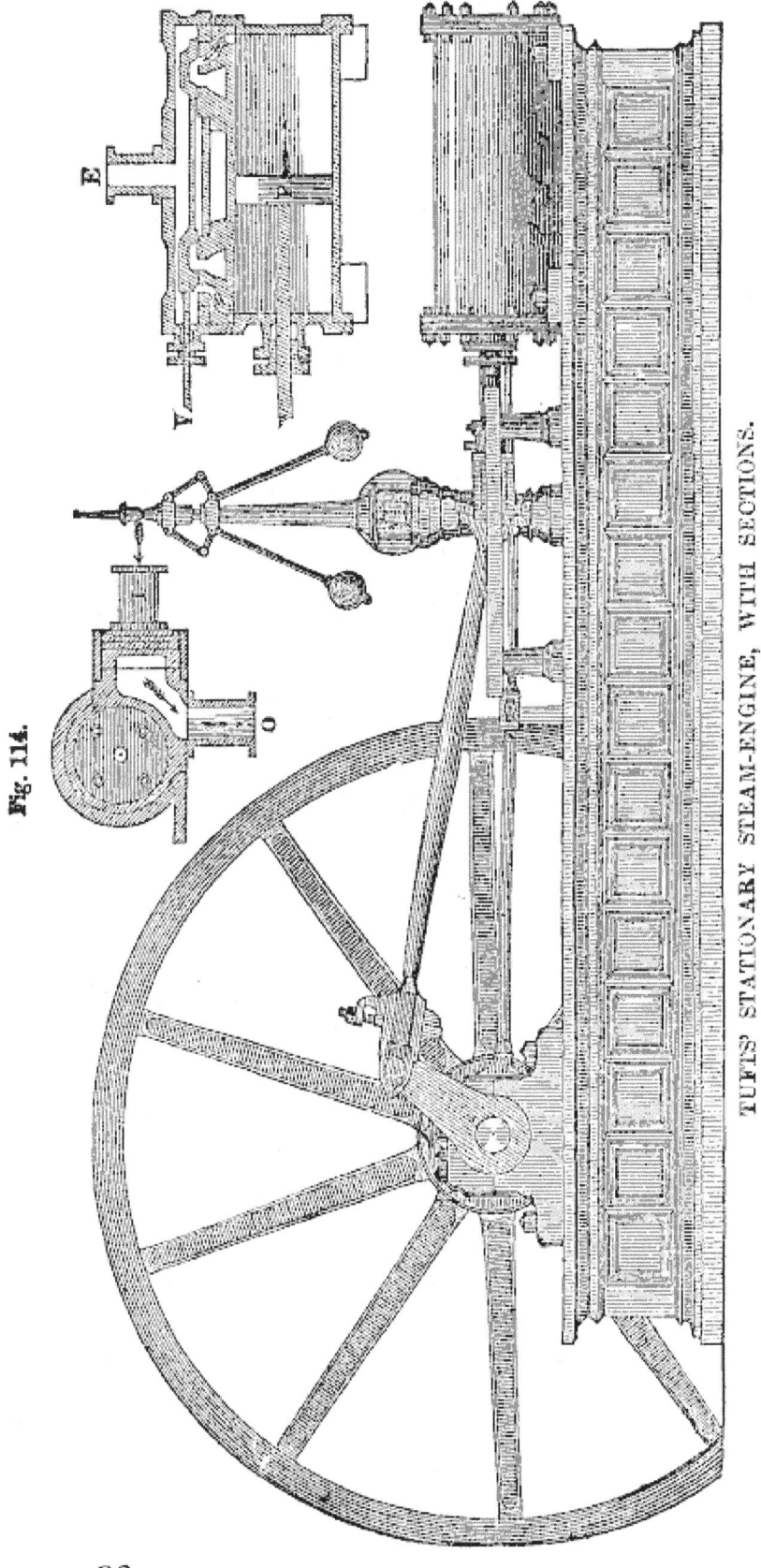

Fig. 114.

TUFTS' STATIONARY STEAM-ENGINE, WITH SECTIONS.

36

422 NATURAL PHILOSOPHY.

Fig. 115.

Fig. 116.

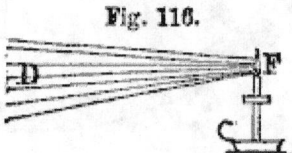

ILLUSTRATIONS. 423

Fig. 117.

Fig. 118.

Fig. 119.

Fig. 120.

Fig. 121.

Fig. 122.

Fig. 123.

Fig. 124.

Fig. 125.

Fig. 126.

Fig. 127.

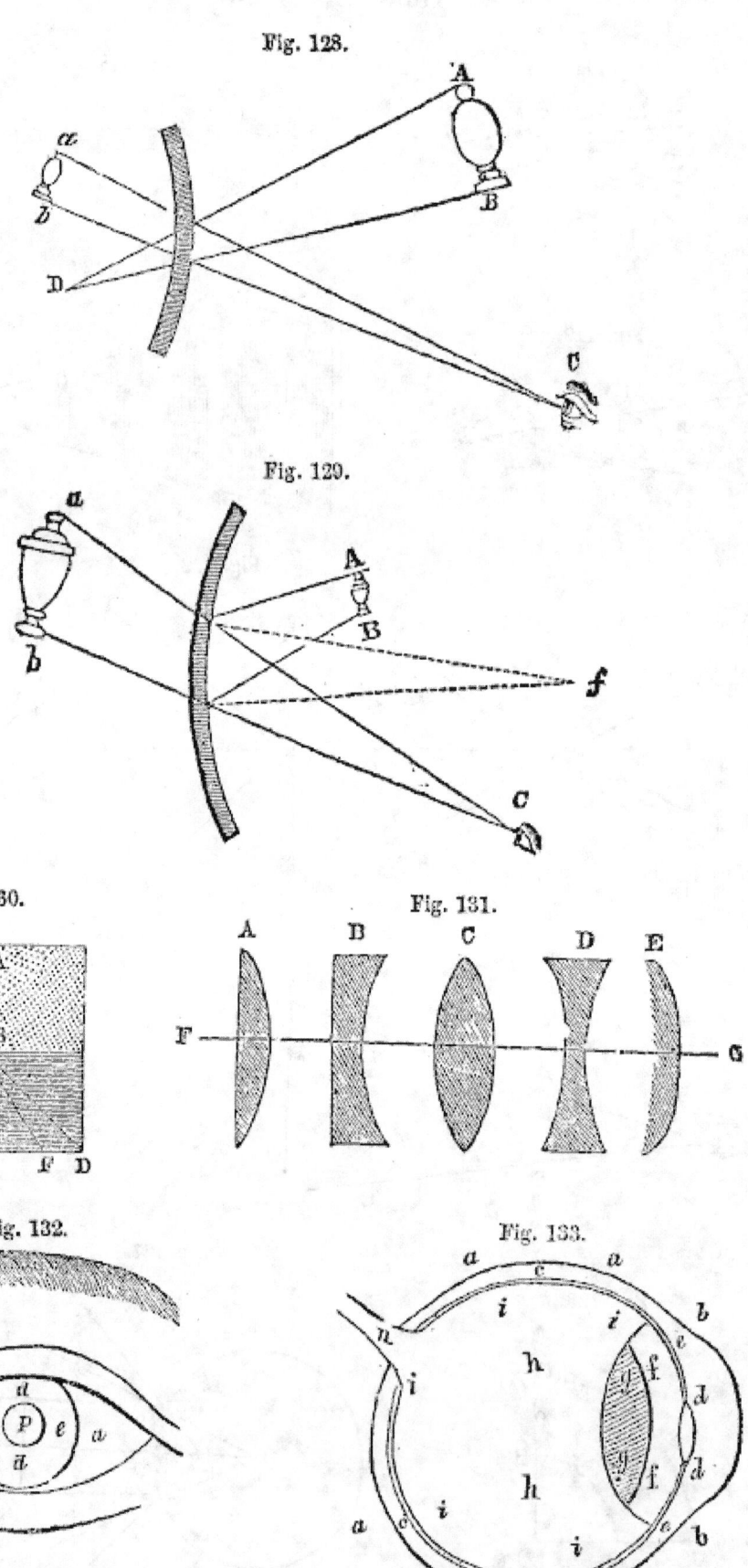

ILLUSTRATIONS.

Fig. 134.

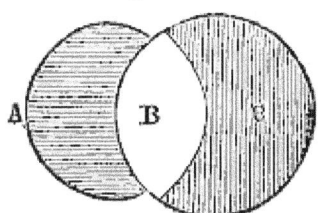

Fig. 135.

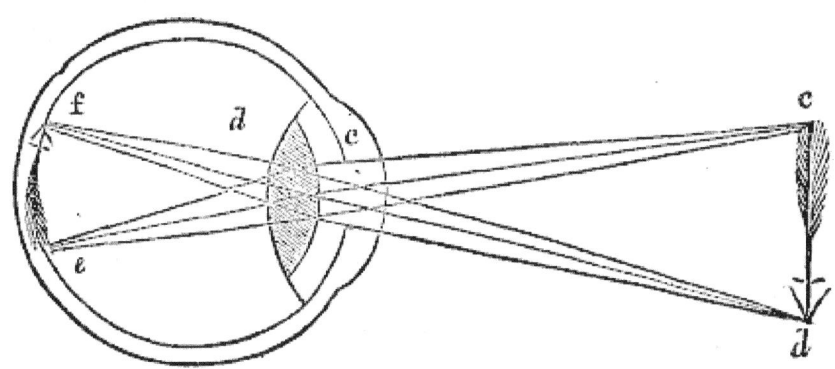

Fig. 136.

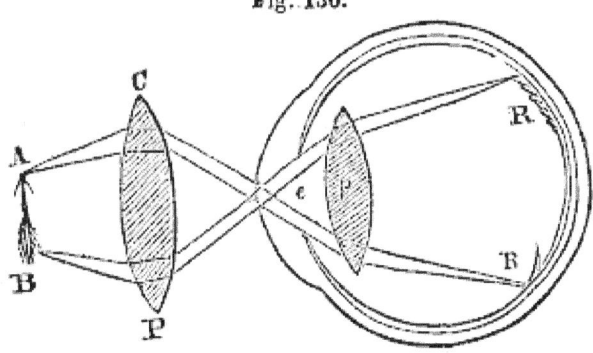

Fig. 137.

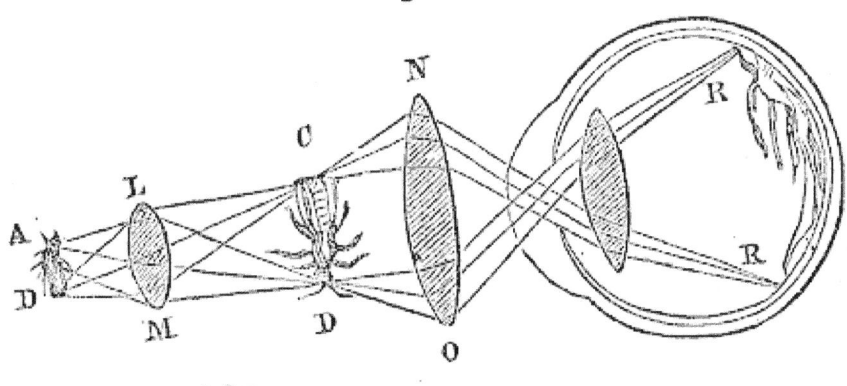

426 NATURAL PHILOSOPHY.

Fig. 138.

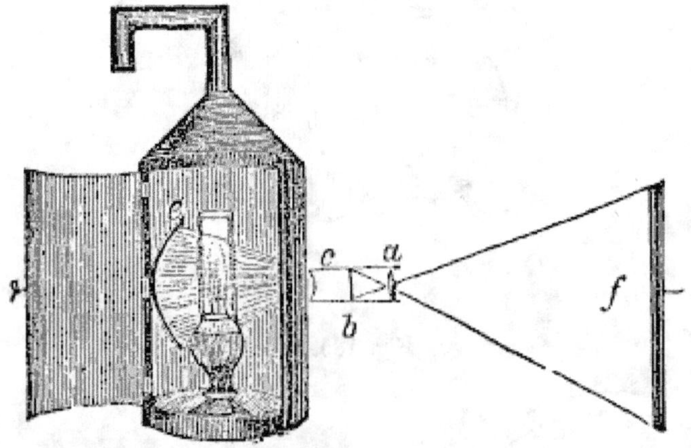

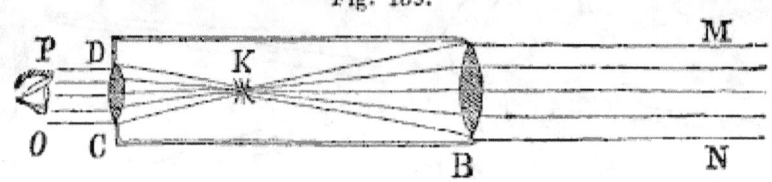

Fig. 139.

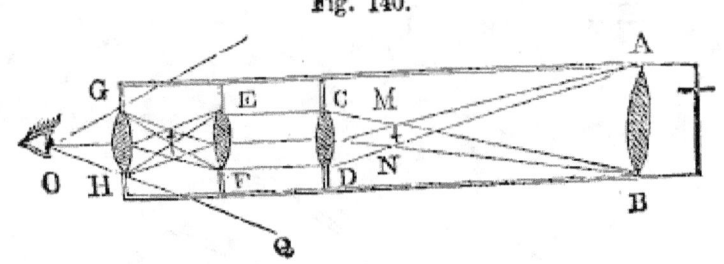

Fig. 140.

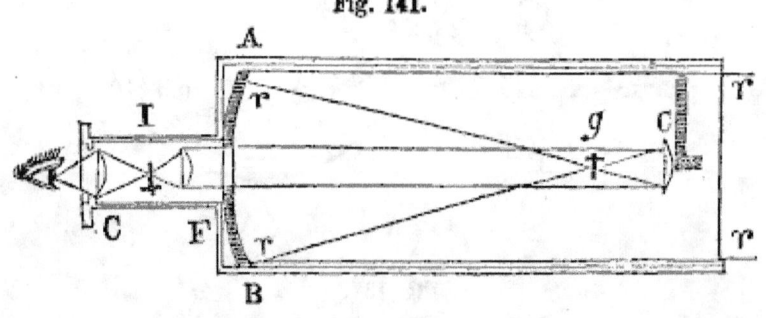

Fig. 141.

Fig. 142.

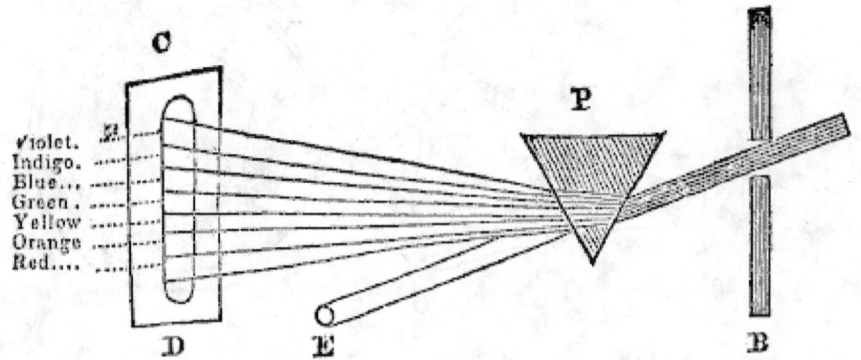

ILLUSTRATIONS. 427

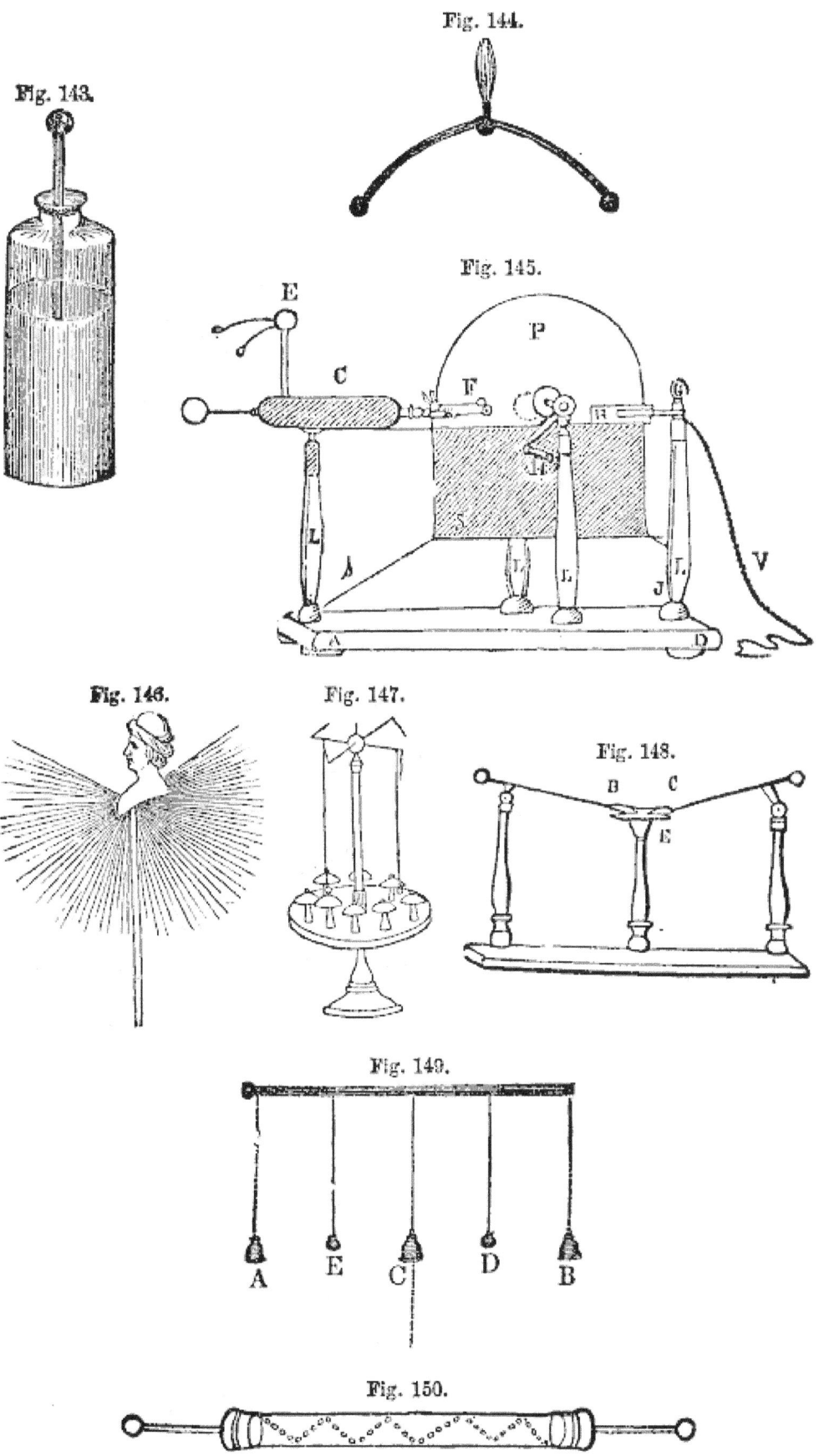

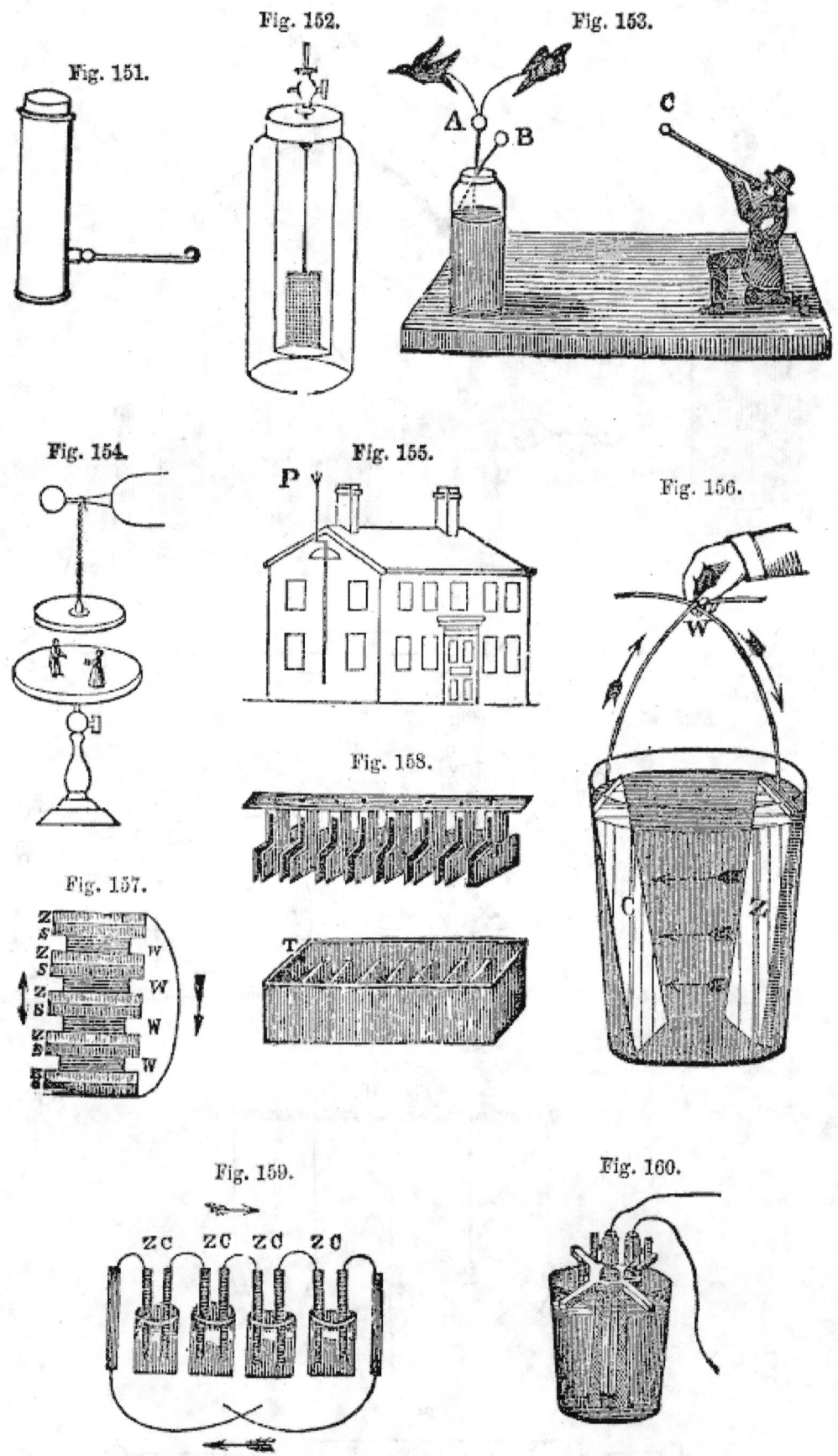

ILLUSTRATIONS.

Fig. 161.

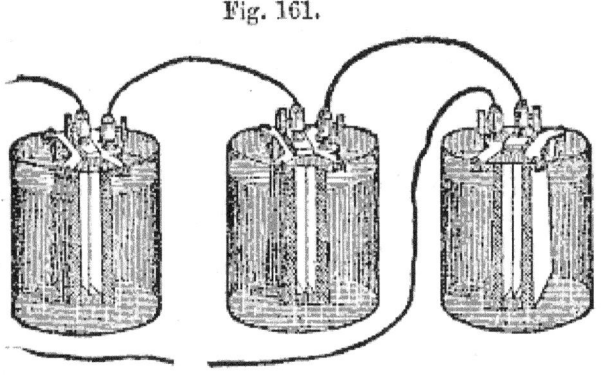

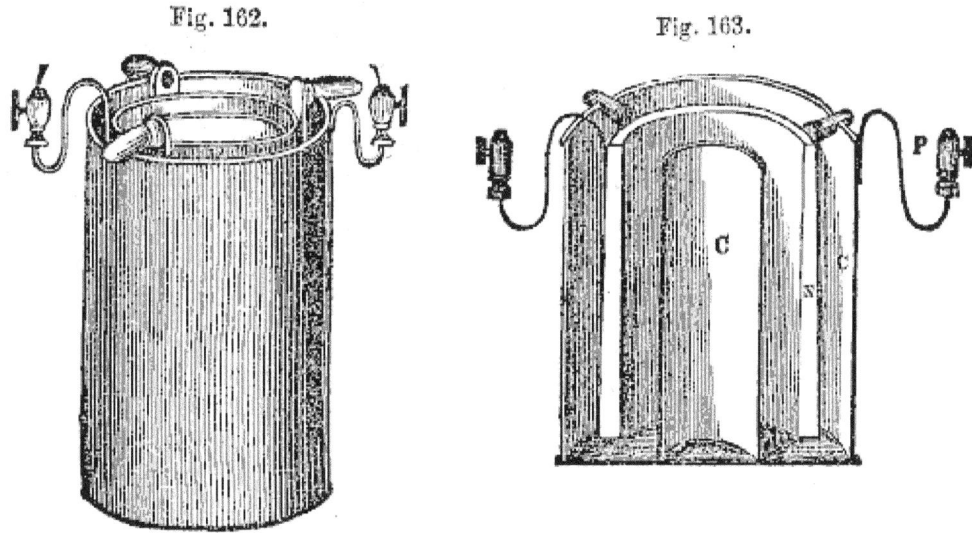

Fig. 162.   Fig. 163.

Fig. 164.

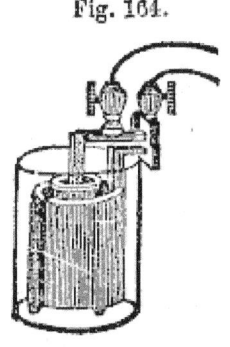

Fig. 165.

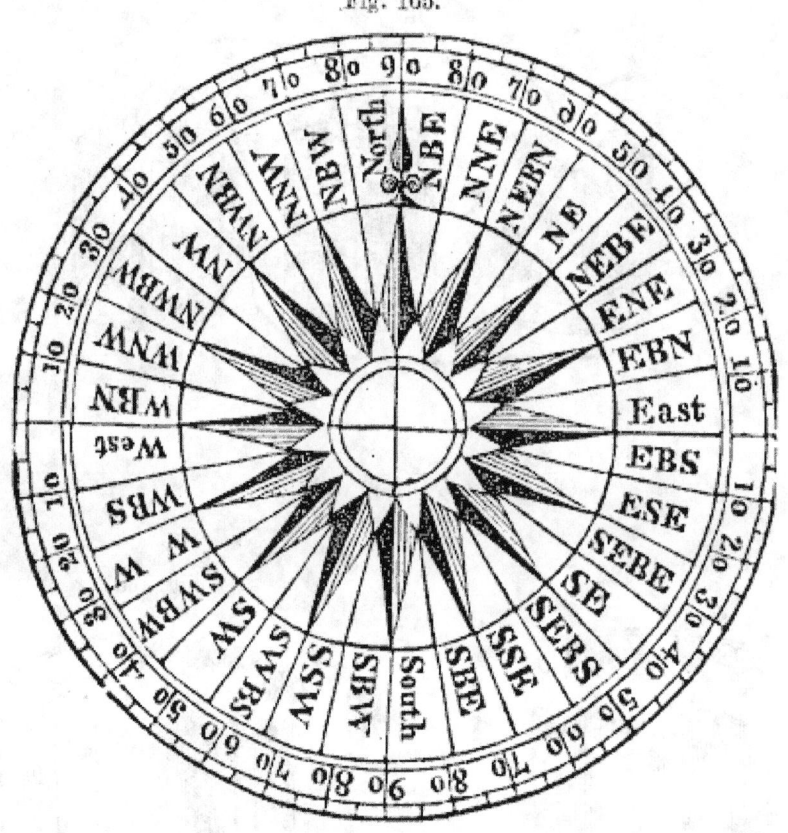

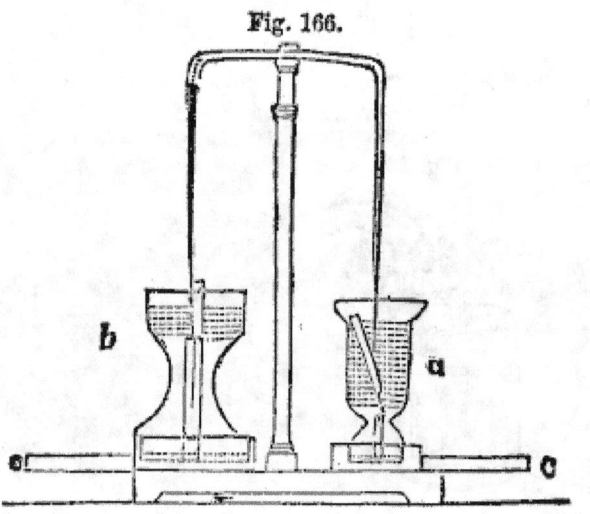

Fig. 166.

Fig. 167.

ILLUSTRATIONS.

Fig. 168.

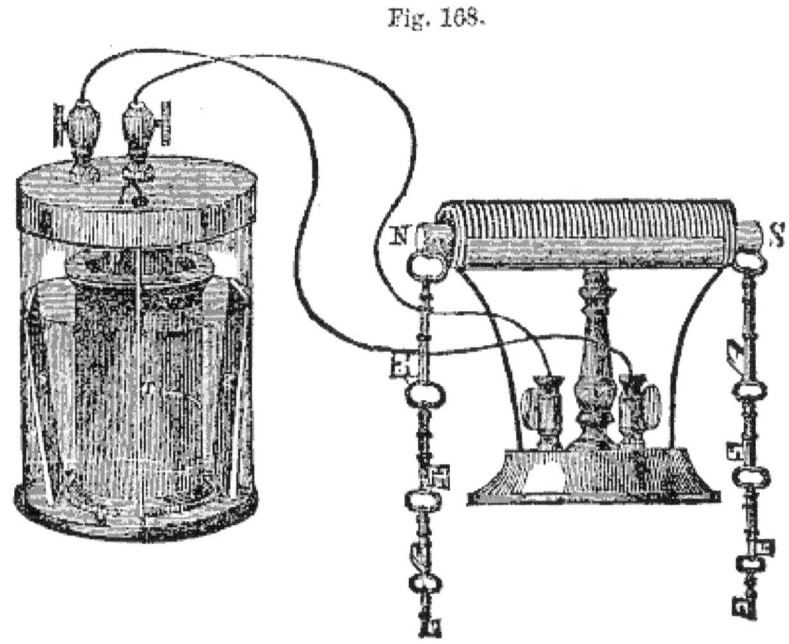

Fig. 169

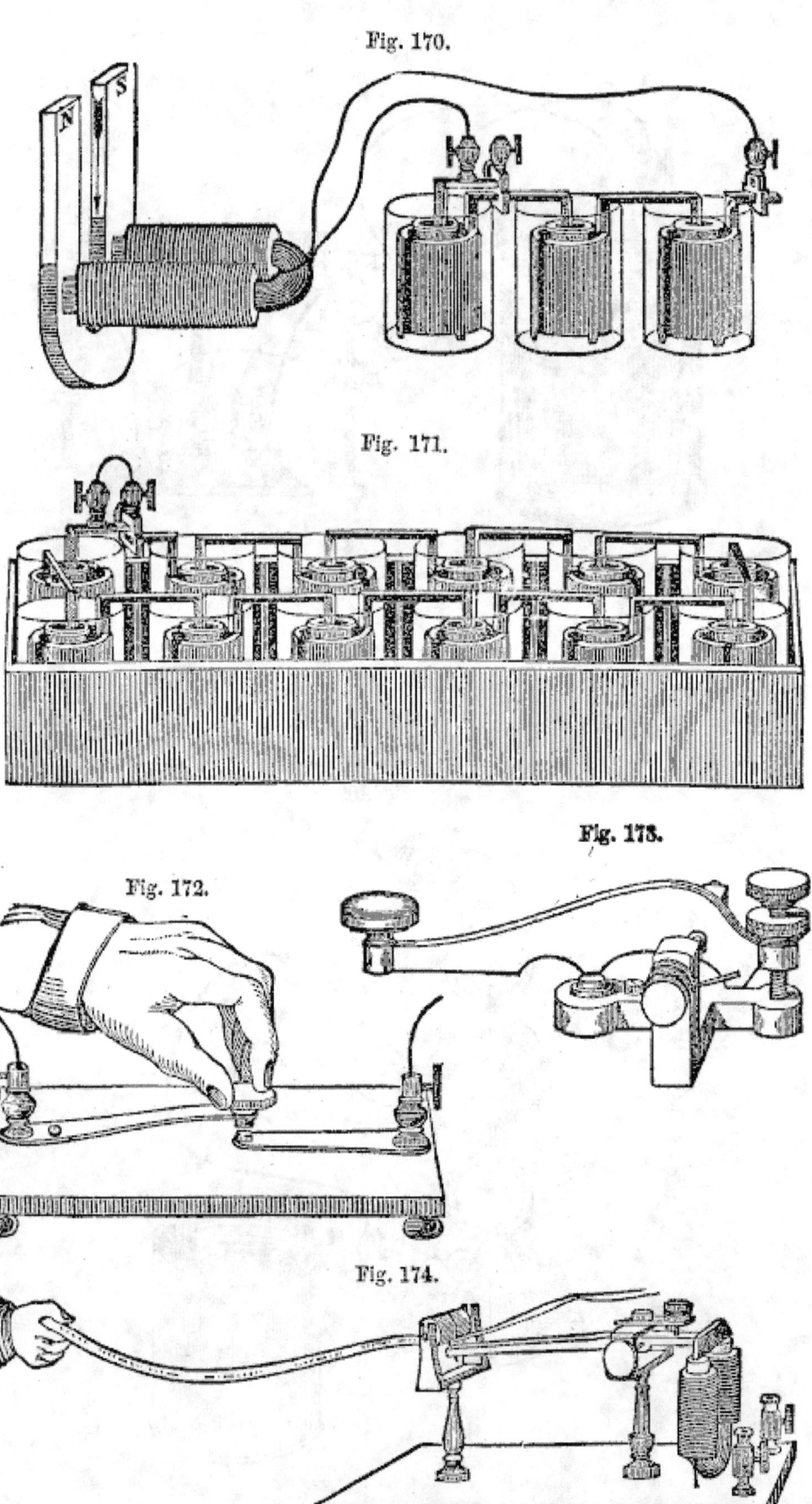

Fig. 170.

Fig. 171.

Fig. 172.

Fig. 173.

Fig. 174.

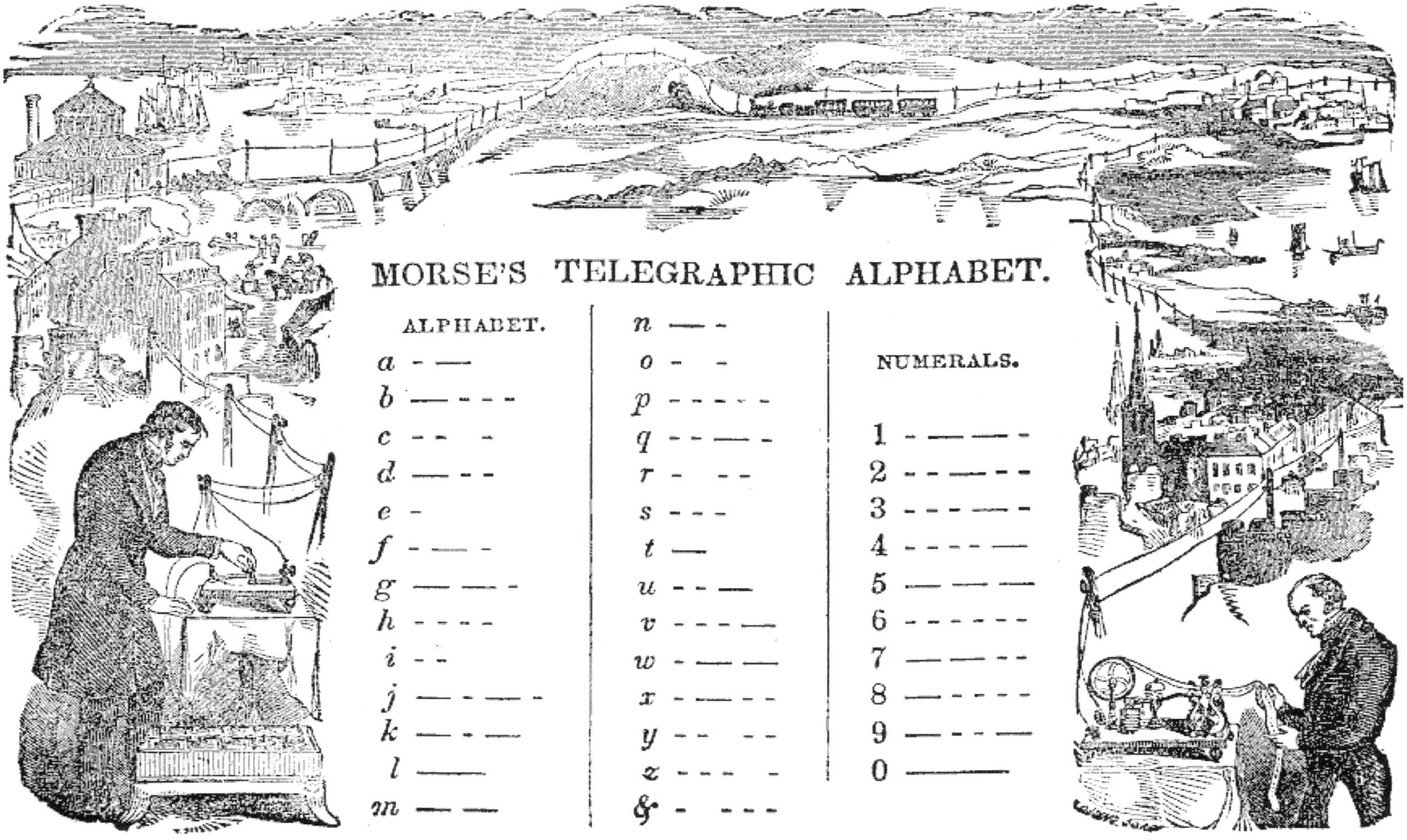

Fig. 175.

MORSE'S TELEGRAPHIC ALPHABET.

ALPHABET.			NUMERALS.	
a ·—		n —·	1 ·———	
b —···		o · ·	2 ··———	
c ·· ·		p ·····	3 ···——	
d —··		q ··—·	4 ····—	
e ·		r · ··	5 —————	
f ·—·		s ···	6 ······	
g ——·		t —	7 ——··	
h ····		u ··—	8 —····	
i ··		v ···—	9 —··—	
j —·—·		w ·——	0 ———	
k —·—		x ·—··		
l ——		y ·· ··		
m ——		z ··· ·		
		& · ···		

Fig. 176

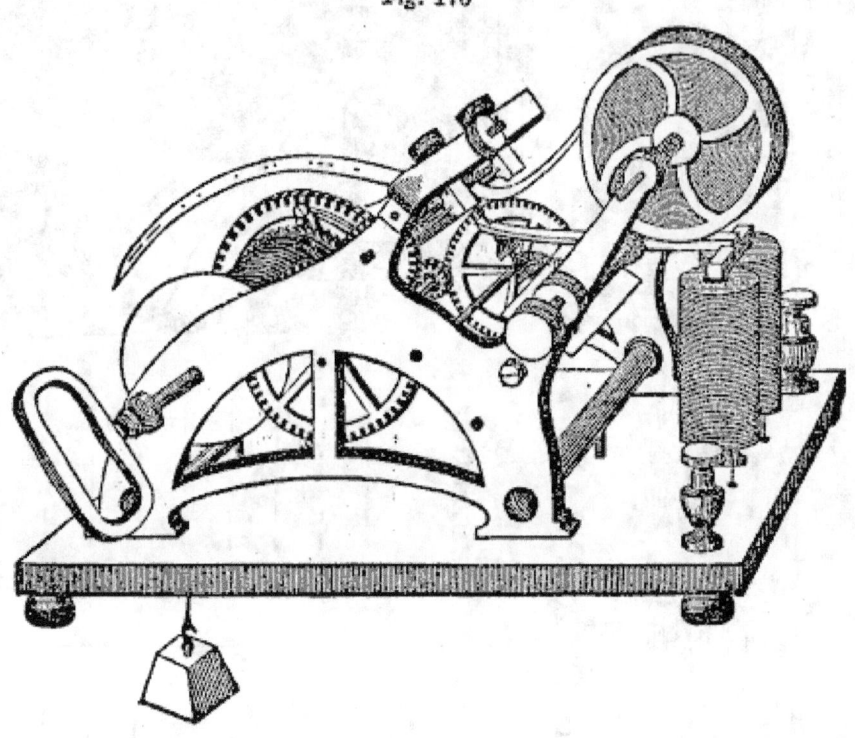

Fig. 177.

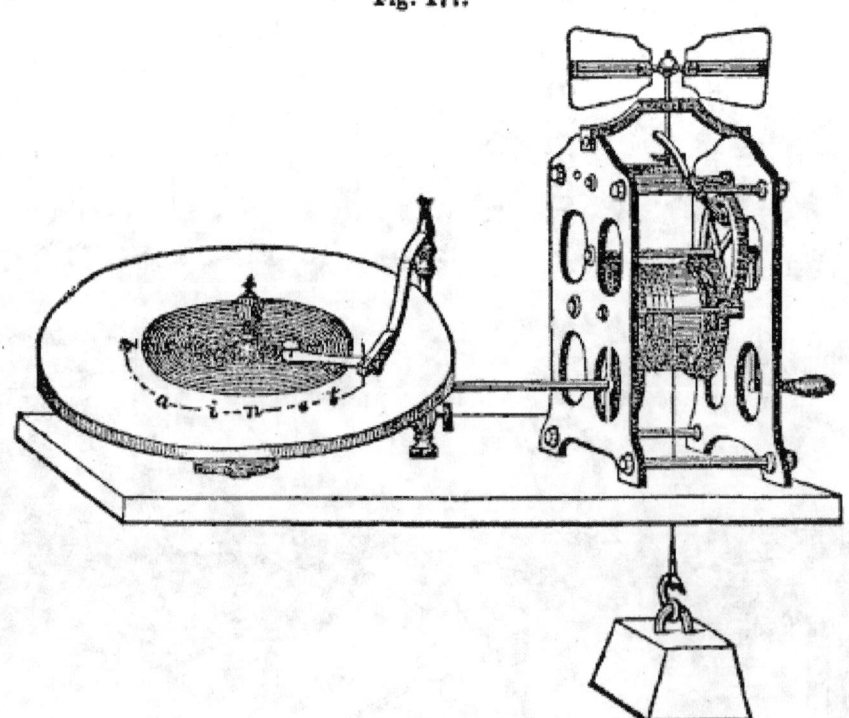

ILLUSTRATIONS. 435

Fig. 178.

Fig. 179.

Fig. 180

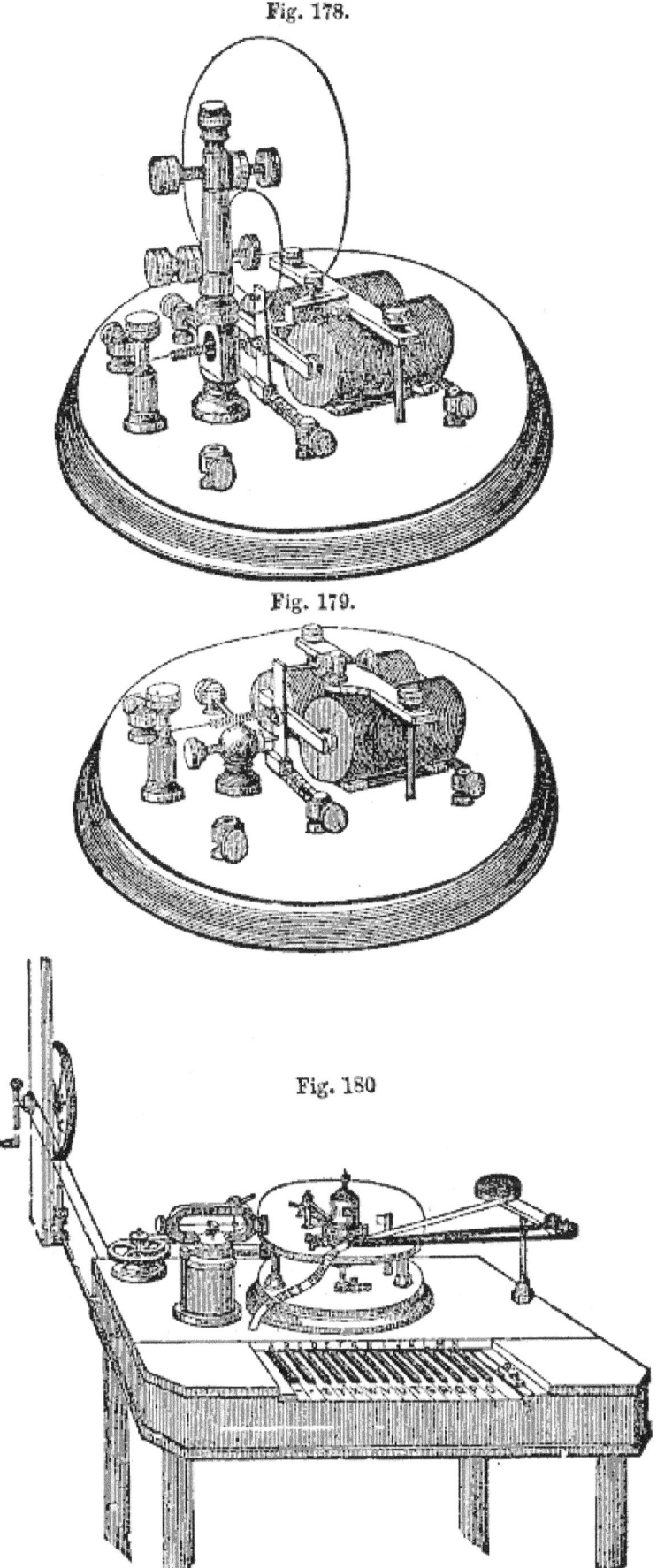

436   NATURAL PHILOSOPHY.

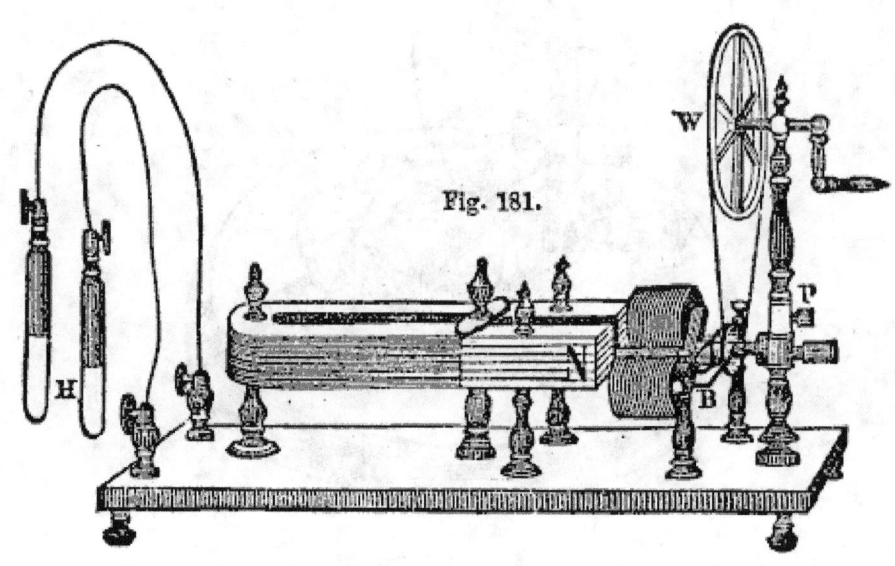

Fig. 181.

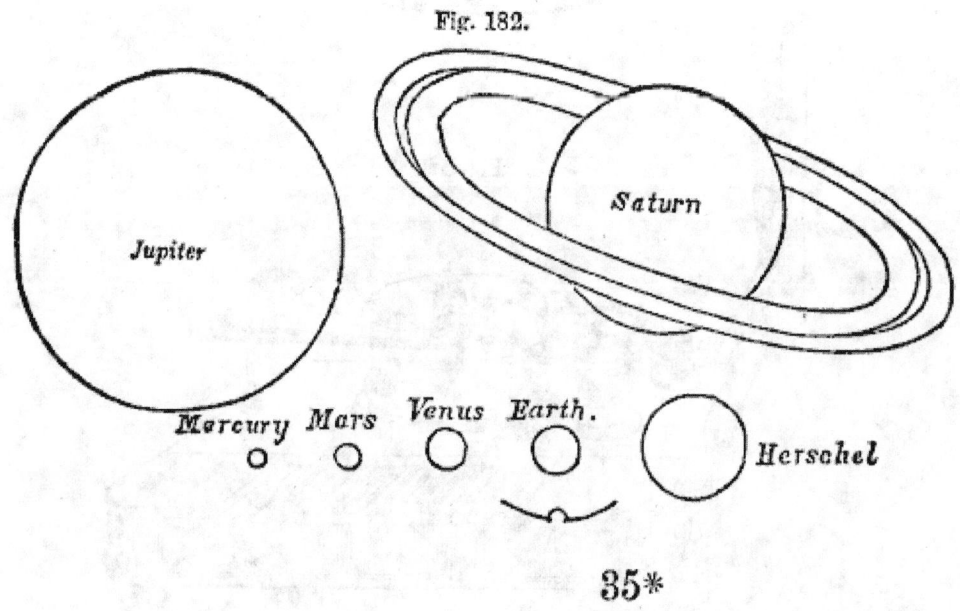

Fig. 182.

35*

ILLUSTRATIONS. 437

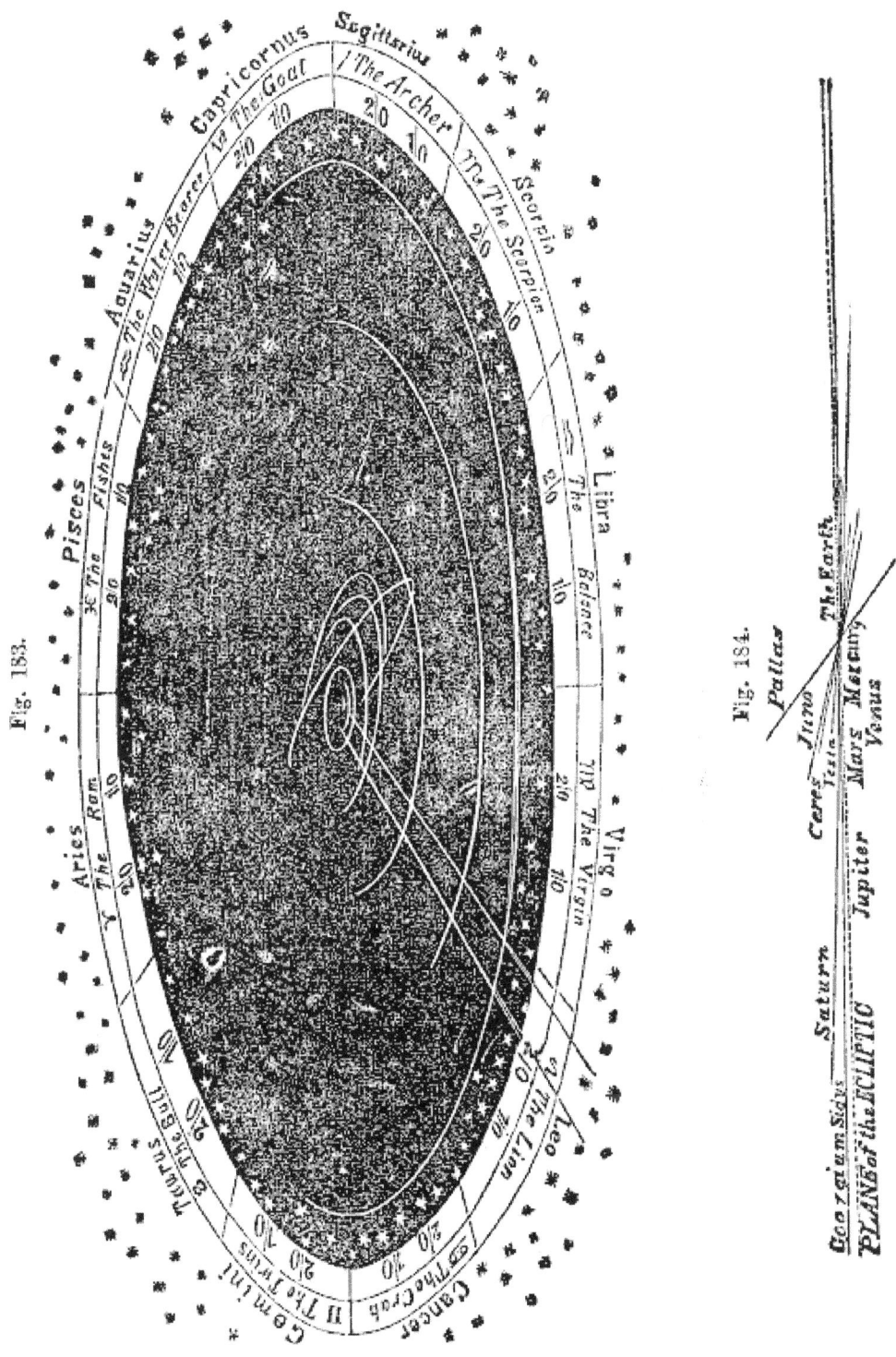

Fig. 183.

Fig. 184.

438 NATURAL PHILOSOPHY.

Fig. 185.

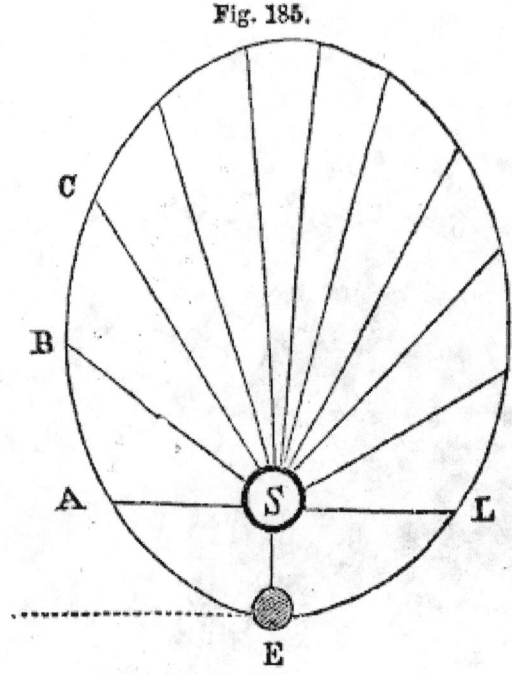

Fig. 186.

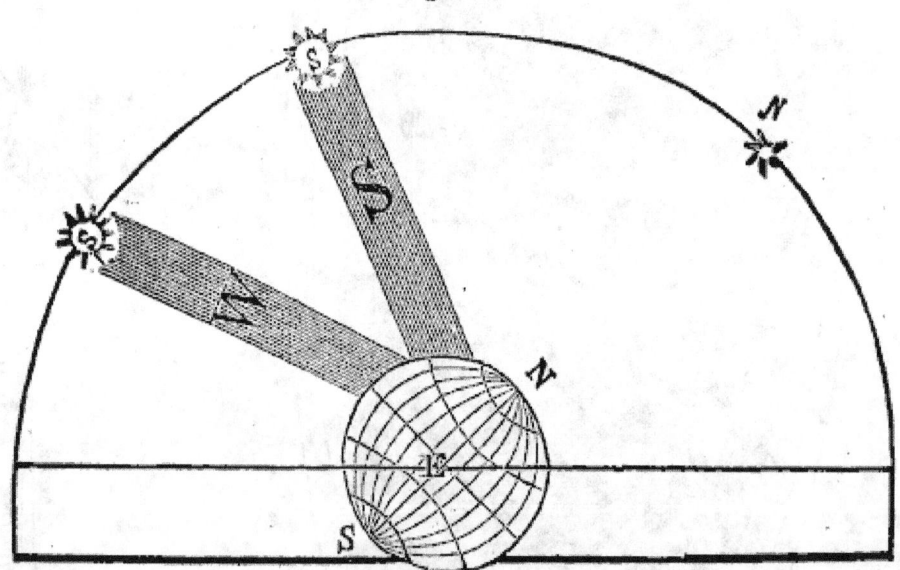

Fig. 187.

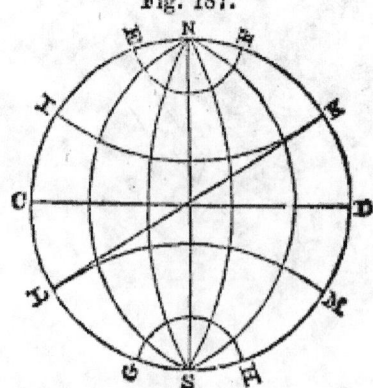

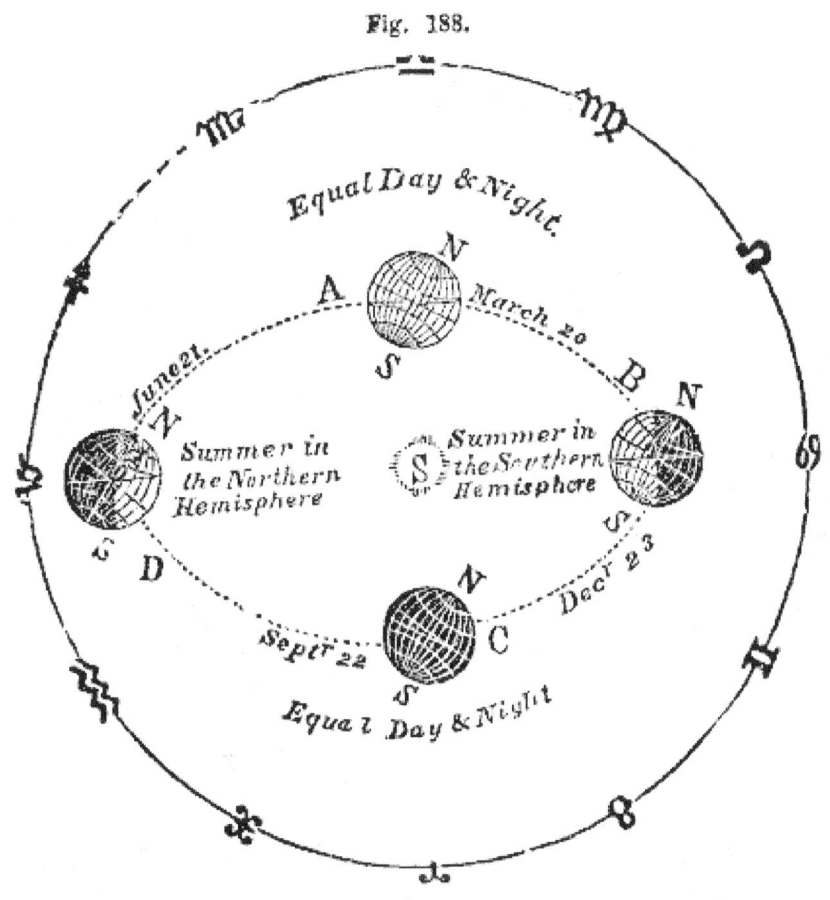
Fig. 188.

Fig. 189

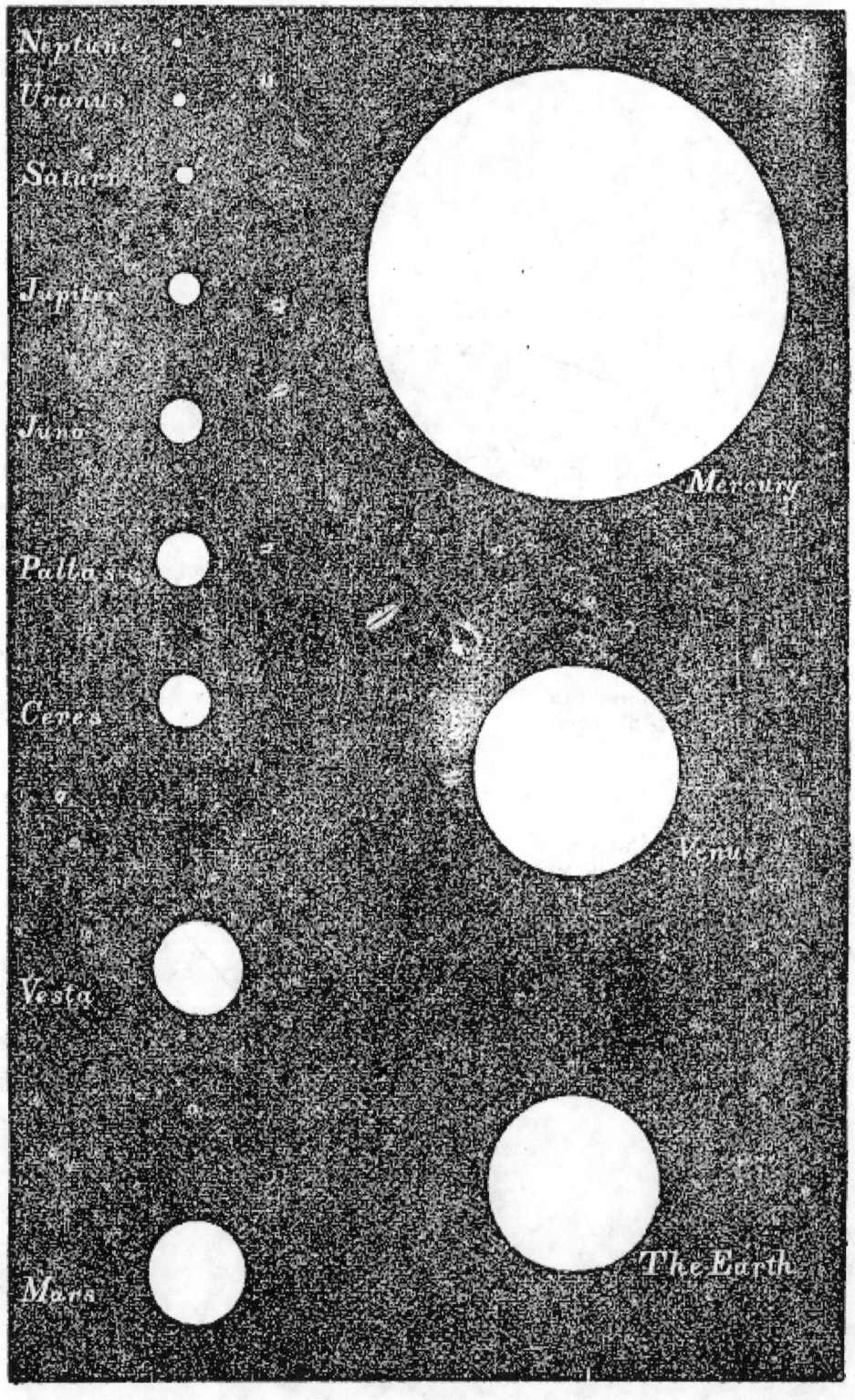

ILLUSTRATIONS. 441

Fig. 190.

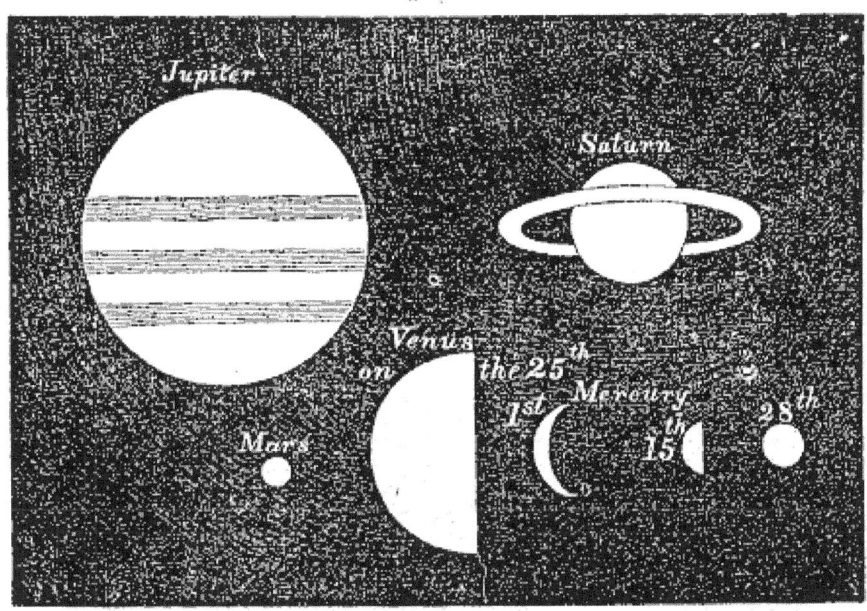

Fig. 191.

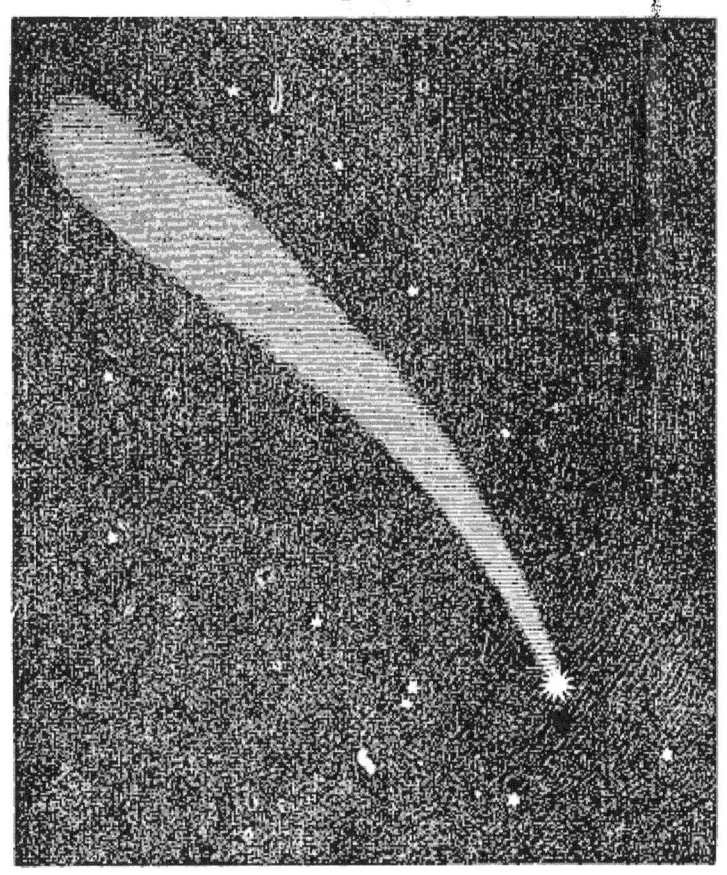

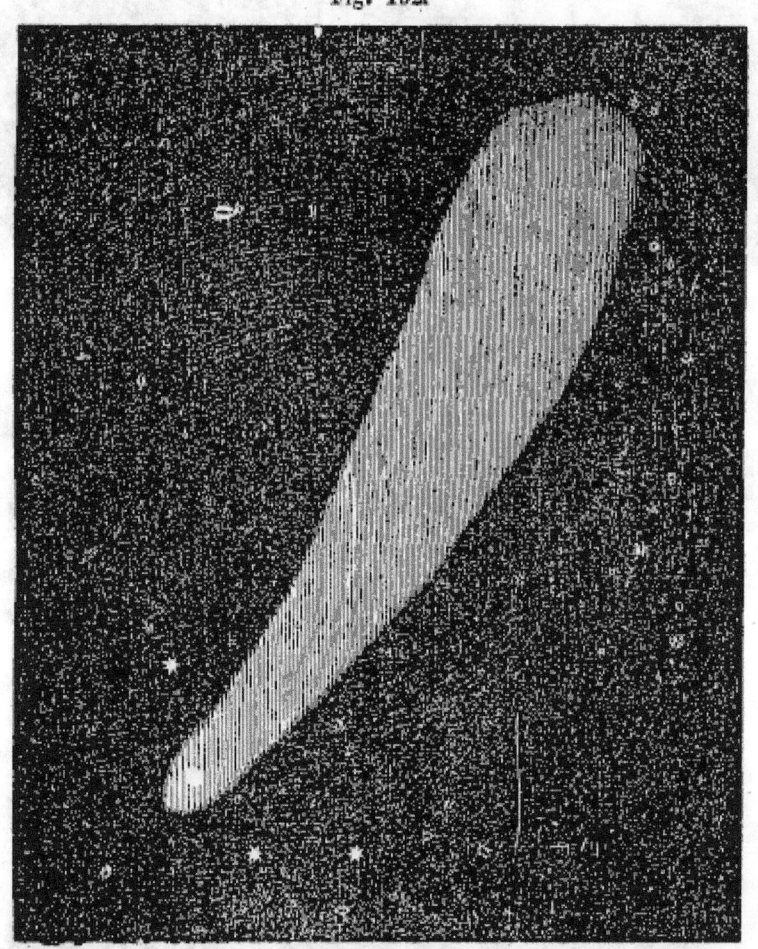

Fig. 192.

Fig. 193.

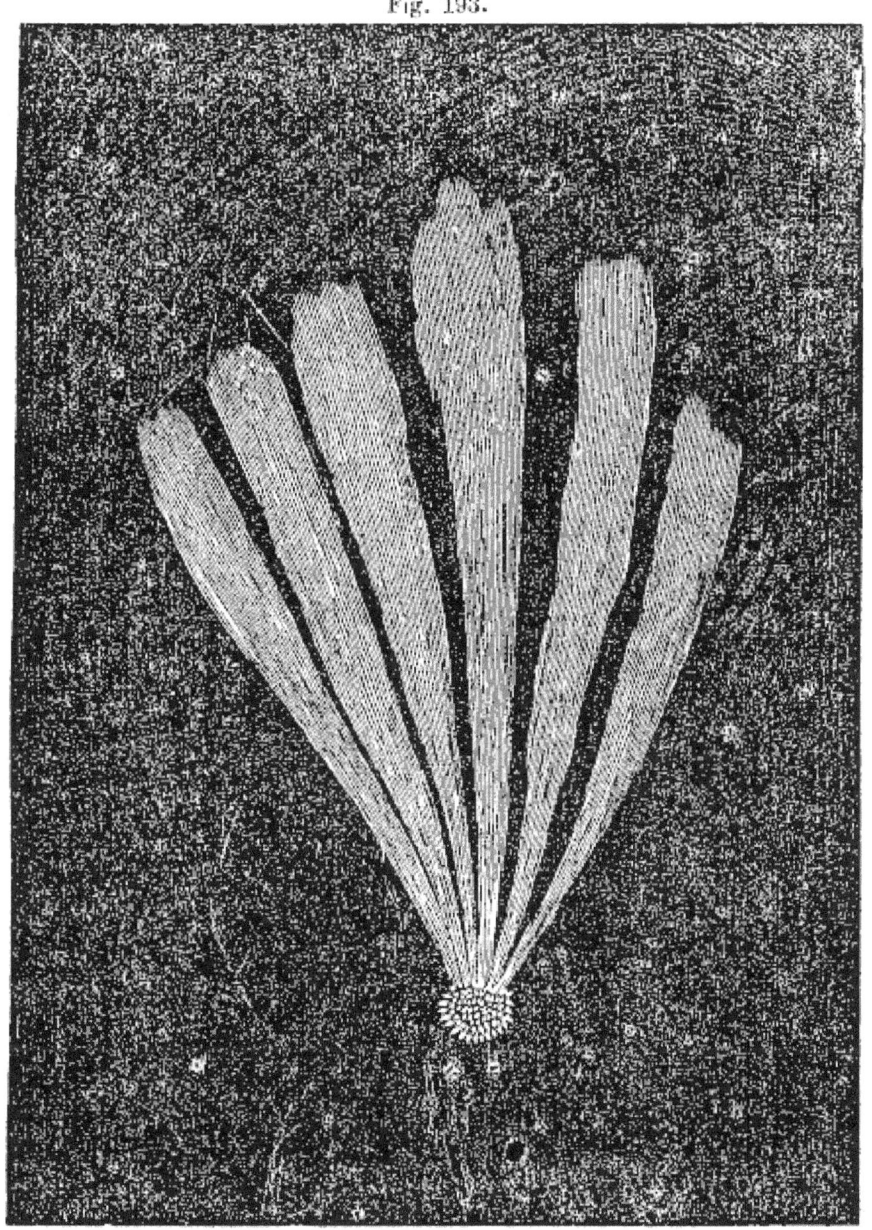

444 NATURAL PHILOSOPHY.

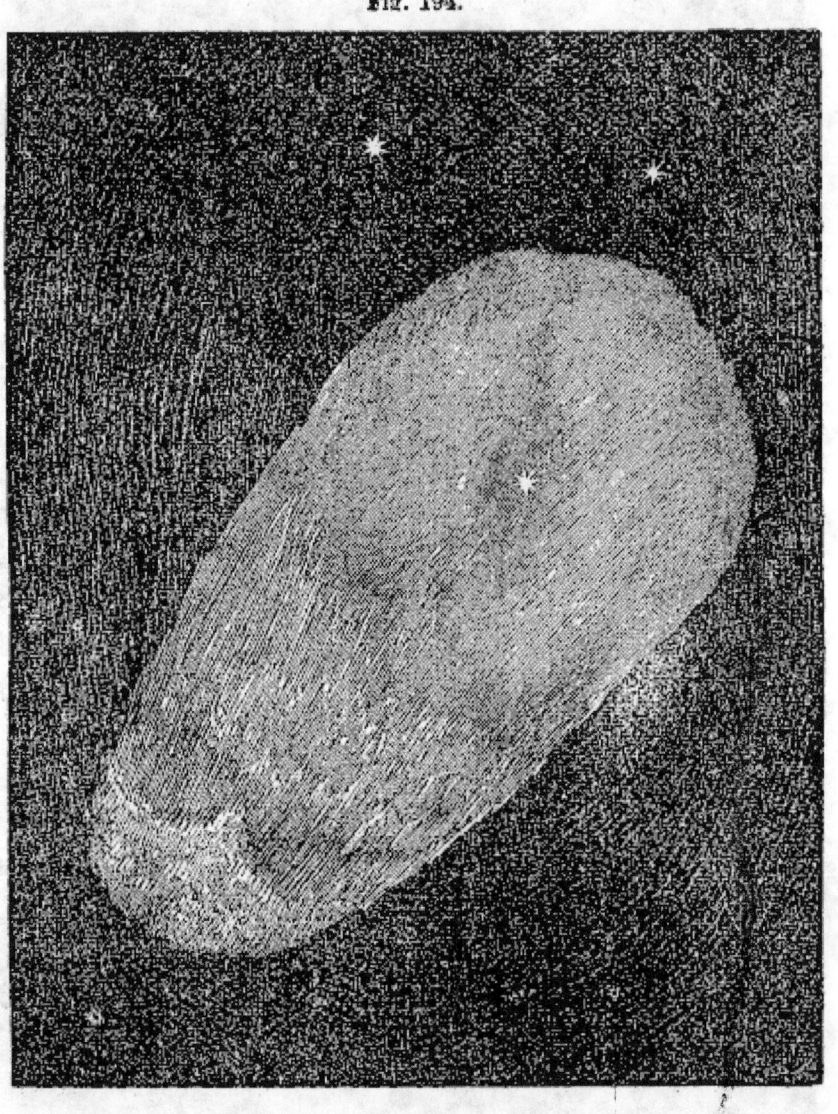

Fig. 194.

ILLUSTRATIONS. 445

Fig. 195.

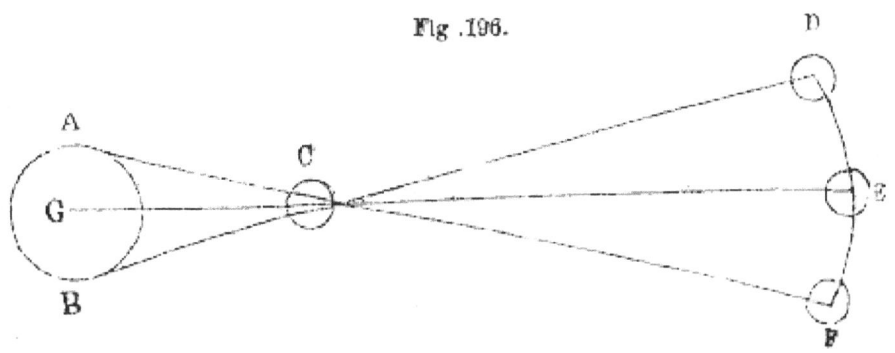

Fig. 196.

38

446 NATURAL PHILOSOPHY.

Fig. 197.

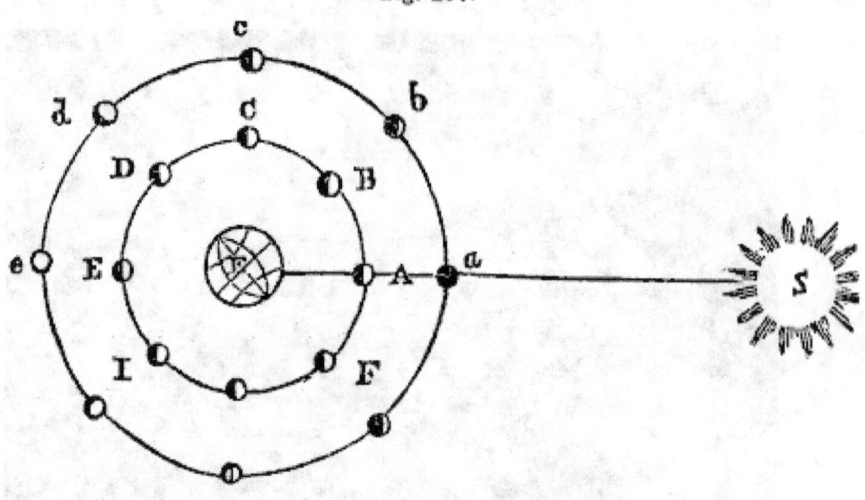

Fig. 198.

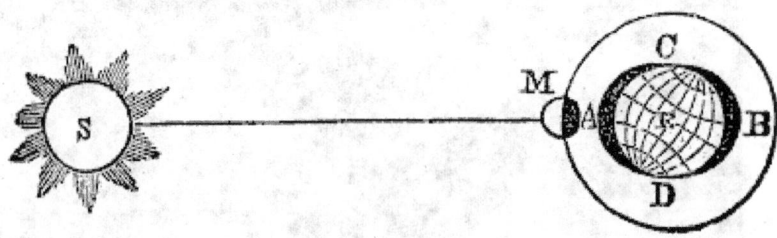

Fig. 199.

Fig. 200.

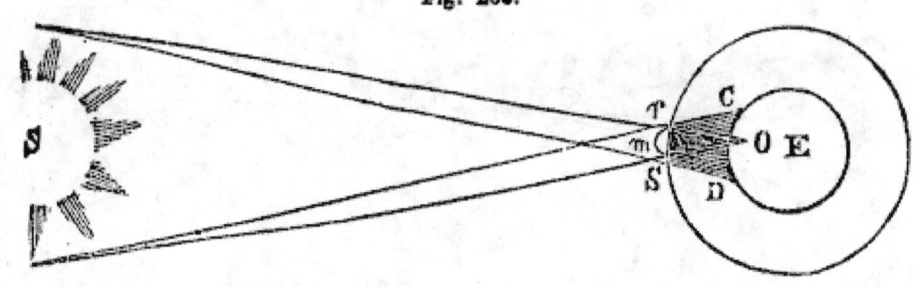

ILLUSTRATIONS.

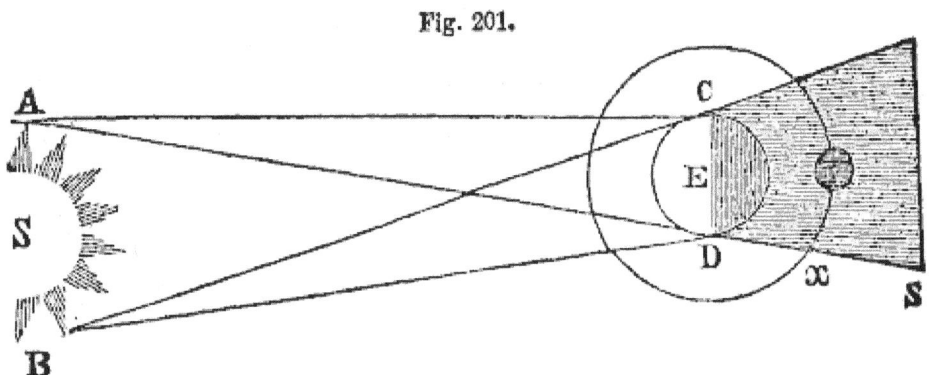

Fig. 201.

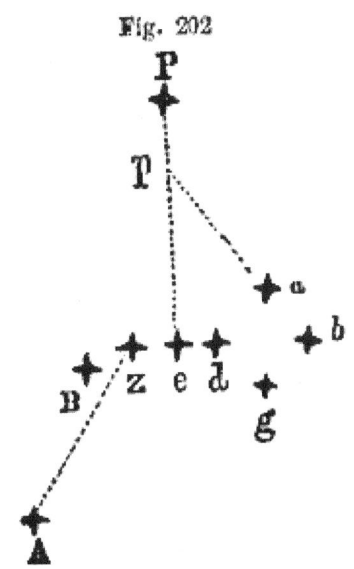

Fig. 202

ELECTRICAL TELEGRAPHIC COMMUNICATION.

See page 323, Parker's Natural Philosophy

INDEX

A.

Aberration of light 384
" spherical . . 247
Accidental colors 252
Achromatic 247
Acid, carbonic 21
Acid, sulphuric, effects of on water 187
Acoustic paradox 177
Acoustics 173
" definition of . . . 18
Acoustic tubes 179
Action 45
Action and reaction, illustration of 46
Action, suspension of 85
Actynolite 21
Aëriform, definition of 19
" fluids 138
Aëriform fluids compressed and expanded without limit . . . 139
Aëriform fluids have no cohesive attraction 139
Aëriform fluids have all the properties of liquids 140
Aëriform fluids have weight . . 139
Aëronaut, how he descends from a balloon 38
Aërolites 387
Affinity, chemical 19, 27
Agents 18
" imponderable 18
" ponderable 18
Air 140
Air, a bad conductor of heat . 191
Air, as an element 19
Air-bladder of fishes 47
Air-chamber 163
Air, component parts of the . . 140
281 note
Air, compression of, caused by gravity 39

Air, compressibility of the . 162
Air, condensation of at surface of the earth 140
Air, condensed, experiments with 163
Air contained in wood and water, experiments to show 161
Air diminishes upwards in density 140
Air, elasticity of the . . . 142, 162
Air, elasticity of the, experiments showing 160
Air, effect of gravity on density of 38
Air essential to animal life, experiment to prove 168
Air essential to combustion, experiments to prove 168
Air, fluidity of 142
Air, fluidity of, experiments showing 165
Air, gravity of the, experiments illustrating 157
Air-gun 164
Air, how a mechanical agent . . 142
" impenetrability of . . . 22, 141
" inertia of 28, 143
Air, inertia of, experiments showing 165
Air, lightness of the 162
" materiality of the 162
Air, miscellaneous experiments with 166
Air necessary to animal life and to combustion 140
Air, of what composed 20
Air, pressure of the as the depth 162
" pressure of in all directions 162
Air, pressure of the on a barometer 140
Air, pressure of the on a square inch 141
Air, pressure of the on the body 141
Air, pressure of the preserves the liquid form of some bodies . . 169

Air, pressure of the retards ebullition 168
Air-pump 154
Air-pump, experiments performed by the 157
Air-pump of steam-engine . . . 201
Air-pump, the double 156
Air, resistance of the 25, 38
Air, resistance of the to a cannon-ball 62
Air, scales for weighing 160
Air, two principles, properties of 139
" when heaviest 146
Air, when the best conductor of sound 176
Air, why not visible 140
Albite 20
Alison, extract from 70
"All's well," how far heard . . 176
Alumina 21
Aluminum 20
Ampère's discoveries in electro-magnetism 309
Ampère's electro-magnetic apparatus 314
Analysis of the motion of a falling body 52
Angle 48
Angles, how measured 48
Angle of vision 219
Angles of incidence and of reflection 48, 49, 216
Angles, right, obtuse and acute . 48
Animal electricity 282
Animals, sagacity of 92
Annealing 31
Antimony 20
" not malleable 31
Aphelion 349
Apogee 349
Apparatus for illustrating the tendency of a body to revolve around its shorter axis . . . 61
Apparition, circle of, perpetual . 385
Apparitions, deceptive 225
Aqueous humor 237, 239
Arago's experiments on velocity of sound 176
Arbor 81
Archimedes' boast to Hiero . . 95
Archimedes, burning mirrors of . 228
Archimedes discovers the method of ascertaining the specific gravity of bodies . . . . . . 127 note
Archimedes, screw of 132
Arc of a pendulum 101
Arcturus 399

Aristotle's opinion of the velocity of a falling body 52
Arsenic 20
" not malleable 31
Asteroids 339
Astræa 339
Astronomy, definition of . 17, 18, 335
Astronomers, distinguished . . . 336
Astronomy, father of 336
Atmosphere, weight of the . . . 141
Atmospheric telegraph 331
Attraction 25, 26, 33
" capillary 111
" chemical 27
" kinds of 27
" law of falling bodies . 51
" mutual 34
" of all bodies 34
" of cohesion 27
" of gravitation 27
" of the earth 33
" on what dependent . . 34
Attwood's machine 52
Augite 21
Austral polarity 302
Axes of the planets, inclination of 350
Axis, exact sense of 81
Axis, longer, a body revolving around 61
Axis of motion 59
Axis of the earth, effects of its inclination 354
Axis of the earth, geological theory of 62
Axis, what bodies revolve around an 59
Axle 81
Azote 20, 140

B.

Babbit's metal 99
Bain's telegraph 326
Baker, the Connecticut 191
Balance-wheel 104
Balance 75
Ballistic pendulum 63
Balloon, how to descend from . . 38
" the pneumatic 161
Ball, thrown in a horizontal direction 64
Balls, force of, how estimated . 63
Bands with one and two centres of motion 83
Banks, Sir Joseph 190
Barber's Grammar of Elocution . 180

INDEX.

Barium 20
Barometer . . . . . . 144 and note
Barometer, the aneroid or portable 145
Barometer, the diagonal 145
Barometer, of the different states of the 148
Barometer, greatest depression of the 147
Barometer, its importance 146 note
" rules of the 147
" the mercurial . . . 145
Base of a body 67
Batteries, thermo-electric . . . 335
" galvanic 287
Battering ram 105
Battering ram, force of, how estimated 105
Battery, electrical 264
Battery, Grove's 293
" how discharged silently 265
Battery of the electro-magnetic telegraph 321
Battery, protected sulphate of copper 293
Battery, Smee's 290
" sulphate of copper . . 292
Beam of light 213
Belgrade, battle of, and the comet 380
Bellows, hydrostatic, how constructed 119
Bell, the diver's or the diving . 150
Bevelled wheels 85
Birds, bodies of 123
" how they fly 47
" muscular power of . . . 47
Bismuth 20
" not malleable 31
Bissextile, meaning of 396
Black 252
Black lead, uses of in overcoming friction 99
Bladder-glass 159
Bladder, inflated, why compressed in water 115
Boats, how propelled 47
Boats, on what principle they float 123
Boats, motion in, why imperceptible 26
Bode's law 342
Bodies 18
" attraction of 33
Bodies of drowned persons, why they sink and afterwards rise . 123
Bodies, what are easily overset . 69
" what stand most firmly . 68
Bodies, what will rise and what will fall in air 40
Body acted upon by three or more forces 57
Body, parts of which move with greatest velocity 60
Bodies, what ones will float and what sink in water 123
Body, when it will fall 66
Bohemia slate, formations of . . 23
Bolt-head, and jar 167
Bomme M. 379
Bones of a man's arm, levers of third kind 77
Borax 20
Boreal polarity 302
Bottle, effect of pressure of the sea upon 115
Boyle 144
Boynton's, Dr., chart of materials which form granite 21
Bramah's hydrostatic press . . 121
Brass, how made brittle 30
Breadth 23
Breast-wheel 82, 83
Brittleness 27, 30
Brittleness, how acquired by iron, steel, copper and brass 30
Bromine 20
Brooks, how formed 124
Buckets of water-wheels 82
Buckets of water, why heavier when lifted from the well . . 126
Bulk of a body, how ascertained from its weight 125
Burdens, how made unequal . . 77
Burning-glasses 228, 235

C.

Cadmium 20
Calcium 20
Calliope 339
Caloric 187
Calorimotor 297
Camera obscura 219, 240
Camera obscura, portable, how made 219
Cannon-ball, greatest velocity that can be given to 63
Cannon-ball, force of the resistance of the air to 62
Cannon, how far heard 176
Caoutchouc, or India-rubber . . 30
" balls, elasticity of . 47
Capillary attraction 111
" " cause of . . 111

Capillary tubes 111
Capstan 80
Capstan and windlass, difference between 80
Carbon 20
Carbonate of lime 21
Carbonate of magnesia 21
Carbonic acid 21
Carriages, high, why dangerous . 68
Carronades 63
Cartesian devil 162
Cask, how burst by hydrostatic pressure . . . . 120 note
Cassegranian telescope . . . 250
Castors, why applied to legs of tables, &c. 85
Catoptrics 215
Celestial bodies, true place of . 384
Celsius' thermometer 149
Central forces 59
Centre of gravity . . 57, 58, 59, 66
Centre of gravity, illustrations of . . . . 66 note
Centre of magnitude . . 58, 59, 66
Centre of motion 58, 59, 71
Centre of sphericity 37
Centre, what bodies revolve around a 59
Centres 58
Centrifugal force 59
Centrifugal force, effect of on a body revolving around its longer axis 61
Centrifugal force, to what proportioned 60
Centrifugal force, where greatest 103
" meaning of 59
Centripetal force 59
" meaning of 59
Ceres 339
Cerium 20
Chain-pump 131
Chaises, tops of, toggle-joint . . 97
Chamfered 91
Chantrey, the sculptor 191
Charged, meaning of 261
"Charlemagne," experiment on board of the 115
Charles V. and the comet . . . 379
Charles' wain or wagon . . . 398
Chart of materials forming the crust of the earth 20
Chemical affinity 19, 27
Chemical attraction 27
Chemical effects of light . 256, 257
Chemical electricity 259
Chemistry 19, 110

Chimneys, glass, how preserved from cracking 192
Chisels, on what principle constructed 91
Chlorine 20
Chlorite 21
Chord, musical, how produced . 182
Choroid 237, 240
Chromatics 251
Chromium 20
Circle 48
Circle of perpetual apparition . 385
Circles 59
Circles, circumference of, how divided 48, 365
Circular motion 58
Circular motion changed to rectilinear by cranks 81
Circular motion, how caused . . 58
Clay 21
Climates, cause of 354
Clock, before and after the sun . 397
" how regulated 102
" moving power of 104
Clock, periods when it agrees with the sun 397
Clocks, why they go fastest in winter 102
Clock, what it is 102
" wheels of, their use . . . 102
Clothing, cause of warmth of . . 189
Clouds 24
" of what composed . . . 186
Cobalt 20, 298
" not malleable 31
Coffee-pots, why with wooden handles 190
Cogs 83, 84
Cohesion, attraction of 27
Cohesion, attraction of, its effects on watery particles 186
Cold 185, 192
Cold, its effects on the density of bodies 192
Colors 254
" accidental 252
Columbium 20
Comets 372
" density of 379
Comet, Halley's, as seen by Sir John Herschel, and by Struve. 377, 378, 379
Comet, Halley's, periodical time of 377
Comets, how regarded formerly . 373
Comets in the solar system, number of 379

INDEX.

Comets, Kepler's opinion of their number 380
Comets, number of 373
Comet of 1680 375
" " 1744 376
" " 1811 373
Comet of 1853, Mr. Hind's account of the 381
Comet of 1856 379
Comets, orbits of 374
Comets, return of, first predicted by Halley, Encke, and Biela . 377
Comets, tails of 374
" velocity of 374
Common centre of gravity of two or more bodies 69
Complex wheel-work 83
Compound battery 290
" lever 75
" motion 55
" " how produced . 54
Compressibility 27, 28, 29
Concave mirrors 222
" " effects of . . 225
Concave mirrors, laws of reflection from 227
Concave mirrors, peculiar property of 224
Concave mirror, true focus of . . 224
" screw 94
Concave surfaces, facts with regard to 236
Condensation 140
Condensed 140
Condenser 198
" of steam-engine . . . 200
Condensing syringe . . . 156, 163
Conduction of heat 190
Conductors of the galvanic fluid . 285
" 258, 260
" of heat 189
Cone 69
Conic sections 341
Conjunction, inferior and superior 349
Connecticut baker 191
Conservatory of arts and trades, how restored to perpendicular . 193
Constellations 383
" of the zodiac . . 347
Contractibility 28
Converging rays 212
Conversation in polar regions heard at great distances . . . 176
Convex mirrors 222
Convex mirrors, laws of reflection from 226

Convex mirrors, effects of . . . 224
Convex screw 94
Convex surfaces, facts with regard to 235
Copernicus 336
Copper 20
Copper and tin, sonorous properties of 30
Copper, how made brittle . . . 30
Cords, tenacity of 32
Cork, how deep it will sink . . . 123
" why lighter than lead . . 34
Cornea 237, 238
Corpuscular theory of light . . . 211
Couronne des tasses 290
Crank, dead point of 81
Cranks 80
Crown-wheel 84
Crust of the earth, materials composing the 20
Crystalline humor, convexity, how increased or diminished . . . 241
Crystalline humor, effect of when too round 242
Crystalline lens 237
Cup of Tantalus 133
Cups, the Magdeburgh 157
Current, velocity of a, how measured 130
Curve of a projectile, on what dependent 64
Curvilinear motion 61
Cutting instruments 91
Cylinder, definition of a 79
Cylinder, how made to roll up a slope 68
Cylinder, wheel substituted for . 79

D.

Daguerreotype proofs 257
Darkness produced by two rays of light . . . . 212 note
Davies' Treatise on Magnetism . 316
Day and night, cause of 358
Days and nights, cause of difference in length of 350
Dead point of a crank 81
Delisle's thermometer 149
Delphi, oracle of 180
Demetrius Poliorcetes 105
Density 27, 28
Density of air, effect of gravity on 38
Depth of a well, how estimated . 53
Descartes 144

Devil, the Cartesian 162
Dew and fog, difference between . 150
Dew, how produced 150
Diagonal 48
" of a parallelogram . . 55
" of a square 55
Diallage 21
Diameter 48
Diameter, equatorial, how lengthened 61
Diameter, equatorial of the earth, longer than polar, and why . . 61
Diameter of the earth, equatorial and polar 102
Diameter of the earth, how ascertained 365
Didynium 20
Digits 395
Dilatability 29
Dionysius, ear of 178
Dionysius, how he overheard his prisoners 178
Dioptrics 230
" laws of 230
Dipping of a magnet 303
Dipping of a magnet, how remedied 303
Direction 41
" line of 66
Discharge, the jointed 264
Dissolving views 246
Distance at which a man is invisible 220
Distance, greatest which can be estimated 382
Distances measured by velocity of sound 177
Distillation 194
Distilled water, why used as standard of specific gravity . . . 123
Diverging rays 212
Divers, limit to the depth of . . 115
Diving bell, or diver's bell . . . 150
Divisibility 21
" extent of 23
" definition of . . . 23
"Dodge," how children 26
Double action of the steam-engine 200
Drowned persons, why they sink and afterwards rise 123
Ductility 27, 31
Dynamics 17
" meaning of 18

E.

Earth 368
Earth, a good conductor of sound 176
" as viewed from the moon . 364
" attraction of the 33
" centre of gravity of . . 37
Earth, consequences of a more rapid rotation of the 366
Earth, constituent elements of the 20
Earth, crust of the, materials composing 20
Earth, diameter of, how ascertained 365
Earth, figure of the 364
" how known to be round . 364
Earth, how much larger than any falling body 33
Earth, motions of its inhabitants 365
" nearer the sun in winter . 352
Earth, parts of which move most rapidly 61
Earth, strata of the 20
Earth, the principal reservoir of electricity 261
Ebullition retarded by pressure of the air 168
Echo 177
" why never heard at sea . . 178
Eclipse 392
" annular 394
Eclipses, greatest number of in a year 396
Eclipse, lunar, to whom visible . 395
" solar, to whom visible . 395
" total of 1806 396
Eclipses, why more of the sun than of the moon 393
Eclipse, why not at every new and full moon 392
Eclipse, partial 394
" total 394
Ecliptic 345
Egeria 339
Ehrenberg's microscopic observations 23
Elastic fluids 139
Elasticity 27, 29, 30
" of air 142
" of gaseous bodies . . 30
" of ivory 46
Electrical battery 264
" bells 273
" fire-alarm 330
" machine 266
Electrical machine, experiments with 270
Electrical sportsman 275
Electric current, direction of, how ascertained 310

Electrical tellurium 272
Electric fluid, velocity of 43
Electricity 17, 18, 258
Electricity acquired by induction 278
Electricity and magnetism, resemblance between 302
Electricity, animal 282
" by induction . . . 266
" circuit of 265
Electricity as excited by galvanism and by friction, difference between 283
Electricity by transfer 266
Electricity, effects of similar states 263
Electricity, frictional . . . 282, 283
Electricity, frictional and chemical, how they differ 294
Electricity, galvanic, quantity of.295
" nature of 259
Electricity, quantity of excited by chemical action . . . . 284 note
Electricity, simplest mode of exciting 262
Electricity, the vitreous or positive, the resinous or negative . 262
Electricity, three states of . . . 335
" voltaic 283
Electrics 258, 260
Electric telegraph, history of the 329
Electro-magnet 317
Electro-magnet, communication of magnetism to steel by means of 318
Electro-magnetic multiplier . . 313
Electro-magnetism . . 17, 260, 308
Electo-magnet, the U or horseshoe 319
Electro-magnetism, definition of . 18
Electro-magnetism, discoveries of Œrsted, Faraday, Ampère, Arago, and Sir H. Davy . . 308, 309
Electro-magnetism, facts of . . 309
Electro-magnetic induction . . 312
Electro-magnetic rotation . 313, 316 note
Electro-magnetic telegraph, signal-key and registering apparatus of the 322
Electro-magnet of Prof. Henry and Dr. Ten Eyck 317
Electro-magnetic telegraph . . 319
Electro-magnetic telegraph, how put into operation 324
Electro-metallurgy 331
Electrometer 268
Electrophorus 269

Electro-plastic process 331
Electro plating and gilding . . . 331
Electroscope 269
Electrotype process 231
Elementary substances, enumeration of 20
Elements, the four 19
Ellipse 341
Elocution, Barber's Grammar of . 180
"Empty," common meaning of . 98
Endosmose 27, 112
Engineer, how enabled to direct his guns 65
Engine, the fire 154
" the steam 196
Equilibrium 74, 75
" of fluids 110
Equilibrium of fluids, exemplified by means of the siphon . . . 133
Equilibrium of fluids, how disturbed by waves 131
Equilibrium of fluids of different densities 113
Equilibrium of mercury, water, oil, air, &c. 113
Equinoxes 358
" precession of the . . . 397
Equivalent, mechanical 58
Erect, why objects are seen . . 241
Erbium 20
Escapement-wheel 104
Essential property, meaning of . 21
Essential properties of matter . 21
Eunomia 339
Evaporation, Dr. Watson's experiment 150
Eye 237
" a camera obscura 240
" different parts of the . . . 237
Eye-glass 248
Eye, imperfections of, how caused 242
" of what composed 237
Eyes, two, why they do not cause double vision 241
Exercises for solution 53
Exhausting syringe 163
Exosmose 27, 112
Expansibility 27, 29
Expansion, how it differs from dilatation 29
Experiments showing inertia of air 165
Extension 21, 23

F.

Fahrenheit's thermometer . . 149

Falling bodies, law of 51
Faraday, announcement of in relation to solar spots and magnetic variation 304
Faraday's discoveries in electromagnetism 308
Faraday's electro-magnetic apparatus 313
Faraday's nomenclature of electricity 259
February, why 29 days every fourth year 397
Feldspar 21
Fire-alarm, the electrical . . . 330
Fire, as an element 19
Fire-engine 154
Fifth 184
" how produced 182
Figure 21, 23
Fishes, how they swim, rise or sink, &c. 47
Fixed pulley, mechanical advantage of 87
Fixed pulley, operation of the . 87
Flavio de Melfi, inventor of mariner's compass 306
Flexibility 27, 31
Float, how heavy bodies can be made to 38
Float-boards of water-wheels . . 82
Flora 339
Florence, experiment made at on impenetrability of water . 22, 109
Fluid and solid bodies, difference between 108
Fluid, definition of 108
Fluidity of air 142, 165
" what constitutes . . . 108
Fluid pressure, law of 115
Fluids, aëriform 138
Fluids, aëriform, expanded and compressed without limit . . 139
Fluids and liquids, how different 109
Fluids, effects of their peculiar gravitation 113
Fluids, equilibrium of . . . 122, 133
Fluids, downward pressure of, how shown 114
Fluids, gravitation of 110
" how different from liquids 109
" how they gravitate . . . 113
" lateral pressure of . 114, 116 117
Fluids, level or equilibrium of . 110
" mechanical agency of . . 138
Fluids of different densities, gravitation of 112

Fluids, particles of, how arranged 114
" pressure of 114
Fluids, pressure of, according to height 119, 120
Fluids, pressure of, on what dependent 118
Fluids, pressure of, to what proportional 115
Fluids, surface of 110
Fluids, upward pressure of . 114, 117
Fluids, why unsusceptible of formation into figures 110
Fluorine 20
Fly 143
Flying of birds, how effected . . 47
Fly-wheels 81
Fly-wheels and the dead points of cranks 81
Fly-wheel in the steam-engine . 203
Focus of concave mirrors . . . 224
Fog and dew, difference between 150
Fog, how produced 150
Force 41
Forces, at an angle 56
" effects of 55
" three or more in action . 57
" unequal at right angles . 56
Forcing-pump 153
Formulæ 44
Fortuna 339
Fountain, glass and jet 159
Fountain, Hero's 138
Fountains, artificial, how constructed 137
Fountains, how formed 137
Fourth 184
Fowling-pieces, length of . . . 63
Franklin, inventor of lightning-rods 281
Free heat 187
Frictional electricity . . . 259, 283
Friction . . . . 90 note, 98
" cause of 99
" how diminished 99
" how increased 99
" loss of power caused by . 99
" important uses of . . . 100
Friction of the beds and banks of rivers 130
Friction, particles of fluids destitute of 108
Friction-wheels 99
Fuel, combustion of 24
Fulcrum 70, 71, 72
" generally a pin or a rivet 76
Fulcrum in levers of different kinds 77

Fulcrum of steelyards 74
Fulton, Robert 200
Fundamental law of mechanics . 71
Fusee of a watch 107

G.

Galaxy 383
Galileo 100, 143, 337
Galileo's experiment at Pisa to prove his law of falling bodies 52
Galileo's law of falling bodies . 52
Galvanic action, three elements necessary for 285
Galvanic batteries 287
" battery 289
" circle 286
Galvanic circle, effects of, how increased 287
Galvanic circle, essential parts of a 286
Galvanic circle, simplest, of what composed 286
Galvanic electricity 259
Galvanic electricity, process for obtaining 286
Galvanic fluid, how excited . . 284
" piles 287
Galvanism 17, 18, 283
" facts explained by . . 296
Galvano-plastic process 331
Galvanotype 331
Garments, light-colored why cool 191
" linen, why cool . . . 189
Garments, to what they owe their strength 100
Garments, woollen, why warm .189
Garnet 21
Gaseous bodies, elasticity of . . 30
Gaseous bodies, to what degree they may be dilated 29
Gases 139
Gases, how prevented from rising from a fluid 168
Gay Lussac's experiments on the velocity of sound 176
Gearing 83
Geology 62
Georgium Sidas 369
Gibbous 388
Glucinum 20
Gold 20
Gold, both ductile and malleable 31
" divisibility of 23
Gold, the most malleable of all metals 31
Glass, its brittleness 32

Glass, the bladder 159
" the fountain and jet . . .159
" the hand 158
" the India-rubber . . . 159
Glass, why easily cracked when suddenly heated 192
Glass, why used in mirrors . . .221
Governor 106, 200
Governor applied to steam-engine by James Watt 106, 203
Governor, explanation of the . .106
" uses of the 106
Grain of hammered gold 23
Grand law of nature . . . . 69 note
Granite 20
Gravitation, attraction of . . . 27
" of fluids . . .110, 112
Gravity 25, 33
Gravity causes pressure of fluids upwards as well as downwards.117
Gravity, centre of 37, 59, 66
Gravity, effect of on density of air 38
Gravity, effects of on different bodies 41
Gravity, force of, not affected by projection 64
Gravity, force of on projectiles . 62
" " where greatest 35
" how it increases and decreases 35
Gravity, law of terrestrial . . . 35
Gravity, specific . . . 40, 126 note
Gravity, specific, scales for ascertaining 126
Gravity, specific, standard of . . 123
" terrestrial 34
Great Bear 398
Green sand 21
Gregorian telescope 250
Gridiron pendulums 103
Grove's battery 293
Gudgeons 80
Guericke, Otto 158
Guinea and feather drop 165
Gunnery, science of 62
Gunpowder, force of 63
Gunpowder, great charges of useless and dangerous 63
Guns, how tested 63
Guns, short ones, why preferable 63
Gun, the air 164
Gymnotus electricus 282

H.

Hail, how formed 124, 150
" how it differs from snow . 124

Hair-spring 104
Hall, Captain Basil 146
Halley's comet as seen by Sir John Herschel 379
Hand-glass 138
Handles of tea-pots, &c., why of wood 190
Hare's calorimotor 297
Harmony 181
" how produced . . . 183
Harmony, science of, on what founded 182
Harvest-moon 389
Heat accompanies all great changes in bodies 110
Heat, application of its expansive power as a mechanical agent . 193
Heat and cold 187
" conductors of 189
" effects of 188
" effects of on bodies . . 185, 188
Heat, effects of on density of substances 192
Heat, effects of on water . . 186, 194
Heat, free 187
" first law of 189
" imperfect conductors of . 190
" its effects on a body . . . 141
" most obvious effects of . . 193
" how propagated 190
" latent 187
" law of the reflection of . . 191
" laws of 185
" nature of 185
" of the sun 188
Heat produced by electrical action 188
Heat, sources of 187
Hearing trumpets 178
Heavenly bodies, motion of the when the most rapid 350
Heavenly bodies, why not seen in their true place 232
Heavens, why bright in the daytime 218
Hebe 339
Height of a building, how estimated 53
Height to which a body projected upward will rise 54
Heliacal ring 318
Heliography 257
Helix 316
Henry's and Dr. Ten Eyck's electro-magnet 317
Hero's fountain 138

Herschel sees stars through a comet 379
Herschel, Sir J. F. W.'s illustration of the size and distance of the planets 344
Herschel, Sir John's opinion of the height of the atmosphere . 38
Herschel's telescope and its power 251, 337
Heterogeneous 19
Hiero employs Archimedes to detect the adulteration of a crown. 124
Hind's account of the comet of 1853 381
Hipparchus, father of astronomy . 336
Homogeneous 19
Hornblende 21
Horizontal motion does not affect that of gravity 65
Horse-power as applied to the steam-engine, meaning of . . 199
Horses, how made to draw unequal portions of a load 77
House's printing telegraph . . . 328
Human voice, powers of the . . 180
Humor, the vitreous 237, 259
" the aqueous 237, 239
Hunter's moon 388
" screw 95
Hydraulics 17, 18, 108, 128
Hydraulic-ram 133
Hydrodynamics 108, 129
Hydro-electric 334
Hydrogen 20
" gas generator 275
" pistol 274
Hydrometer 128
Hydrostatic bellows, how constructed 119
Hydrostatic paradox 118
Hydrostatic press, Bramah's . . 121
Hydrostatic pressure, as a mechanical power 121
Hydrostatic pressure, caused by height, not by quantity . . . 119
Hydrostatics 17, 18, 108
Hygeia 339
Hygrometer 149, 150
Hyperbola 341
Hypersthene 21

I.

Ice formed under a receiver . . 169
" how made to melt rapidly . . 191

INDEX.

Ice, why wrapped in woollen or packed in shavings 190
Ice, why wooden spoons and forks are used for 190
Image from concave mirrors . . 225
" " convex mirrors . . 223
" inverted 218
Impenetrability 21, 22
Imponderable agents 18
Incidence, angle of 48
Incident motion. 47
Incident rays 216
Inclination of earth's axis, effects of 354
Inclined plane 90
" " advantage of . . 91
" " application of the 91
" " principle of the . 90
Incombustible bodies 188
Indestructibility 21, 23
India rubber 30
" " balls, elasticity of . 47
" " glass 158
Induction, electricity by . . 266, 278
" electro-magnetic . . 312
Inertia 21, 24, 26, 41
" experiment to illustrate . 25
" of air 38, 143, 165
" of a fluid, effects of the . 134
" of fly-wheels 81
" of water 98
Inferior conjunction 349
" planets 343
Infusoria 23
Instruments for raising water . . 131
Insulated, meaning of . . . 261, 270
Intensity as applied to electricity, meaning of 295
"In vacuo" 98
Iridium 20
Iodine 20
Irene 339
Iris *of the eye* 237, 238
Iris, *the planet or asteroid* . . . 339
Iron 20
Iron, a knowledge of the uses of the first step towards civilization 31
Iron, ductile but not malleable into thin plates 31
Iron, how made brittle 30
" oxide of 21
" when most malleable . . . 31
Ivory, elasticity of 30, 46

J.

Jansen 337
Jerusalem, siege of 106
Jet, the straight and revolving 163
Jointed discharger 264
Juno 339
Jupiter 367, 368
Jupiter, a prolate spheroid, and why 62
Jupiter's belts 368
Jupiter, satellites of 367

K.

Kaleidoscope 229
Kepler 337
" laws of . . . 337, 350, 352
Kepler's opinion of the number of comets 380
Klinkenfues 381
Knee-joint 96

L.

Ladder a lever 77
Lakes, why more difficult to swim in 126
Lamp, defects of, how remedied . 112
Lamps, why they will not burn . 111
Lamp, wick of, how it supplies the flame 111
Lantanium 20
Latent heat 187
Lathes 80
Law, Bode's 342
Law, fundamental of mechanics, pyronomics, acoustics and optics 49
Law, Mariotte's 142
" of falling bodies 51
Laws of heat 185
" of reflected sound 178
Laws of reflection from concave mirrors 226, 227
Law of the heavenly bodies . . 340
Lead 20
" not ductile 31
" why heavy 34
Le Verrier 371
Leap-year 396
Leaves of a wheel 84
Length 23
Lens, axis of a 233

Lens, concavo-convex 233
" convex as a burning-glass . 235
" double concave 233
" double convex 233
Lenses 232
Lens, effect of how estimated . . 234
" focal distance of a . . . 234
Lenses in spectacles 236
Lens, single concave 233
" single convex 233
" the crystalline 237
Level, how ascertained 113
" or equilibrium of fluids . 110
Levels, spirit or water 113
Lever 93
" advantage in use of . . . 73
Lever, force of the, on what dependent 76
Lever, how used 72
" kinds of 72
" many forms of the . . . 75
" of first kind 73
" of second kind 76
" of third kind 78
" perpetual, the 80
Lever, power of not dependent on its shape 76
Lever, principle of the 71
" the bent 76
Lever, things to be considered in the 72
Leyden-jar 263
" how charged . . . 271
Leyden-jar, how discharged silently 265
Light, aberration of 384
" absorbed by all bodies . . 217
" beam of 213
" color of 251, 252
Light, corpuscular and undulatory theories of 211
Light, heat and chemical action of 254
Light, how projected 213
Light, intensity of, law of decrease 212
Light, passing into different mediums 230
Light, polarization of 256
" reflected 215
" " laws of . . 216
" reflection of 211
Light, Sir Isaac Newton's opinion of 211
Light, theories of 211
Light, thermal, chemical and non-optical effects of 256

Light, velocity of 43
" zodiacal 360
Lightning, how caused 278
Lightning-rods 265
" by whom invented 281
Lightning-rods, the best, how constructed 280
Lime 21
Lime, carbonate of 21
Linen garments, why cool . . . 189
Line of direction 66
Liquid, how it differs from a fluid 109
Liquids have a slight degree of cohesion 109
Liquids not easily compressed . 29
Liquid, quantity of discharged from an orifice 129
Lithium 20
Load-stone 298
Locomotive steam-engine . . . 208
Looking-glasses 221
Looking-glass, length of to reflect the whole person 222
Lucifer 363
Luminous bodies 210
Lutetia 339

M.

Machine 71
Machinery, propelled by electricity 279
Machine, Attwood's 52
Machines, velocity of, how regulated 106
Magazine, magnetic 307
Magdeburgh cups 157
Magnesia 21
" carbonate of . . . 21
Magnesium 20
Magnet, attraction and repulsion of 300, 301
Magnet, attractive power of, where greatest 300
Magnet, broken 302
Magnet communicates its properties 301
Magnet, dipping of a 303
" effect of heat upon . . 302
Magnet, horse-shoe or U, how armed 308
Magnetic influence, all bodies susceptible of 301
Magnetic magazine 307
Magnetic needle 304
Magnet, keeper of a 302, 308
Magnet, properties of 299

Magnetic poles 300
" power on surface . . . 302
Magnetism 17, 18, 298
Magnetism and electricity, resemblances of 302
Magnet, modes of supporting . . 300
Magnetic poles, where strongest 304
Magnet, north and south poles of, where most powerful . . . 306
Magneto-electricity 17, 332
Magneto-electricity, most powerful effects of, how obtained . . 332
Magneto-electric machine . . . 333
Magnet, polarity of 299
Magnet, poles of changed by electricity 303
Magnet, powers of, how increased 301
" kinds of 299
" artificial, how made 306, 307
" the receiving 327
" U or horse-shoe . . . 301
" variation of . 303, 304 note
Magnitude, centre of . . . 59, 66
Main-spring of a watch . . 104, 107
Major third 184
Malleability 27, 31
Malleability dependent on temperature 31
Manganese 20
Marco Paolo 306
Mariner's compass 304
Mariner's compass, inventor of the 306
Mariner's compass, needle of, how placed 305
Mariner's compass, how mounted 305
" " points of the 305
Mariotte's law 142
Mars 366
Massila 339
Materials, strength of 95
Materials which compose the crust of the earth 20
Materials, tenacity of 32
Matter, attractive 34
" definition of 19
" essential properties of . 21
" gaseous form of 19
Matter, its different states or forms 19
Matter, liquid form of 19
Matter, quantity of, how estimated 40
Materiality of air 162
Matter, solid form of 19
Mechanical agency of fluids . . 138
" equivalent 58

Mechanical operations always attended by heat 198
Mechanical paradox 68
" power 70
" powers 71
Mechanical powers, enumeration of the 72
Mechanical powers, on what principle constructed 71
Mechanical powers, principal law of the 89
Mechanical powers, reducible to three classes 72
Mechanical properties of gases, vapors, &c. 139
Mechanics 17, 41
" fundamental law of . 71, 91, 118
Mechanics, fundamental law of, its application to hydrostatic pressure 119
Media 97, 229
Medium 97
Mediums 97, 229
Medium in optics 230
Melpomene 339
Meniscus 233
Mercurial pendulum 103
" tube 160
Mercury 20
" the planet, transit of . . 363
Mercury, the planet, why not often seen 362
Metallic points 265
Metals, good conductors of heat . 190
" names of the 20
Metals, order of their conducting power of heat 190
Metals, tenacity of 32
Meteoric stones 387
Meteoric stones, Dr. Brewster's opinion of 367
Metes 339
Mica 21
Microscope, a double 243
" a single 242
Microscope, compound magnifying power of, how ascertained 244
Microscope, magnifying power of, how ascertained 244
Microscope, the solar 244
Microscope, the solar, magnifying power of 244
Microscopes, what have the greatest magnifying power . . 247
Milk, why affected by thunder and lightning 281

Milky-way 383
Minor third 184
Mirror 221
    " concave 222
    " convex 222
    " plain 221
Mirrors of half the height show a whole-length figure . . . 217
Mirrors reverse all images . . 222
    " use of glass in 221
Miscellaneous experiments with air 166
Mobility 27
Molybdenum 20
Momenta 50
Momentum 41, 50
Momentum of a body, how ascertained 50
Monochord 182
Moon 386
    " as cause of tides 391
    " as seen through a telescope 389
Moon, common errors in respect to the 386
Moon, density of the 387
    " difference in daily rising 389
    " gibbous 388
    " harvest and hunter's . . 389
    " horned 388
    " in quadrature 388
Moon-light, objects seen by, why faint 217
Moon, surface of the 386
    " uninhabitable 364
Morienne 144
Morse's telegraph 320
    " telegraphic alphabet . . 323
Motion 41
Motion, accelerated, retarded and uniform 44
Motion, axis of 59
    " centre of 59
Motion, how transmitted by hydrostatic pressure 121
Motion, incident and reflected . 47
Motion impelled by two or more forces 55
Motion of the heavenly bodies, cause of the 34
Motion, perpetual 45
    " regulators of 100
    " reversed 83
Motion, slow or rapid produced at pleasure by machinery . . 84
Motion, when imperceptible . . 220
Moving power in machines, how stopped 85

Mountain, how burst by hydrostatic pressure 120
Musical scale 183
    " sounds 181
Multiplier, electro-magnetic . . 313
Multiplying-glass 235
Musical chord, how produced . . 182
Musical instruments, why affected by the weather 182
Music of a choir dependent on the uniform velocity of sound . 176
Music of strings, how caused . . 181
Mutual attraction 34

N.

Natural Philosophy, definition of 17
Neap tides 391
Needle, the magnetic 304
Needle, how placed in a mariner's compass 305
Negative electricity 259, 262
    " (galvanic) pole 287
Neptune 371
Newcomen and Savary's steam-engine 197
Newton, Sir Isaac 23, 337
Newton, Sir Isaac, discovery of gravitation 100
Newton's discoveries, on what based 352
Newton's (Sir Isaac) opinion of light 211
Newton, Sir Isaac's, opinion of the earth's compressibility . . 29
Nickel 20, 298
Niobium 20
Nitrogen 20
Non-conductors 258, 260
Non-electrics 258, 260
Nut and screw 93

O.

Oars, on what principle constructed 77
Object, apparent size of, on what dependent 220
Objects, when invisible . 218, 220
Octave 184
    " how produced 182
Œrsted's discoveries in electro-magnetism 308
Oil, effects of in smoothing the surface of water 131
Oil, glutinous matter in 111
Oil-mills 92

INDEX.

Oil, why it floats 39
Olber's, Dr., opinion on lunacy . 386
Opaque bodies 217
Opera-glasses 249
Opposition 350
Optical paradox 212
Optic-nerve 237, 240
Optics 17, 210
" definition of 18
Oracles of Delphi, Ephesus, &c. . 180
Orbit, meaning of 340
Orbits of the planets, inclination of 347
Orbits of the planets, not circular 343
Otto Guericke 158
"Out of beat," meaning of . . . 104
Overshot-wheel 82
Osmium 20
Oxyde of iron 21
Oxygen 20

P.

Pails, why two can be carried more easily than one 69
Palladium 20
Pallas 339
Parabola 62, 341
Parachute 38
Paradox 118
" acoustic 177
" hydrostatic 118
" mechanical 68
" optical 212
" pneumatic 169
Paradox, optical, pneumatic, acoustic, &c., no paradox . . 212 note
Parallax 385
Parallel motion, appendages for 200
Parallelogram 48
Parthenope 339
Pascal 144
Pelopium 20
Pendulum 100
Pendulum, cause of slowness and rapidity of vibrations 102
Pendulums, continuous motion of, how preserved 103
Pendulum, how lengthened or shortened 102
Pendulum, how to be suspended 103
" its motion, how caused 101
Pendulums, length of, proportion of 103
Pendulum, length of to vibrate seconds 102

Pendulum, length of to vibrate two seconds 103
Pendulum, length of varies with the latitude 102
Pendulums, table of the lengths of to beat seconds in different latitudes 104
Pendulum, the ballistic 63
" the gridiron . . . 103
" the mercurial . . . 103
Pendulums, to what variations subject 103
Pendulum, use of the ball of. 101 note
Penumbra 394
Percussion, force of 93
Perigee 349
Perihelion 349
Permanent magnets 301
Perpendicular 48
Perpetual lever 80
" motion 45
Perpetual motion, approximation to 288
Phocea 339
Phosphor 363
Phosphorus 20
Photography 257
Physical spectra 228
Physics, definition of 17
Piazzi 343
Pincers 75
Pinions 83
Pipes, tones of, on what dependent 181
Pivots 81
Plane, the inclined 90
Planet, meaning of 339
Planet and star, difference between 339
Planets, characters by which they are represented 346
Planets, inferior and superior . 343
" minor 339, 367
" " how discovered . 342
Planets, minor, by whom discovered 343
Planets, minor, size of 344
" names of the . . 338, 339
Planets, relative appearance of, as seen through a telescope . 372
Planets, the primary 338
Planets, when in a particular constellation 349
Platinum 20
Platinum, both ductile and malleable 31, 32
Plough, constellation of the . . 398
Plumb-line 37

Pneumatics 17, 18, 138
Pneumatic balloon 161
Pneumatic paradox 169
" shower-bath . . . 166
" scales 160
Pointers 398
Poker 75
Polarity 299
" boreal and austral . . . 302
Polarization of light 256
Polar or pole star 384, 398
Poles, magnetic 300, 304
Poles, magnetic, where strongest 304
Ponderable agents 18
Pope Callixtus and the comet of
  Halley 378
Pores 28
Porosity 27, 28
Positive electricity 259, 262
Positive (galvanic) pole . . . 287
Potash 21
Potassium 20
Power 72
Power, how gained by use of the
  lever 76
Power, how to be understood 73 note
Powers, mechanical 70, 72
Power that acts 7
Power, weight and velocity, pro-
  portion of 90
Precession of the equinoxes . . 397
Press, Bramah's hydrostatic . . 121
Presses, screws applied to . . . 95
Pressure at any depth, how esti-
  mated 115
Pressure, fluid, law of 115
Pressure, hydrostatic, as a me-
  chanical power 121
Pressure, hydrostatic, caused by
  height, not by quantity . . . 119
Pressure of fluids 114
Pressure of fluids in proportion to
  height of column 120
Pressure of the air . . . 141, 162
" of water at great depths 109
Pressure on hydrostatic bellows,
  how estimated 119
Primary planets 338
Principle of all machines . . . 72
Principle of the mechanical pow-
  ers 71
Prism 252
Projectiles 62
Projectile, random of 65
Projection, force of 62
Projection, force of, has no effect
  on gravity 64

Propeller 204
Properties, essential and acciden-
  tal, of matter 21
Properties, essential and unessen-
  tial 23
Prussian blue 327
Psyche 339
Ptolemy 336
Pulley 86
" fixed and movable . . . 86
" fixed, use of 87
Pulleys, mechanical principle of
  same as that of levers 88
Pulley, movable, how it differs
  from a fixed 87
Pulley, movable, principle of the 89
Pulley, power of, how ascertained 88
Pulleys, practical use of 89
Pump, the chain 131
" the common, for water . . 152
" the forcing 153
" the air 154
Pupil 237, 238
Pyramid, why the firmest of struc-
  tures 68
Pyrometer 193
" Wedgewood's . . . 193
Pyronomics 17, 18, 185, 187
Pythagoras 336

Q.

Quadrature 388
Quartz 21
Questions for solution 36, 42, 43, 50
  53, 54, 78, 86, 90, 96, 106, 116, 127,
  184

R.

Radiation of heat 190
Radii 48
Radius 48
" vector 350
Rain, how formed . . . 124, 150, 186
Rainbow, how produced . . . 255
Ram, the battering 105
" the hydraulic 133
Random of a projectile 65
Rarefaction 140
Rarefied 140
Rarity 27, 28
Ray of light 212
Rays of light absorbed . . . 215
" " converging . . . 212
Rays, converging and diverging,
  laws of 227

Rays of light, diverging 212
Rays of light from terrestrial objects 213
Reader, The Rhetorical 180
Reaumur's thermometer . . . 149
Receiver 154
Rectangle 48
Rectilinear motion converted to circular 81
Reflected motion 47
Reflecting substances 211
" telescope 246, 249
Refraction 229
Refracting substances 211
" telescope 246
Refrangibility 230
Registering apparatus of the telegraph 322
Regulators of motion 100
Rein, F. C., hearing trumpets or cornets . . . . 178 note
Repulsion 28
Resinous electricity 262
Resistance 41
Resistance of a medium, to what proportioned 97
Resistance of the air 38
" to be overcome . . . 71
Resultant 58
" motion 57
" of two forces . . . 56
Resultant of two forces, how described 58
Retarded motion of bodies projected upwards 54
Retina 237, 240
Reversed motion 83
Revolving-jet 163
Revolution of the planets, length of 341
Rhetorical Reader 180
Rhodium 20
Rhodes, siege of 105
Rifles, how tested 63
Rivers, how formed 124
Rivers, why difficult to swim in . 126
Rivulets, how formed . . 124, 136
Roads, inclined planes 91
Rolling friction 98
Romans, the ancient, how they conveyed water 137
Rope-dancer, how enabled to perform his feats 67
Ropes, strength of, on what dependent 100
Rosse's telescope 251
Rotation, electro-magnetic . . 313

Rudders, on what principle constructed 77
Rules relating to musical strings 184
Rules by which changes of the weather may be prognosticated by means of the barometer . 147
Rules relating to musical pipes . 184
Rush's Treatise on the Voice . . 180
Ruthenium 20

S.

Safety-valve 199
Sagacity of animals 92
Sap, ascent of, to what due . 112
Satellites, general law of . . . 370
Saturn 368
Saturn's rings 368
Scales for ascertaining specific gravity 126
Scale, the musical 183
Scales, the pneumatic 160
Schorl 21
Science of harmony, on what founded 182
Scissors 75
Sclerotica 237
Screw 93
" a compound power 94
" advantage of the 94
" convex and concave . . . 94
" power of, how estimated . 94
" Hunter's 95
" of Archimedes 132
" uses of the 95
"Sea-Eagle," experiment made on board of the 109
Seasons, cause of the 350
" explanation of the cause 355, 356
Sea-water, cause of its increased specific gravity 126
Seebeck, Professor, discoveries of in thermo-electricity . . . 334
Selenium 20
Serpentine 21
Shadow 213
Shadows, darkest, how produced 214
Shadows from several luminous bodies 215
Shadows, increasing and diminishing 214
Shadow of a spherical body, form of 214
Shadows, why of different degrees of darkness 213
Shaft 81

Shepherds, balancing of in south of France 67
Ships, on what principle they float 123
Sidereal time 396
" year 396
Silence produced by two sounds 177
Silica 20, 21
Silver, best conductor of heat . . 190
Simple motion 55
Sidereal year, how measured . . 397
Signal-key of the electric telegraph 322
Signs of the zodiac 346
Signs used in almanacs . . . 389
Silurus electricus 282
Silver 20
Siphon 132
Siphon, equilibrium of fluids exemplified by means of the . . 133
Siphon, experiments with the . 167
" principle of the 133
Sky, why blue 253
Slate formations in Bohemia . . 23
Slaves in West Indies, how they steal rum 122
Steel, how made brittle 30
Sliding friction 98
Smee's battery 290
Smoke, why it ascends 39
Snow, how formed . . . 124, 150
" how it differs from hail . 124
Snow and ice, how made to melt rapidly 191
Snuffers 75
Soap-bubble, thickest part of . . 23
Soda 21
Sodium 20
Solar microscope 244
Solar system, account of the 337, 338
" time 396
" year, how measured . 396, 397
Solstices 358
Sonorous bodies 174
Sonorous property of bodies, to what due 175
Sound 174
Sound affected by the furniture of a room 179
Sound, by what laws reflected . . 178
Sound, by what reflected and dispersed 179
Sound, focus of 179
Sound, how communicated most rapidly 175
Sound of the human voice . . . 179
" of strings, how caused . . 181
" rapidity of 176

Sounds, distance to which they may be conveyed 176
Sounds, musical 181
" producing silence . . . 177
Sound, velocity of . . . . 176 note
Sounds, what pleasing to the ear 183
" when loudest 174
Sources of heat 187
Space 41
" how estimated 43
Speaking-trumpets 178
Specific gravity . . . . 40, 126 note
Specific gravity of bodies, how ascertained 125, 127
Specific gravity, scales for ascertaining 126
Specific gravity, standard of . . 123
" gravities, table of . . . 124
Sphericity, centre of 37
Spectacles 236
Spectrum of a prism 254
Spherical aberration 247
Spherical body, how made to roll down a slope 68
Spider's web 23
Spiral tube 274
Spirit level 113
Spirit or water level, with what filled 113
Spots in the sun 304
Sportsman aiming at a bird . . . 57
Spring, how high it will rise . . 137
Springs, how formed 136
Spring-tides 391
Spur-gear 84
Spur-wheel 84
Square 48
Square rods, why better than round as conductors of electricity . . 279
Standard of specific gravity . . 123
Stars, distance of the 382
Stars, distance of the, Sir John Herschel's opinion of 383
Stars, how distinguished from planets 339
Stars, the fixed 381
Stars, why not seen in the daytime 363
Stars, why not seen in their true place 384
Statics 17, 18
Stationary steam-engine 209
Steam 195
Steamboats 203
Steam, cause of the ascent of . . 124
" dry and invisible . . . 196
Steam-engine applied to boats . 203

Steam-engine, power of, how estimated 199
Steam-engine, the 196
" improvers of the . 200
Steam-engine, Newcomen and Savary's 197
Steam-engine, Watts' double acting, condensing 197
Steam-engine, Watts' improvements of the 197
Steam-engine, the locomotive 204, 208
Steam-engine, the stationary . . 209
Steam-engine, Tufts' stationary 207
Steam, foundation of its application to machinery 30
Steam, how condensed into water 195
" how made to act 196
Steam, on what its mechanical agency depends 195
Steam, pressure of, on what dependent 195
Steam-ship 203
Steam, space occupied by . . . 196
" temperature of 195
" why it ascends 39
Steatite 21
Steelyards 75
" how to be used . . . 74
Steelyards, mechanical principle of the 73
Stereo-electric current 334
Stethoscope 175
Still 194
Stilts used in south of France . 67
Straight jet 163
Strata of the earth 20
Stream, velocity of, how measured 130
Strings, musical sounds of, how produced 181
Strings, musical quality of the sounds of 181
Strontium 20
Struve's opinion of the distance of the stars 382
Substance, heterogeneous . . . 19
" homogeneous 19
Sucker 160
Sulphate of copper battery . . . 292
Sulphate of copper battery (protected) 293
Sulphur 20
Sun, as cause of tides 391
" as viewed from the planets . 360
" its size, &c. 359
Sun, moon and planets, relative size of the 343

Sun, planets and stars, inhabited 359
Sun, red appearance of the, how caused 253
Sun's heat, effect of on the earth. 150
Superior conjunction 340
" planets 343
Surinam eel 282
Suspension of action 85
Synchronous tickings of a clock . 104
Syracuse, King of, employs Archimedes to detect the adulteration of a crown . . . . . 127 note
Syringes for striking fire . . . 168
Syringe, the condensing . . 156, 163

T.

Table of specific gravities . . . 124
Table of the lengths of pendulums 104
" of velocities 42
Tackle and fall 89
Talc 21
Tangent 48, 60
Tantalus . . . . 133 note
Tantalus' cup 133
Tantalize, origin of the word . . 133
Tapestry of Bayeux 380
Tea-pots, why they have handles of wood 190
Teeth 83
Telegraph, atmospheric 331
" Bain's 326
" electric, history of the 329
" electro-magnetic . . 319
Telegraph, electro-magnetic, representation of the 323
Telegraph, electric, principles of its construction 320
Telegraph, House's printing . . 328
Telegraphic battery 321
Telegraph, meaning of . . . 319 note
Telescopes 246
Telescope, achromatic 247
" Cassegrainian . . . 250
" day and night . . . 248
" Gregorian 250
" Herschel's 251
" " power of . 337
" Lord Rosse's . . . 251
" reflecting 246
" refracting 246
" simplest form of the . 247
Tellurium 20
Tenacity 27, 32
" of cords 32
" of the metals 32
" of metals, how increased 33

Tenacity of various substances . 32
Tender of a steam-engine . . . 204
Terbium 20
Terrestrial gravity 34
Thermal effects of light 256
Thermometer 149
" Celsius' 149
" Delisle's 149
" Fahrenheit's . . 149
Thermometer, on what principle
   constructed 29
Thermometer, Reaumur's . . 149
Thermo-electric 334
" batteries . . . 335
Thermo-electricity 260, 334
Thetis 339
Thorium 20
Threads of a screw 93
Thunder-clouds, distance of, how
   measured 177
Thunder-house 277
Thunder-storm, safest position in. 281
Tides 390
" neap and spring 391
Time, apparent and true, difference between 397
Time as kept by clock and by the
   sun 397
Time employed in the ascent and
   descent of a body equal . . . 54
Time, how estimated 43
" sidereal and solar . . . 396
Time of ascent and descent of a
   body 45
Tin 20
Tin and copper, sonorous properties of 30
Tin, not ductile 31
Tissue figure 270
Titanium 20
Toggle-joint 96
" operation of the . . 97
Tones of the voice, how varied . 180
Tonic 183
Tonnage of vessels, how estimated 123
Torpedo 282
Torricelli 143
Torricellian vacuum 143
Towns and fortifications, attacks
   on 63
Transfer of fluids 167
Transit of Mercury and Venus . 363
Translucent bodies 211
Transparent bodies 211
Tropic 356
Trumpet 178
Trumpets, hearing 178

Trumpet, speaking 178
Tubes, capillary 111
" mercurial 160
Tufts' stationary steam-engine . 207
Tune 41
Tungsten 20
Tycho Brahe 336

U.

Umbrella, use of in leaping from
   high places 38
Undershot wheel 82
Undulations of light 211
" of water, effects of . 131
Undulatory theory of light . . 211
Universal discharger 272
Urania 339
Uranium 20
Uranus 369
" moons of 370
Ursa Major 398

V.

Vacuum 98, 143
Vacuum, a perfect, not to be procured by means of the air-pump 156
Vacuum, the Torricellian . . . 143
Valve 152
Vanadium 20
Vapor, cause of ascent of . . . 124
Vapors 139
Vegetables, why white or yellow
   when growing in dark places . 256
Vehicle in motion, cause of accidents from 25
Velocities, table of 42
Velocity 41, 71
" absolute and relative . . 42
" how estimated 42
Velocity of balls thrown by gunpowder 63
Velocity of light and of the electric fluid 40
Velocity of parts of a body, how
   diminished 60
Velocity of sound . . . 176 and note
Velocity of sound, distances
   measured by the 177
Velocity of sound, experiments of
   Arago, Gay Lussac and others 176
Velocity of a stream, how measured 130
Velocity of the surface of a
   stream, greatest 129

Velocity required in machines, how regulated 106
Ventriloquism 180
Venus 363
" transit of 363
Venus, why never seen late at night 363
Vertical line 37
Vesicular form of matter, definition of 19
Vespasian, battering-ram of . . 106
Vesper 363
Vessels, tonnage of, how estimated 123
Vesta 339
Vision, angle of 219
Victoria 339
Vitreous electricity 262
Vitreous humor 237, 239
Vitriol, effects of on water . . . 187
Voice, Dr. Rush's Treatise on the 180
" sound of the 179
Voice, the human, imitative power of the 180
Voice, tones of the, how varied . 180
Voltaic battery 289
" electricity . . . 259, 283
" pile 288

W.

War, how it has been elevated to a science 63
Warmth of clothing, cause of . . 189
Watch, how it differs from a clock . 104
" how regulated 105
" moving power of a . . . 104
Water 21
Water, converted into steam, space occupied by 30
Water, distilled, the standard of specific gravity 123
Water, elasticity and compressibility of 24
Water expands when freezing . . 192
Water-fowl, buoyancy of 123
Water frozen under the air-pump . 169
Water, how applied to move machinery 83
Water, how converted into steam . 195
Water, how high raised by means of common pump 153
Water, how much diminished in bulk by pressure 29
Water, instruments for raising . 131
Water-level 113
Water, motion of, how retarded . 129

Water, not destitute of compressibility 109
Water, of what composed . . . 20
Water, pressure of at great depths 109, 116
Water, pressure of at any depth, how estimated 115
Water, pressure of at different depths 115
Water-pump 152
Water-spouts 172
Water, weight of a cubic foot of . 126
" weight of a cubic inch of 115
Water when falling, why less injurious than ice 114
Water, when perfectly pure . . 124
Water, why it appears more shallow than it is 231
Water-wheels 81
" most powerful . . 82
Watson, Dr., experiment of, to show degree of evaporation . . 150
Watt, James 106
Watt, James, his improvements of the steam-engine 197
Waves, how caused 130
Waves of light, laws of . . 212 note
Wedge 92
" advantage of the . . . 92
Wedge, effective power of, on what dependent 92
Wedge, power of the 92
Wedges, use of 92
Wedgewood's pyrometer . . . 193
Weight 34, 72
" cause of 34
" lifter 165
Weight, loss of in bodies weighed in water 126
Weight of any body, how ascertained by its cubical contents . 125
Weight raised by wheel and axle, how supported 79
Weight, what bodies have the greatest 34
Welding 31
Wheel and axle 78
" " advantage of . 79
" " construction of . 79
" " how supported . 81
" " principle of the . 80
Wheel, escapement 104
Wheels, friction 99
Wheels in machinery acting as levers 78
Wheels, large and small, advantages of each 85

Wheels, locked, how and why . . 85
Wheels of a clock, their use . . 101
" power of 80
" size of limited by what . 85
" tires of how secured . . 193
Wheels, toothed, method of ascertaining power of 85
Wheels, use of on roads 85
Wheel with teeth, of three kinds 84
Whirlwinds 172
Whispering-gallery 179
Whispering-gallery in Newburyport 179
Whisper, motion of a, rapidity of the 176
White 251
Whitefield 179
Wick of a lamp, principle of the 111
Width 23
Wightman's apparatus for inertia 25
William, Duke of Normandy . . 380
William the Conqueror 380
Winch applied to wheel and axle 79
" double 80
Wind 170
Wind, cause of the different directions of the 171
Wind, east, cause of at the equator 171
Wind instruments, sound of, on what dependent 181
Winds quality of the, how affected 171
Wind, why it subsides at sunset . 171
Windlass 80
Windlass and capstan, difference between 80
Wind-mills 80
Window, where the hand should be applied to raise 77
Wollaston, experiments of . . . 254
Wooden spoons and forks, why preferred for ice 190
Woollen garments, why warm . 189
Worcester, Marquis of 200
Worm of a still 195

Y.

Year 341
Year, leap 396
Year, sidereal and solar 396
Yttrium 20

Z.

Zodiac 345
Zodiacal light 360
Zodiac, constellations of the, change of 347
Zodiac, signs of the 346
Zinc 20
Zinc, at what temperature malleable 31
Zirconium 20

www.ingramcontent.com/pod-product-compliance
Lightning Source LLC
Chambersburg PA
CBHW081233180526
45171CB00005B/407